INTRODUCTION TO ANTENNA PLACEMENT AND INSTALLATION

Aerospace Series List

Principles of Flight Simulation	Allerton	October 2009
Aircraft Fuel Systems	Langton, Clark, Hewitt, Richards	May 2009
Computational Modelling and Simulation of Aircraft and the Environment: Volume 1 – Platform Kinematics and Synthetic Environment	Diston	April 2009
The Global Air Transport System	Belobaba, Barnhart, Odoni	April 2009
Handbook of Space Technology	Ley, Wittmann & Hallmann	April 2009
Surrogate Modelling in Engineering Design: A Practical Guide	Forrester, Sobester & Keane	August 2008
Aircraft Performance Theory and Practice for Pilots	Swatton	August 2008
Aircraft Systems, 3rd Edition	Moir & Seabridge	March 2008
Introduction to Aircraft Aeroelasticity and Loads	Wright & Cooper	December 2007
Stability and Control of Aircraft Systems	Langton	Sept 2006
Military Avionics Systems	Moir & Seabridge	February 2006
Design and Development of Aircraft Systems	Moir & Seabridge	June 2004
Aircraft Loading and Structural Layout	Howe	May 2004
Aircraft Display Systems	Jukes	December 2003
Civil Avionics Systems	Moir & Seabridge	December 2002

INTRODUCTION TO ANTENNA PLACEMENT AND INSTALLATION

Thereza M. Macnamara BSc, MSc (London)

WILEY

A John Wiley and Sons, Ltd., Publication

Registered office
John Wiley & Sons Ltd, The Atrium, Southern Gate, Chichester, West Sussex, PO19 8SQ, United Kingdom

For details of our global editorial offices, for customer services and for information about how to apply for permission to reuse the copyright material in this book please see our website at www.wiley.com.

Library of Congress Cataloging-in-Publication Data
Macnamara, Thereza.
Introduction to antenna placement and installation / Thereza MacNamara.
p. cm.
Includes bibliographical references and index.
ISBN 978-0-470-01981-8 (cloth)
1. Airplanes–Radio antennas–Installation. 2. Antenna radiation patterns. I. Title.
TL694.A6M28 2010
621.382′4 – dc22
2010000252

A catalogue record for this book is available from the British Library.
ISBN: 9780470019818 (Hbk)

Typeset in 10/12 Times by Laserwords Private Limited, Chennai, India.
Printed and bound by CPI Group (UK) Ltd, Croydon, CR0 4YY
C9780470019818_040124

This book is dedicated to my three granddaughters Shanti Jasmine (AKA Cheeky), Sophia Jane Zahra (AKA Softie) and Teya Mared (AKA Twinks) who will hopefully read it when they are older.

Contents

About the Author

Thereza M. Macnamara attained her first degree in applied physics and her master's degree in microwaves at London University. After two years of teaching physics up to Advanced level, she worked as a microwave engineer at G&E Bradley. She then worked as a research physicist for Morganite Research and Development before returning to work as a microwave engineer, working on a standard field facility, waveguide components, thermistor detectors, and calibration at Wayne-Kerr Laboratories, Flann Microwave Instruments and Bradley Electronics. After a short break to have a family, she returned to work as an examiner at the British Patent Office and lecturing in mathematics and physics whilst her children were growing up. She then took up a post as a senior RF engineer at ERA Technology, where she worked on antennas, feed networks and as an EMC engineer, before taking up the post of an electromagnetics specialist at BAE Systems where she worked for 17 years. Whilst at BAE Systems she worked in the R&D department and on Nimrod, Typhoon, Harrier, Tornado and Jaguar aircraft, and then became the technical coordinator of the EU funded research project IPAS (Installed Performance of Antennas on AeroStructures). Apart from many technical papers, she has also authored a reference book on EMC, entitled "Handbook of Antennas for EMC".

Preface

This book has been written as a reference book in a tutorial style. Each chapter is designed to be fairly self-contained, and the reader does not to read through the chapters sequentially, although sections have been cross-referenced. The reader also does not have to read a chapter from the beginning in order to find the definition of a particular symbol, for instance, since the details of each symbol are given after each equation. Although this may seem repetitive to the reader reading the book from the beginning, this approach is invaluable to the reader seeking to refer to a particular topic, and not wishing to spend time looking up the meaning of symbols used in equations and formulas. Practical details are given and the use of mathematical equations is kept to a minimum since most engineers who prefer the mathematical approach will find an abundance of textbooks using Maxwell's equations but will be unable to find simple explanations of the physical phenomena underlying the mathematics.

Where appropriate, reference to documents and books is given with the particular page or section number. This enables the reader to quickly access the particular topic, and the author considers this a welcome departure from references made to entire books, with the reader required to find the relevant section.

This book is intended as a background and reference book primarily for antenna and integration engineers involved in the integration of a single antenna or the entire antenna layout on the airframe of an air vehicle. However, the techniques could equally be applied to antennas on any structure such as a land or sea vehicle as well as spacecraft.

Engineers in other disciplines such as electromagnetic health/hazard (EMH), systems and aerodynamic engineers seeking specific or general information on aircraft antennas as well as those involved with measurement, certification and qualification phases will also find this book very useful. The treatment is essentially practical and even experienced antenna design engineers will find it very useful, since they may not be familiar with distortion of the antenna pattern when installed on an airframe. The pitfalls as well as the benefits of different sites on the fuselage of an aircraft are demonstrated by the use of measured radiation patterns.

The reader is expected to have attained an academic level of about undergraduate degree or Higher National Diploma (HND) standard or have had some practical experience as a systems, EMH or antenna engineer.

The book uses SI units throughout, and lists all the abbreviations and acronyms commonly used. SI units avoid the use of the solidus (forward slash or division sign) and instead the use of negative indices is recommended. However, the solidus is commonly used in some cases, and thus has been retained in these cases.

A large part of the material used in this book is based on Installed Performance of Antennas on AeroStructures (IPAS) and the production of this book is part of the exploitation plan for the European Union (EU) funded Project IPAS.

The radiation pattern of an antenna is distorted when it is installed on an airframe and the installed pattern is peculiar to each airframe. Thus the radiation pattern on one airframe is completely different from that on another airframe and the pattern is also dependent on the electrical dimensions (the dimension in terms of wavelengths) of the surface and the electrical distance of the antenna from obstacles.

Chapter 1 explains the main properties of electromagnetic waves as applied to antennas in a qualitative manner, and is useful for engineers not familiar with antennas or the jargon used.

Chapter 2 provides an overview of most common systems connected to antennas used on commercial and military aircraft, and outlines the salient points of the systems with regard to the antenna. It does not delve into any detail concerning the receiving and processing equipment.

Chapter 3 outlines the antenna siting process and the interfaces with the other disciplines, showing the trade-offs that must be considered and iterations usually required before arriving at the final antenna layout. It demonstrates the difficulty of obtaining the optimum layout due to its low priority compared with the aerodynamic considerations. It also shows some typical antenna layouts.

Chapter 4 shows how the interactions between waves of the same frequency (but with different amplitudes and phases) produce resultants of different amplitudes. This provides the reader with an understanding of the complex nature of installed antenna patterns. The chapter explains qualitatively the effect of obstacles, frequency, wings and the curved ground plane on the radiation pattern of an antenna on an aircraft, by demonstrating how the antenna pattern is affected with the change in frequency and at different positions on the fuselage.

Chapter 5 describes the most common antennas used on aircraft but does not elaborate on antenna theory. It also includes the effect of the electrical size of the ground plane on monopoles, comparison between the different types of spiral antennas, and the effect of the aperture illumination on the radiation pattern of aperture antennas.

Chapter 6 on RF interoperability is particularly pertinent to systems and will provide background information to systems engineers. This chapter is equally applicable to radiated emissions, sometimes undertaken as part of the EMH clearance programme. It also demonstrates the difficulty of predicting the coupling between antennas that are not within line of sight, and provides empirically derived values for antenna gains used in calculating the coupling. The derivations were developed after a measurement programme undertaken in IPAS.

Chapter 7 qualitatively describes the different software commonly used for predicting the radiation patterns of antennas on structures without delving into the mathematics used, so that the reader does not have to understand the complexities of Maxwell's equations in order to understand the aspects of computational antenna modelling on structures. Work undertaken under the IPAS research project involved computational predictions and measurements on the same scaled models by different partners using their own in-house facilities so that direct comparisons could be obtained between the different codes, as well as between measured and predicted results. This has provided an unique comparison between the different types of computational modelling software available, since normally predictions are only performed by each aircraft manufacturer with a single suite of computational software and one set of measurements for each type of airframe.

Chapter 8 provides a basic understanding of the measurement sites for antenna pattern measurements and for radiated emissions as well as practical details of measurements. It discusses the trade-offs to be considered in the choice of a scaled model and gives a qualitative description of near and far field facilities as well as compact ranges. Ground and in-flight tests are also described and some results shown.

The final chapter contains reference data such as conversion tables, conductivities, dielectric constants, conversion between dBm and watts, electrochemical electromotive forces (EMFs), EM spectrum with frequency wavebands and wavelengths, common formulas, the periodic table (listed alphabetically by symbol), preferred scientific prefixes and definitions.

Every care has been taken in the preparation of the manuscript. However, the author would appreciate any comments on the topics covered or errors in the text, be they typographical, technical or factual.

Acknowledgements

I would like to thank Enrique Escolano, Jaco Verpoorte, Pat Foster, Harmen Schippers, Torsten Fritzel and Carlo Rizzo for their help and useful comments on reviewing parts of this book.

Thereza M. Macnamara

Series Preface

In many of the books in this Series to date, the functions of the aircraft systems, mission systems and avionics have been described for both commercial and military aircraft types. One thing was omitted in most descriptions of communications, navigation, identification and surveillance systems – the role of the antenna.

That omission has been corrected in this important reference book on the care and attention needed to select the right antenna and to define its location on the airframe. That airframe is a crowded piece of real estate where systems engineers compete for the best site for optimum performance of their system. Antennas must be placed to ensure system performance and satisfy aerodynamic considerations, whilst ensuring interoperability and preventing mutual interference.

This book shows how antennas designs are analysed, modelled and tested throughout the aircraft design process to ensure the optimum results for systems performance. This is a book for antenna, systems engineers and for specialist practitioners in radio frequency applications for both commercial and military aircraft.

Allan Seabridge

1

Basic Antenna and Propagation Theory

1.1 Introduction

This chapter explains the principles of antenna theory and propagation qualitatively, without going into the complex mathematical equations that are usually found in textbooks and reference books dealing with these subjects. Although this results in explanations of a simplistic nature, it enables the reader with a basic physics background to understand electromagnetic (EM) theory.

EM waves are transverse waves, unlike sound and ultrasonic waves, which are longitudinal waves that require a medium. By analogy transverse waves are like the waves one would obtain by moving a rope up and down to transmit a sine wave, whereas a longitudinal wave is like a series of train wagons being shunted along, so that each wagon moves horizontally back and forth whilst the wave also moves horizontally along the whole train.

Sonic waves cannot be transmitted in a vacuum, whereas EM waves do not require a medium and can be transmitted in a vacuum, such as deep space. This is why we can see the stars but do not hear the sound of meteors, and so on.

The EM spectrum extends from direct current (DC) that has no/zero frequency to cosmic radiation.

Above DC, we commonly encounter low frequencies from 3 Hz up to around 300 Hz used for communications with submarines.

Alternating current (AC) frequencies are used to transmit mains power. In Europe 50 Hz is used, whereas in North America and some other countries 60 Hz is more common.

The mains power on aircraft is usually 400 Hz. There are various reasons for the choice of this frequency for powering aircraft systems, one of them being that this frequency was selected as a compromise between weight, size and efficiency of the aircraft power units.

Above these frequencies, there are many applications such as communications with mines, broadcasting, and so on, until the frequency used in aircraft systems which extends into the microwave and millimetre wave region. The details of the aircraft systems are covered in Chapter 2.

Introduction to Antenna Placement & Installation Thereza M. Macnamara

We then have the higher frequencies used for radio astronomy before the infrared (IR) region where we have the heating effect. Then we move into the optical or visible region of the spectrum which is actually a very small range of frequencies.

Above these frequencies, we have the ultraviolet (UV) region which causes sunburn at its high frequency end. Much of the UV light is absorbed by the atmosphere, and thus the higher levels are only encountered in hills and mountains. It is worth noting that photochromic lenses darken on exposure to UV light and therefore do so much more quickly at higher altitudes, whereas they take a lot longer to darken in a car, where most of the UV is absorbed by glass. UV is not absorbed by plastics and thus can penetrate the Plexiglas of aircraft windows/portholes. EM waves up to these frequencies are non-ionizing.

Above ultraviolet the EM waves are ionizing and penetrate materials such as tissue, in the case of X-rays, and even concrete, in the case of gamma rays. Then we have cosmic rays that originate in space.

All these forms of radiation are in the EM spectrum; however, when engineers refer to Electromagnetics (in the case of antennas) or electromagnetic health (EMH) they usually are only referring to the EM spectrum from about 1 kHz to about 300 GHz.

The term antenna is used as a generic term for wire antennas such as dipoles as well as for aperture antennas such as horns, reflectors, and so on.

In some cases however, the term 'antenna' is restricted to aperture antennas in the upper radio frequencies (RFs) and microwave regions – above about 500 MHz – and the term aerial is used at the lower frequencies.

In guided circuits, such as wire and printed circuit tracks that have resistive and reactive components (inductors and capacitors), the electric and magnetic fields are 90° out of phase, but in free space the electric and magnetic fields are in phase.

1.2 Characteristics of Electromagnetic Waves

In order to understand the propagation of EM fields, all the properties of radiation have to be considered. The main phenomena that affect the propagation are reflection, refraction and diffraction.

At a macroscopic level we can think of EM radiation as travelling in straight lines, but at a microscopic level we have to consider the wave properties of the radiation as well.

If we think of the waves as spherically emanating from the source (like the waves obtained by throwing a pebble into the water), then the spherical shells are the wavefronts and the rays are the radii from the source and therefore perpendicular to the wavefronts. In Figure 1.1 the wavefronts and rays are shown in two dimensions for clarity.

Reflection and refraction can be explained by the rectilinear propagation (light travelling in straight lines) of EM waves, but diffraction can only be explained by the wave theory. Geometric optics deals with rectilinear propagation, whereas physical optics deals with light as waves. Figure 1.2 shows the rays produced in a plane normal to the cylinder axis, for an antenna located off the body of the cylinder. The stronger the colour, the greater is the intensity of the ray.

Figure 1.3 is a graphic indication of the rays reflected and diffracted off an airframe.

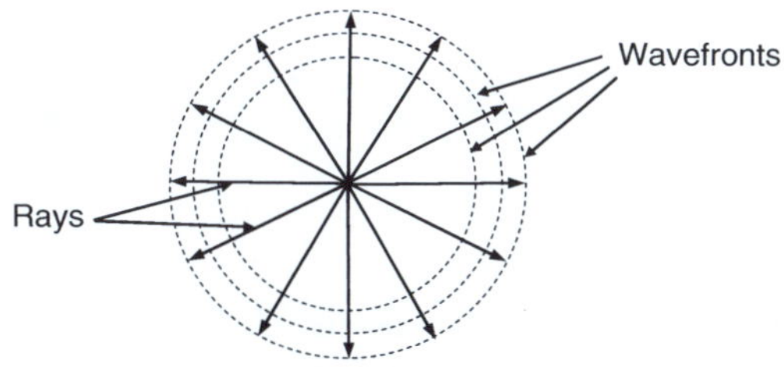

Figure 1.1 Relationship between wavefronts and rays.

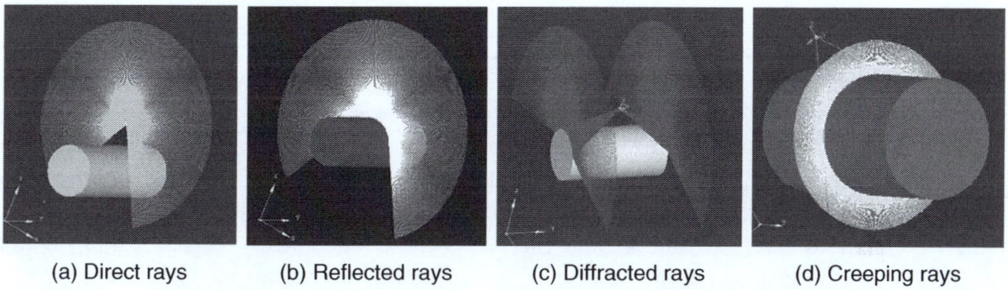

Figure 1.2 Graphic illustration of the direct, reflected, diffracted and creeping rays obtained for an antenna located above the surface of a cylinder. Figure 4 of [1]. Reproduced by kind permission of EADS. See Plate 1 for the colour figure.

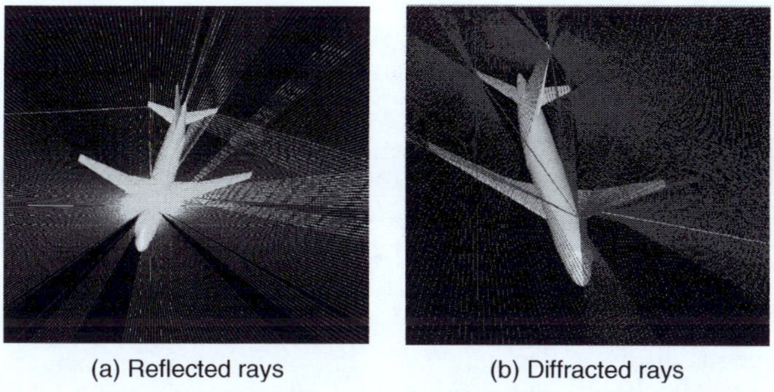

Figure 1.3 Rays obtained using the EADS ASERIS-HF GTD code. Figure 27 of [2]. Reproduced by kind permission of EADS. See Plate 2 for the colour figure.

1.2.1 Reflection

Reflection is explained by Snell's first law, which states that if an incident ray strikes a planar reflecting surface, it is reflected at the same angle to the normal as the incident angle and that the incident, normal and reflected rays are all in the same plane. Snell's laws are named after the Dutch mathematician Willebrord Snellius (1580–1626).

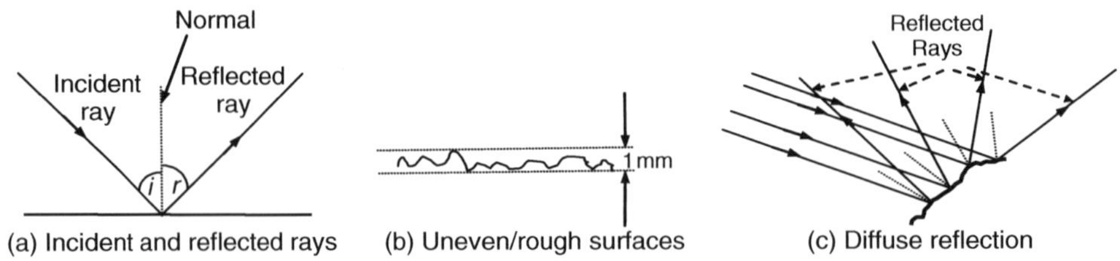

Figure 1.4 Snell's law, uneven surfaces and diffuse reflection for uneven surfaces.

The incident angle i is the angle between the incident ray and the normal to the surface, and the reflected angle r is the angle between the reflected ray and the normal to the surface, as shown in Figure 1.4a.

In the case of light waves, we can understand this phenomenon quite easily since we encounter it when looking in a plane mirror. However, if we consider a surface that only partially reflects the light, for instance parts of the dashboard reflected in the windscreen, then we are more likely to understand the EM waves reflected by surfaces that do not reflect most of the EM wave. In the case of the surface of an aircraft the complex reflected waves are akin to the images seen in the 'crazy' mirrors of a fairground.

In the case of a flat, perfectly conducting surface, the reflected ray would be similar to that obtained from a mirror, and in RF terminology this is called specular reflection.

If the surface is uneven, as in the case of most airframes, then there will be diffuse reflection, with waves scattered in a number of directions, in the same way as a sheet of white paper would diffuse the light falling on it.

When we say that a surface is flat or uneven, we mean relative to the wavelength of the radiation falling on it. For instance, if a surface roughness varies by 1 mm as shown in Figure 1.4b, this surface would be smooth/flat to a EM wave of 500 MHz that has a wavelength of 60 cm, whereas it is rough to visible light of frequency 450 THz (450×10^{12}) that has a wavelength of 0.7 μm.

This unevenness also explains diffuse reflection obtained, for instance, when light falls on a sheet of white paper. A parallel beam of light falling on an arbitrarily uneven surface is shown in Figure 1.4c, and we can see that the reflected rays are scattered in several directions, known as diffuse reflection.

Apart from the direct wave that is propagated into free space, the first order reflected wave (i.e. the first reflected wave that is the result of a direct wave striking a reflecting surface) contributes the greatest to the resultant in the far field. If the first order reflected wave is in phase with the direct wave, then the resultant would a maximum, whereas if the first order reflected wave is in antiphase (180° out of phase) with the direct wave then the resultant would a minimum.

1.2.2 Refraction

In real life we encounter the phenomenon of refraction when looking into a swimming pool where the floor of the pool appears less deep than it actually is. Refraction is the bending of rays when an incident ray strikes a medium with a different refractive index. In general, the denser media have higher refractive indices. Spectacle lenses that are thinner usually have higher refractive indices, to attain the same power (dioptre).

Refraction is explained by Snell's second law, which states that when an incident ray enters a more optically dense medium (i.e. with a higher refractive index), it is bent towards the incident ray. Conversely, if it enters a less dense medium, it is bent away from the incident ray, as shown in Figure 1.5.

Snell's law is expressed as

$$\mu_1 \sin i = \mu_2 \sin r, \quad (1.1)$$

where

μ_1 is the refractive index of the first medium
i is the angle of incidence at the interface between the first and second medium
μ_2 is the refractive index of the second medium
r is the angle of refraction.

For instance, if the first medium has a refractive index of 1 (like air), the second medium has a refractive index of 1.4 and the angle of incidence is 36°, using Equation 1.1 the refracted angle is 24.8°.

However, if the refractive indices were reversed and the wave was travelling from a medium with a refractive index of 1.4 to one with a refractive index of 1, then the refracted angle would be 55.4°, as shown in Table 1.1.

If the second medium is in the form of a plate where the top and lower surfaces are parallel, then the ray emerging from the plate will be parallel to the incident ray but displaced from it by a distance that is directly proportional to the thickness of the plate.

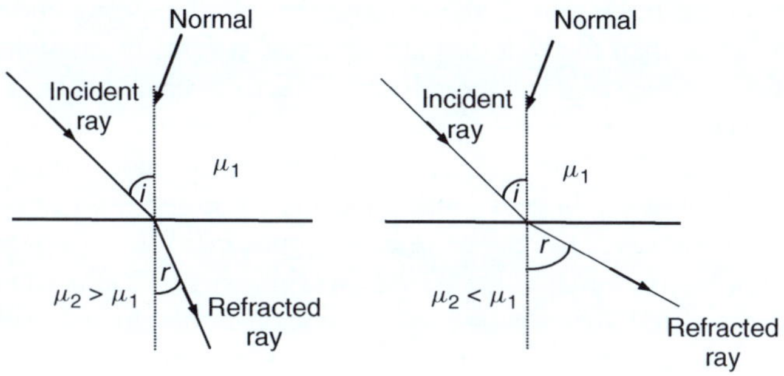

Figure 1.5 Snell's law of refraction.

Table 1.1 Change of refracted angle with refractive index.

μ_1	μ_2	Angle i	Angle r
1	1.4	36	24.8
1.4	1	36	55.4

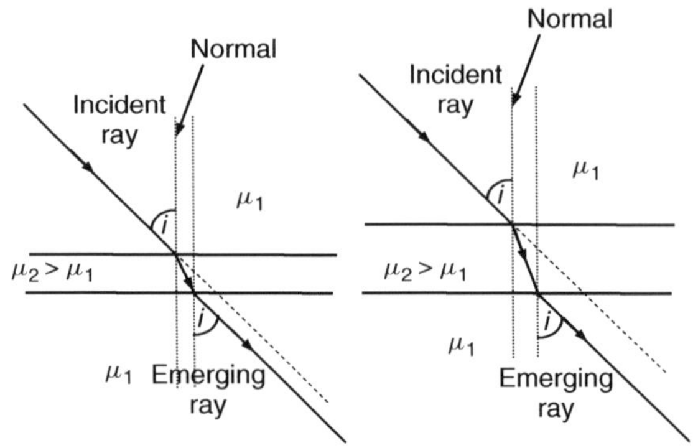

Figure 1.6 The effect of increasing the thickness of the plate.

In the case of aircraft antennas the materials that would subject the EM waves to refraction are composite fibreglass dielectric surfaces like those used in radar domes (radomes) that protect the antennas whilst still allowing the radiation to pass through them.

In Figure 1.6 we can see that as the thickness of the plate is increased the displacement of the emerging ray is also increased. The thickness that we have to consider is the electrical thickness, that is, the thickness in terms of wavelength. Thus if we keep the physical thickness the same but double the frequency the wavelength will have halved, and so that is tantamount to doubling the thickness of the plate. The displacement will be *double* the amount for the higher frequency (assuming that the refractive index has also doubled with frequency). This accounts for the fact that the fibreglass covering used for low frequency antennas such as 'blades' can be quite thick, but in the case of the nose cones the radomes have to be thin. If the radomes have to be strong enough to withstand birdstrike, for instance, then the radomes are made of several layers, where the effect of some of the layers compensates for the adverse effect of other layers. These are known as sandwich radomes.

We must also consider the effect of increasing the refractive index. If the refractive index is increased, the refracted ray is bent more and thus the emerging ray is more displaced compared to the incident ray. This can be seen in Figure 1.7. When the refractive index μ_2 of the plate is increased to a larger value μ_3 the emerging ray is displaced to a larger extent.

It should also be noted that the refractive index of the same material varies with frequency, and in general it increases with increasing frequency.

1.2.2.1 Total Internal Reflection

We have seen how a ray entering a medium of lower refractive index is bent away from the normal according to Snell's law. At a certain angle of incidence i_c known as the critical angle, the angle of refraction is 90° so that the ray travels along the interface between the two media, as shown in Figure 1.8.

At angles of incidence greater than i_c, the ray is refracted back into the first medium. Because the ray now behaves like a reflected ray, with the angle of reflection equal to the angle of incidence, this effect is known as total internal reflection.

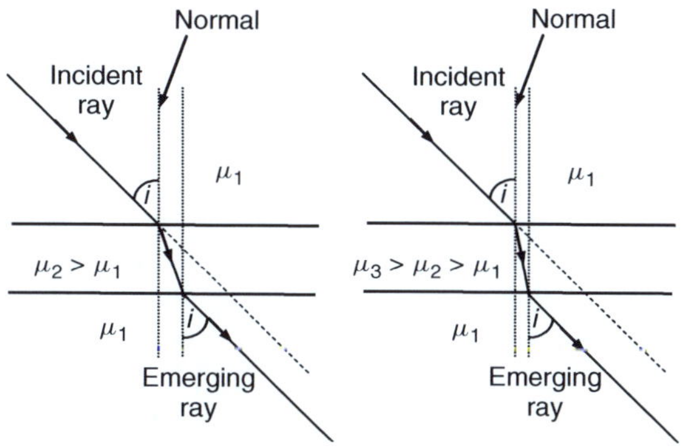

Figure 1.7 The effect of increasing the refractive index of the plate.

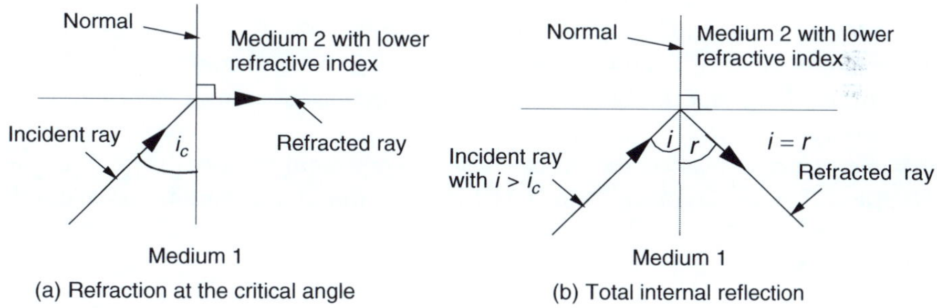

Figure 1.8 The effect of increasing the angle of incidence for a wave travelling to a medium of lower refractive index.

In the earth's atmosphere the refractive index decreases with height above the ground. If we think of the atmosphere as layers of air with decreasing density (and hence decreasing refractive index) we can see from Figure 1.9 that as the altitude is increased, a wave travelling upwards is gradually bent further away from the normal, so that it eventually undergoes total internal reflection and is reflected back towards the earth. Rays that are

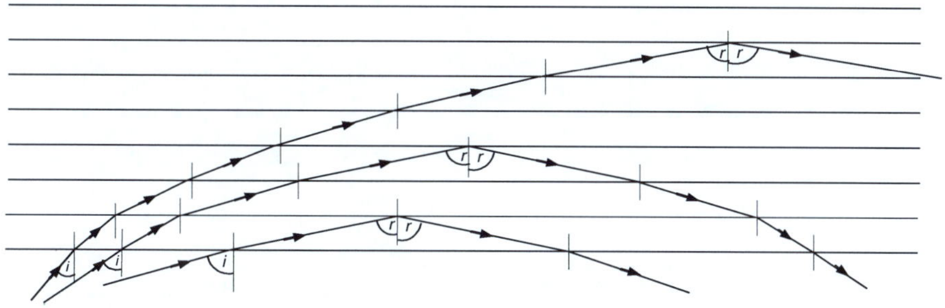

Figure 1.9 Total internal reflection for rays incident at different angles.

incident at large angles undergo total internal reflection at lower altitudes than those that are incident at small angles.

1.2.3 Diffraction

Diffraction is the bending of light waves when they strike an edge. In Isaac Newton's time, when light was considered as travelling in straight lines, this was called rectilinear propagation. However, when it was discovered that the edges of shadows were not sharp, Newton put forward his corpuscular theory to explain this, by postulating that light consisted of particles like miniature golf balls and when they passed over an object, some fell into the shadow region and blurred the edge of the shadow. In Figure 1.10 the corpuscles are shown striking the edge of a plate and most fall to the right of the extended line joining the source to the edge of the plate, and some fall in the semi-lit region, but none fall in the shadow region. Only the corpuscles near the edge are shown for clarity.

Later, when interference fringes were discovered, the corpuscular theory could not explain how combinations of particles could form light as well as dark areas. This could only be explained by considering light as waves. The wave theory explained this phenomenon by showing that the two waves could either cancel or sum, depending on their relative phases.

The wave theory explained the propagation of light until the discovery of the photoelectric effect. It was discovered that EM radiation falling on certain surfaces released

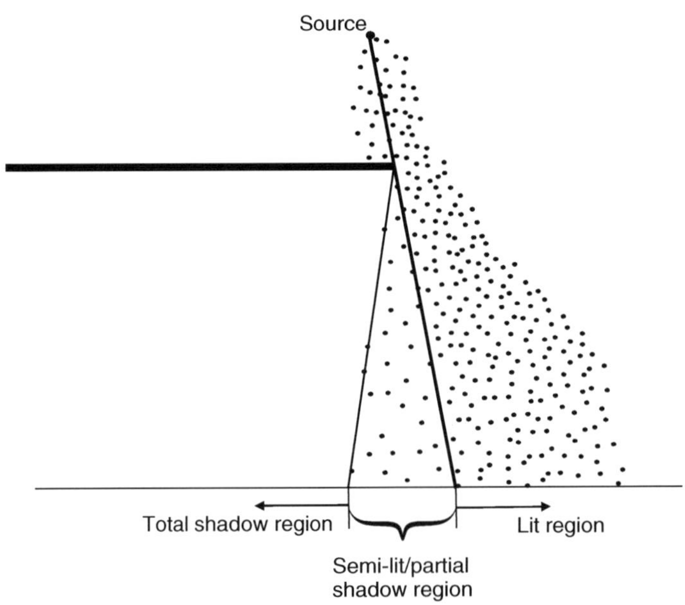

Figure 1.10 Newton's corpuscular theory of light.

particles of a different frequency from those surfaces. However, when the frequency of the irradiating radiation was decreased the photoelectric effect did not occur. The wave theory of light could not explain this, and thus it was concluded that the EM radiation consisted of quanta and the energy of each quantum was h/λ (where h is Planck's constant and λ is the wavelength). At low frequencies the wavelength is large and thus the energy of each quantum is small.

If light did not consist of quanta then the photoelectric effect would occur by just increasing the intensity of the radiation, regardless of the frequency.

However, in order to still explain the interference fringes it was concluded that light had a dual nature, consisting of waves and quanta.

Under the wave theory of light, when light falls on a surface, secondary sources of light or wavelets, are formed (commonly called Huygens sources). Each of these secondary sources emit spherically, which accounts for the fact that an illuminated edge can be seen at angles away from the normal, whereas the original source cannot be seen at these angles.

This phenomenon that results in the 'bending' of light rays at an edge is known as diffraction and applies to all forms of EM radiation, such as RF, IR, UV, and so on.

A good example of diffraction can be seen with the appearance of the diamond rings when the sun goes into and comes out of a total eclipse.

1.2.3.1 Creeping Waves

EM waves also tend to follow a curved surface and 'creep' along its surface. This effect is due to the EM wave travelling along smooth surfaces, and the diffracted rays being emitted tangential to the surface into the surrounding space. These creeping waves account for the fact that when, for instance, a monopole is installed on the top of a cylinder a significant part of the radiation occurs in the lower hemisphere.

Creeping waves that travel around the earth are also sometimes called ground waves or Norton waves. They are guided along the ground in the same way as an EM wave is guided along a transmission line. As they travel over the ground they undergo attenuation. The energy lost from the wavefront into the ground is replaced by energy in parts of the wavefront higher up above the ground. This results in a continuous movement of energy towards the ground.

Creeping waves are particularly pertinent to aircraft, since most fuselages are of circular or elliptical cross-section. The waves travel around the curved surface, taking a number of paths known as geodesics. A geodesic is defined as the shortest route between two points on a mathematically derived surface that includes the points, and in the case of the earth's surface, this would be a segment of a great circle cut. A great circle has the same circumference as the equator. Geodesics are sometimes defined as curves whose tangent vectors remain parallel if they are transported along it.

In the case of a monopole on a cylinder, if we consider the circular cross section, there are two geodesics between diametrically opposite points. These travel along the circumference of the circular cross-section, one in each direction. However, in the case of elliptical cross-sectional cylinders, there may be as many as six geodesics.

1.3 Interaction between Two Waves

Because light consists of waves, we have to consider the phase as well as the amplitude. If we consider the interaction between two waves, we have to consider their relative amplitudes and their relative phases. The interaction and the resultant produced can be depicted by looking at the waveforms or considering the waves as vectors using phasors. The resultant of the phasors can be obtained by drawing the individual phasors to scale or by calculation using simple trigonometry.

When we look at the waveforms in the time domain, the resultant is also in the time domain, so we can see the actual waveform of the resultant. However, when we use phasors we get the amplitude and the phase of the resultant, but we do not see its actual waveform. Also, phasors can only be used for waves of the same frequency, whereas time domain addition can be used for waveforms of different frequencies.

1.3.1 Waveforms in the Time Domain

If the two waves are in phase and of equal amplitude, as shown in Figure 1.11a,b, and if we add the magnitudes at each phase angle together, we will get a resultant wave of double the amplitude and in phase with both the waves, as shown in Figure 1.11c. This is known as constructive interference. The same effect occurs in the case of sound waves in organ pipes and is known as resonance in that case.

If the two waves are 180° out of phase (i.e. in antiphase) and of equal amplitude, as shown in Figure 1.12a,b, the resultant is a wave of zero amplitude. This is known as destructive interference, and the resultant is shown in Figure 1.12c.

If the two waves are in antiphase (180° out of phase) and the second wave has half the amplitude of the first one, as shown in Figure 1.13a,b, then the resultant wave will have half the amplitude of first wave (i.e. the same amplitude as the second wave) and be in phase with the first wave, as shown in Figure 1.13c.

If the two waves have equal amplitude and are out of phase by 90°, as shown in Figure 1.14a,b, then the resultant wave will have 1.41 times the amplitude of either wave and be 45° out of phase with each wave, as shown in Figure 1.14c.

If the second wave, as shown in Figure 1.15b, is 1.93 times the amplitude of the first wave and 135° out of phase with it, the resultant is a wave of amplitude 1.41 times

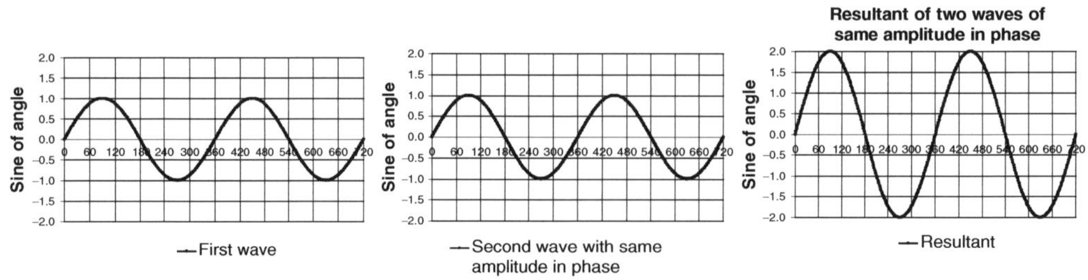

(a) First wave (b) Second wave (c) Resultant wave

Figure 1.11 Resultant of two waves that are in phase and of equal amplitudes. The resultant is a wave of double the amplitude and in phase with both the waves.

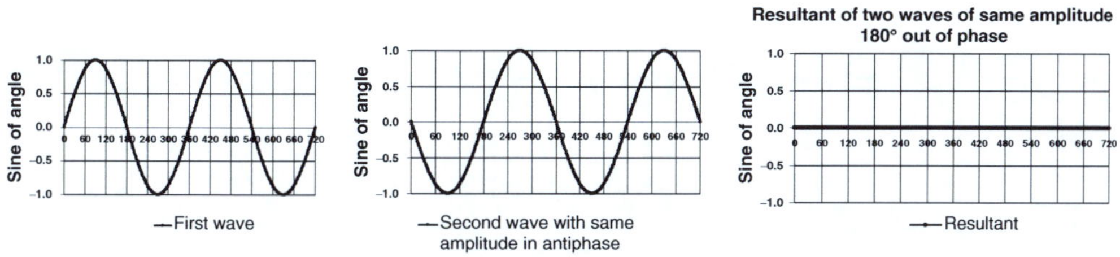

(a) First wave (b) Second wave (c) Resultant wave

Figure 1.12 Resultant of two waves that are 180° out of phase and of equal amplitudes. The resultant is a wave of zero amplitude.

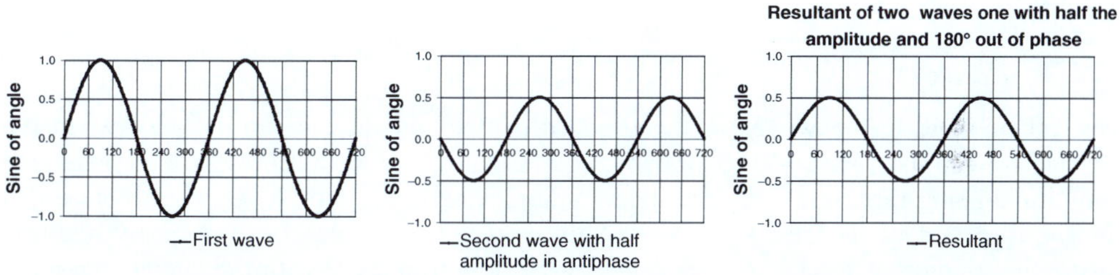

(a) First wave (b) Second wave (c) Resultant wave

Figure 1.13 Resultant of two waves that are 180° out of phase and where the amplitude of the second wave is half that of the first wave. The resultant is a wave in phase with the first wave and of half the amplitude of the first wave.

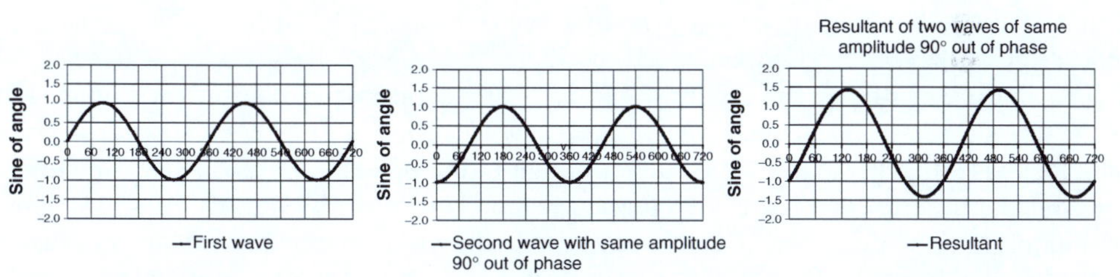

(a) First wave (b) Second wave (c) Resultant wave

Figure 1.14 Resultant of two waves of equal amplitude that are 90° out of phase. The resultant is a wave in phase with the first wave and of 1.41 times the amplitude of the first wave.

the amplitude of the first wave and about 105° out of phase with it. This can be seen in Figure 1.15c. Note that this is the same resultant (in amplitude) as obtained in Figure 1.14, although the second waves are not the same in either amplitude or phase.

The same occurs in the case of the radiation pattern of an antenna. It gives the resultant of the interaction(s) of usually two or more waves, but we cannot identify the source(s) of the individual waves that give us the resultant and more than one combination can have the same resultant. Thus we do not have an unique solution in these cases.

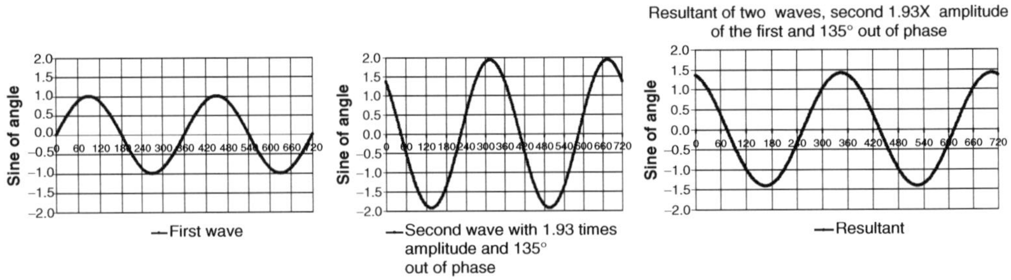

(a) First wave (b) Second wave (c) Resultant wave

Figure 1.15 Resultant of two waves, the second being 1.93 times the amplitude of the first wave and 135° out of phase with it. The resultant is a wave of amplitude 1.41 times the amplitude of the first wave and about 105° out of phase with it.

1.3.2 Phasors

Instead of plotting the waves in the time domain to obtain the resultants between the waves, we can use phasors. Phasors are vectors that are used to represent the waves, the amplitude being represented by the magnitude or length and the phase being represented by the direction of the vector. The resultant of two vectors can be derived by addition using the triangle of forces. The phasors can be drawn to scale with the angles measured using a protractor, or trigonometry can be used to calculate the resultant.

1.3.2.1 Phasors Drawn to Scale

The first vector is drawn to a suitable length and the second one is drawn to the same scale at an angle to represent the phase between the two. The resultant is obtained by joining the line between the start of the first vector and the end of the second vector. For a description of vector addition see [3], p. 52.

The five cases shown in Figure 1.11–1.15 are depicted as phasors in Figure 1.16. Note that the first and second vector are drawn cyclically (i.e. both are drawn in an anticlockwise direction) and the resultant drawn in the opposite or clockwise direction. If we consider case (e) of Figure 1.16 that corresponds to the addition of the waves shown in Figure 1.15, we can see that the angles are drawn as mathematical angles, that is, anticlockwise from the first quadrant. The following steps describe in detail the process for case (e):

1. Draw a horizontal line to a convenient length with an arrow at the right-hand end. This is used as the scale unit 1 for the first phasor.
2. Using a protractor mark out 135° from the arrow end of the first phasor.
3. Mark out the length of the second phasor to be 1.93 times the length of the first phasor, with an arrow at its top end.
4. Join the start of the first line and the arrow end of the second phasor.
5. The new line represents the resultant phasor.

The resultant is of length (and hence amplitude) 1.41 times the first vector and its phase is 105° compared with the first vector.

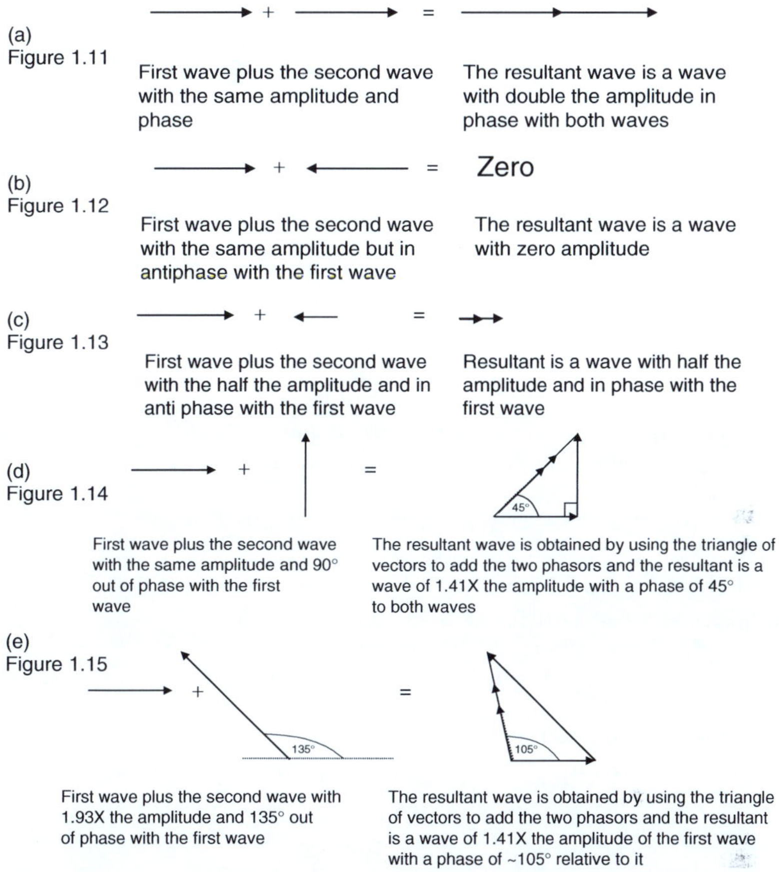

Figure 1.16 Interaction between two waves drawn as phasors to scale.

1.3.2.2 Phasors Used in Calculations

Simple trigonometry can be used to obtain the resultants of Section 1.3.2.1 above.

The cosine and sine formulas (see [3], pp. 39–42) are used in this case. In the triangle the sides are denoted *a, b* and *c*, and the angles opposite these sides are denoted *A*, *B* and *C*, respectively. Referring to Figure 1.17a:

1. The first phasor corresponds to side b and the angle opposite it is angle B.
2. The second phasor corresponds to side c and the angle opposite it is angle C.
3. The resultant phasor corresponds to side a and the angle opposite it is angle A.
4. The angle ϕ is the phase difference between the first and second waves. Note that angle A in the cosine formula is $180 - \phi$.
5. The angle C gives us the phase between the resultant and the first wave.

Thus we need to find the values of length a and angle C to find the magnitude and phase of the resultant, respectively.

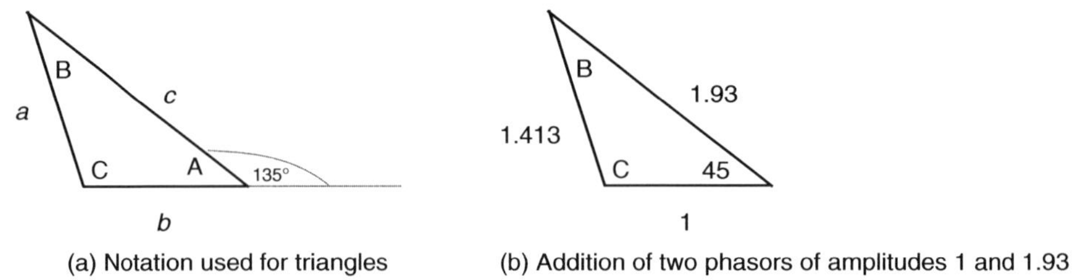

Figure 1.17 Interaction between two waves using trigonometry to calculate the resultant amplitude and phase.

The cosine formula can be used for any type of triangle – acute, right-angled or obtuse. In the case of a right-angled triangle, the cosine of angle $A(=90^\circ)$ is zero and the formula reduces to Pythagoras' theorem. The cosine formula is given by

$$a^2 = b^2 + c^2 - 2bc\cos A. \tag{1.2}$$

In the case of (e) in Figure 1.16, we have $b = 1$, $c = 1.93$, angle ϕ is 135° so angle A is 45°. Using these values in Equation 1.2 we get

$$a^2 = 1 + 1.93^2 - 2 \times 1 \times 1.93\cos(45^\circ).$$

Thus $a^2 = 1.995$, giving $a = 1.41$, which is the amplitude of the resultant.

The sine formula is now used to calculate the angle C. The sine formula states that

$$\frac{a}{\sin A} = \frac{b}{\sin B} = \frac{c}{\sin C}. \tag{1.3}$$

We only need to use

$$\frac{a}{\sin A} = \frac{c}{\sin C},$$

that is,

$$\sin C = \frac{c\sin A}{a} = \frac{1.93\sin 45^\circ}{1.41}$$

so that angle $C \approx 75^\circ$ or 105°. In order to decide which is the correct angle, we need to look at the value of the third angle B.

The exterior angle 135° is equal to the sum of the interior angles B and angle C. Thus if angle C is 75° then angle B is 60°, giving $b/\sin B = 1.1547$. If angle C is 105° then angle B is 30°, giving $b/\sin B = 2$. Since $b/\sin B = a/\sin A$, and since $a/\sin A$ is 1.9954, the second result for $b/\sin B$ is the correct one, and angle B must be 30. Therefore the correct value for angle C is 105°. Thus we get the same value as that obtained by drawing the phasors to scale in Section 1.3.2.1 above.

Using trigonometry gives us a more accurate result for the phase and magnitude of the resultant, but it can be seen that it has to be used carefully, since more than one value could result in the case of the phase angle.

1.4 Polarization

In the case of light waves, the polarization is random. Polarizing filters work on the principle of eliminating the polarization in all planes except one. Thus if we were to put one polarizing filter (a piece of Polaroid) behind the other and then rotate one, there would be positions where there is maximum light passing through both and a minimum (darkness) at right angles to the maximum position. We can think of the polarizing filter as a set of vertical slots so that only light with the plane of polarization parallel to these slots can pass through the slots. When the two lots of slots are at right angles to each other no light can get through.

In Section 1.2.1 we considered light as rays and we ignored the polarization. However, the polarization is altered when a wave is reflected off a surface. This is easier to see if we look at light diffusely reflected off a flat surface – not a mirror. If we look at light reflected at different angles through Polaroid filter, at one particular angle (depending on the surface) only plane polarized light would be reflected, and this is known as the Brewster angle. The reflected wave does not have any polarization in the plane containing the incident, normal and reflected waves. The Brewster angle was named after Sir David Brewster (1781–1868). The angle between the refracted and reflected rays is 90° at the Brewster angle. We can check that it is plane polarized light by using a second polaroid filter and rotating it relative to the first until it is perpendicular to first, at which point no light would be visible.

In the case of RF waves the polarization could be linear, elliptical or circular, although theoretically all polarizations could be considered to be variations of elliptical polarization.

The plane wave EM field in free space consists of a magnetic field at right angles to an electric field. Both fields vary sinusoidally in space as well as time. Thus if we were to consider the field at a fixed point in space it would vary in amplitude sinusoidally and if we were to capture the field at a fixed point in time, the fields would vary as a sine wave in space. Each of the electric and magnetic fields is represented by a vector and has an amplitude as well as a direction.

Figure 1.18 shows the orientations of the electric and magnetic field for a plane wave. The electric field varies in the yz plane, and the magnetic field varies in the xz plane, whilst the wave travels along the z axis.

When we refer to polarization, by default this refers to the polarization of the electric field vector, rather than the magnetic field vector. The polarization of an EM signal refers to direction of the electric field vector during at least one full cycle.

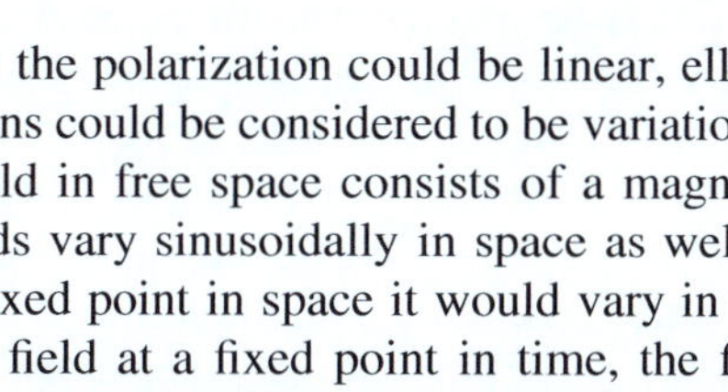

Figure 1.18 Orientation of the electric and magnetic fields for a plane wave.

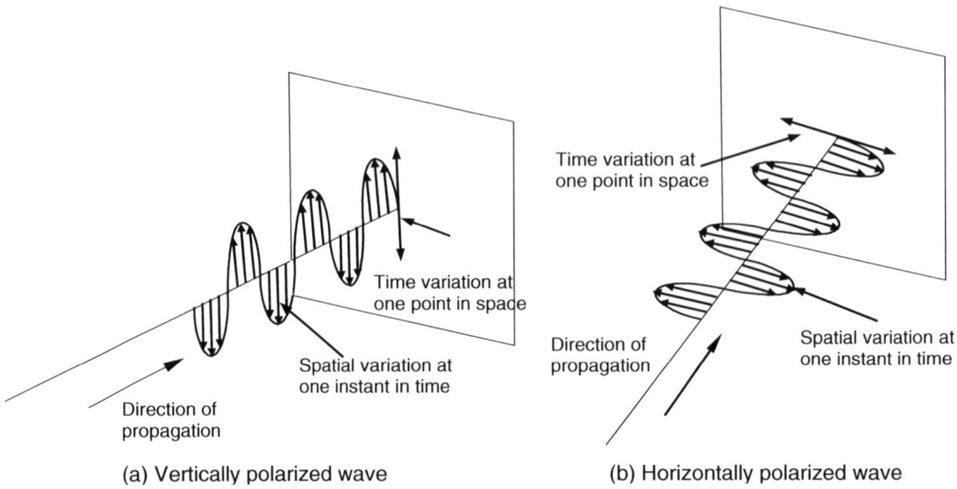

Figure 1.19 Linearly polarized waves.

1.4.1 Linear Polarization

The wave could be linearly polarized in any plane. In the case of a linearly polarized plane wave, the wave progresses in the direction of propagation and the electric field would appear as shown in Figure 1.19. If we were to look at the wave at one point in space, the electric field would move up and down (i.e. vary sinusoidally with time) in the case of the vertically polarized wave as shown on the square. The total excursion of the electric field would be from the positive maximum (amplitude of the sine wave) down to zero and then down to the negative maximum, and back again to the positive maximum over one period of time – that is, when the wave has completed one cycle. Thus one period is the time taken to complete one total excursion and is the reciprocal of the frequency. In the spatial domain the distance for one complete cycle gives us the wavelength.

The other most common types of linear polarizations used for RF propagation are left slant and right slant polarizations. The right slant is usually defined as making an angle of $+45°$ (clockwise like navigational angles) with respect to the vertical when looking from the transmitted radiation, as shown in Figure 1.20a. The left slant is usually defined as making an angle of $-45°$ with respect to the vertical when looking from the direction of the transmitted radiation, as shown in Figure 1.20b.

1.4.2 Circular and Elliptical Polarization

Circular polarization is a form of elliptical polarization. In the case of elliptical polarization, the magnitude as well as the direction of the electric field will vary during a cycle, whereas in the case of circular polarization, the magnitude of the electric field vector remains constant.

In the case of elliptical polarization, the tip of the electric field vector would appear like a helix with an elliptical cross-section in space if we were to freeze it at one instant in time. In addition, if we were to look at the electric field vector at one point in space,

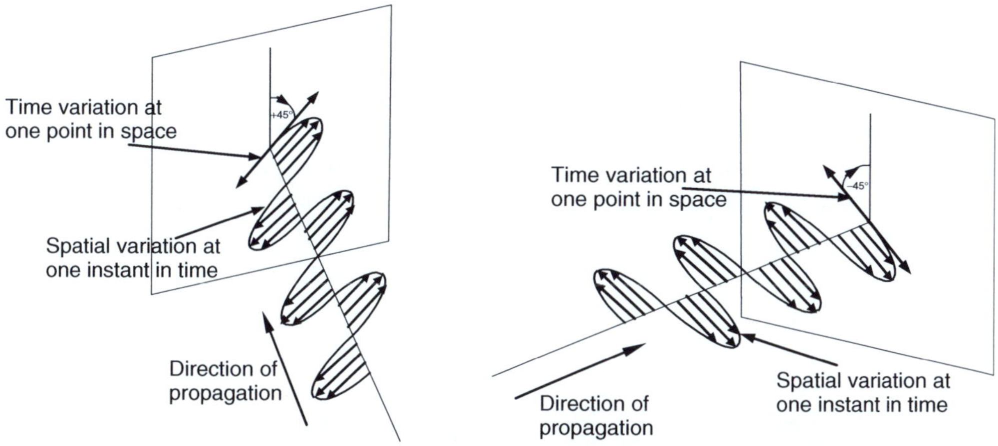

(a) Right slant polarized wave (b) Left slant polarized wave

Figure 1.20 Linearly polarized slant waves.

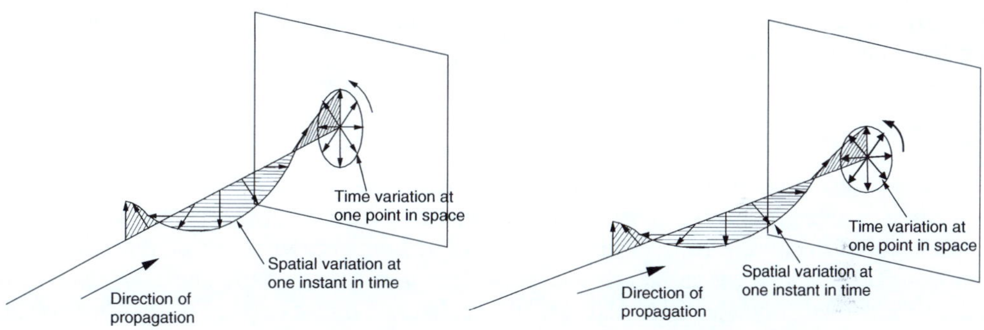

(a) Left hand elliptical polarization (b) Left hand circular polarization

Figure 1.21 Elliptical and circular polarization.

it would appear like an elliptical disc over one period or cycle in time, as shown in the square of Figure 1.21a.

In the case of circular polarization, although the magnitude of the electric field vector remains constant, its direction changes as the wave propagates in space so that its tip appears like a helix at one instant in time; and at one point in space, a circular disc is swept out, as shown Figure 1.21b. The direction in which the vector rotates determines the hand of polarization. Of course, the direction has to be defined in relation to the viewpoint, since the same vector would appear as either clockwise or anticlockwise, depending on the direction from which it is viewed.

The Institute of Electrical and Electronics Engineers (IEEE) defines right hand circular polarization (RHCP) as a clockwise rotation of the electric field vector when looking along the direction of propagation from the transmit antenna, and left hand circular polarization

(LHCP) as an anticlockwise rotation. Thus in Figure 1.21 we have left hand elliptical and circular polarization, since the electric field vectors are rotating anticlockwise in both cases.

In the EM field the ellipticity e of circular polarization is defined as the ratio of the minor to major perpendicular components of the electric fields E_{minor} and E_{major}, and is a measure of the polarization purity or the cross-polar discrimination. E_{minor} and E_{major} are the semi-major and semi-minor axes of the ellipse as shown in Figure 1.22. The ellipticity is given by

$$\rho_e = \frac{E_{\text{minor}}}{E_{\text{major}}}. \tag{1.4}$$

The ellipticity in dB, $\rho_{\text{e(dB)}}$, is given by

$$\rho_e \text{ (dB)} = 20 \log \left(\frac{E_{\text{minor}}}{E_{\text{major}}} \right) \tag{1.5}$$

For circular polarization the ellipticity is one, that is, the two electric field components are equal. In dB the ellipticity would be 0 dB for circular polarization. If, in the case of Figure 1.22a, we take the vertical component as E_{major} and the horizontal component as E_{minor}, then when E_{minor} is zero, the ellipticity is zero and we have linear vertical polarization. Similarly, when E_{major} is zero in Figure 1.22b, the ellipticity is infinite and we have linear horizontal polarization.

1.4.2.1 Tilt Angle

There is a difference between the tilt angle and tilt axis. The tilt angle is defined as the angle of the major axis, whereas the tilt axis is the angle of the polarization ellipse as defined in Section 1.4.2.2. The plane of polarization is the plane of the ellipse (or circle), that is, the time variation at a point in space, and the tilt angle is defined as the angle α that the major axis makes with a reference direction such as the horizontal. The angle α is measured looking from the direction of propagation and is measured clockwise from a reference axis (such as the horizontal) to the major axis of the ellipse as shown in Figure 1.22c. However, sometimes the angle is measured anticlockwise from the horizontal, giving an acute angle α' as shown in Figure 1.22c.

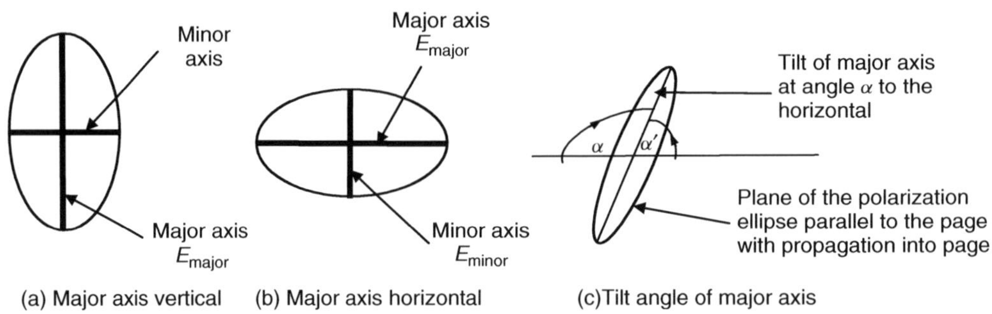

Figure 1.22 The electric fields for elliptical polarization.

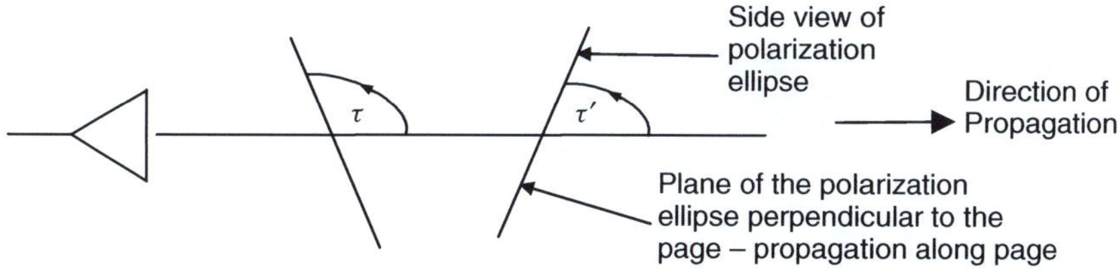

Figure 1.23 Tilt axis for elliptical/circular polarization.

1.4.2.2 Tilt Axis

The tilt axis τ is defined as the angle that the plane of polarization makes with the direction of propagation. This angle is measured from the axis and direction of propagation to the plane of the polarization ellipse. In Figure 1.23 the tilt axes are shown for two different cases, where the tilt axis is an obtuse angle τ and an acute angle τ'.

1.4.3 Axial Ratio

The axial ratio is sometimes defined by the reciprocal of the ellipticity ρ_e. However, the more commonly used definition of the axial ratio χ_{r} is given by

$$\chi_{\mathrm{r}} = \frac{1+\rho_e}{1-\rho_e}, \tag{1.6}$$

where ρ_e is the ellipticity, and since

$$\rho_e = \frac{E_{\mathrm{minor}}}{E_{\mathrm{major}}},$$

we get

$$\chi = \frac{1+E_{\mathrm{minor}}/E_{\mathrm{major}}}{1-E_{\mathrm{minor}}/E_{\mathrm{major}}}. \tag{1.7}$$

The axial ratio χ_{dB} in dB is given by

$$\chi_{\mathrm{dB}} = 20\log\left(\frac{1+E_{\mathrm{minor}}/E_{\mathrm{major}}}{1-E_{\mathrm{minor}}/E_{\mathrm{major}}}\right). \tag{1.8}$$

or

$$\chi_{\mathrm{dB}} = 20\log\left(\frac{1+\rho_e}{1-\rho_e}\right) \tag{1.9}$$

The ellipticity in terms of the axial ratio can be derived since

$$\chi(1-\rho_e) = 1+\rho_e$$

$$\chi - \chi\rho_e = 1+\rho_e$$

$$\rho_e = \frac{\chi-1}{\chi+1}. \tag{1.10}$$

Similarly, the ratio of the perpendicular components of the electric fields can be calculated if the axial ratio χ_{dB} is known, by using the formula

$$\rho_e = \frac{E_{minor}}{E_{major}} = \frac{10^{\chi_{dB}/20} - 1}{10^{\chi_{dB}/20} + 1}. \tag{1.11}$$

1.4.4 Measurement of Polarization Purity

An antenna designed to receive one type of linear polarization (or one hand of polarization) will in the ideal case not receive a signal with polarization that is orthogonal or of the opposite hand. However, a practical antenna will receive a signal that is orthogonal or of the opposite hand at a level equal to its cross-polar level. Typical cross-polar levels are −14 dB, although levels as low as −40 dB are possible. Section 9.13 (in Chapter 9) shows the level received by antennas of different polarizations.

The ability of an antenna to reject the cross-polar radiation is known as cross-polar discrimination (XPD). The ellipticity or polarization purity, which is a measure of XPD, can be measured when undertaking the measurement of the radiation patterns. The antenna under test (AUT) is illuminated by a rotating antenna, such as a horn, so that the polarization angle is continuously varied over 360°. This is known as the spinning technique. The typical radiation pattern is shown in Figure 1.24, and the axial ratio χ_r in dB is the distance between a trough and peak.

For instance, if the axial ratio on the radiation pattern is 0.2 dB, the linear value of χ_r is $10^{0.2/20} = 1.0233$. Then

$$\chi - \chi\frac{E_{minor}}{E_{major}} = 1 + \frac{E_{minor}}{E_{major}}$$

$$\chi - 1 = \chi\frac{E_{minor}}{E_{major}} + \frac{E_{minor}}{E_{major}}$$

$$\chi - 1 = \frac{E_{minor}}{E_{major}}(\chi + 1)$$

$$\frac{E_{minor}}{E_{major}} = \frac{\chi - 1}{\chi + 1} = \frac{1.0233 - 1}{1.0233 + 1} = 0.0114.$$

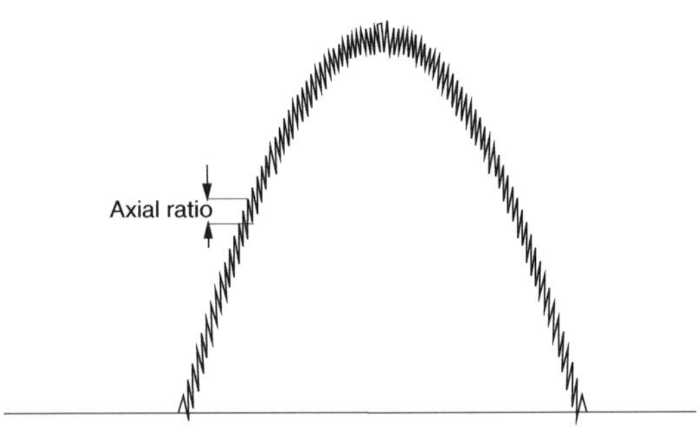

Figure 1.24 A typical radiation pattern obtained for investigation of the polarization purity of an antenna.

Polarization purity or XPD in dBs is $20\log(0.0114) = -38.8\,\text{dB}$. This means that the difference in reception between the copolar and cross-polar radiation is 38.8 dB. This is a very good level. Most blade antennas have XPDs of around -15 dBs.

Note, however, that the sense of polarization, tilt angle and tilt axis cannot be determined from this measurement.

1.5 Characteristics of an Antenna

A transmitting antenna transfers a guided EM wave from a transmission line into free space. In the case of an aircraft the antennas provide the interface between the systems inside the airframe and the outside world. Note that the plural of antenna (used in engineering) is *antennas* and not antennae.

It is important to know the exact spatial distribution of power (i.e. the radiation pattern in 3D space) provided by antenna in order to deduce the performance of systems connected to the antenna and to enable suitable measures to be taken for any deficiency in a particular direction. For instance, if an aircraft has to communicate with the ground it is important for a reasonable amount of power to be radiated/deflected towards the ground. On the other hand, if the prime aim of an antenna is satellite communications or navigation then the radiated power and/or the boresight of the transmit/receive antenna should be oriented towards the satellite. Antenna manufacturers provide the radiation pattern of the antenna on a standard ground plane. However, when an antenna is installed on a structure the spatial power dispersion is very different from that obtained when the antenna is on a standard ground plane.

An antenna can have the following characteristics, not all of which are meaningful to all antenna types:

1. radiation pattern
2. directivity, gain and efficiency
3. beamwidth and gain of the main lobe
4. position and magnitude of the lobes
5. bandwidth
6. polarization of electric field that it transmits or receives
7. handling power.

There are two principal planes in which the antenna characteristics are measured. These are known as the azimuth plane and elevation plane, and can be considered as the horizontal and vertical planes respectively for land-based antennas. The angles in the azimuth plane are conventionally denoted by the Greek letter phi (ϕ), and in the elevation plane they are denoted by the Greek letter theta (θ).

On an aircraft the azimuth plane is known as the yaw plane, and there are two elevation planes. The elevation plane that is transverse to the aircraft (i.e. wing to wing) is known as the roll plane, and the plane that is longitudinal to the aircraft (i.e. nose to tail) is known as the pitch plane, as shown in Figure 1.25.

In the yaw plane the elevation angle θ is zero and the azimuth angle ϕ varies from 0 to 180°.

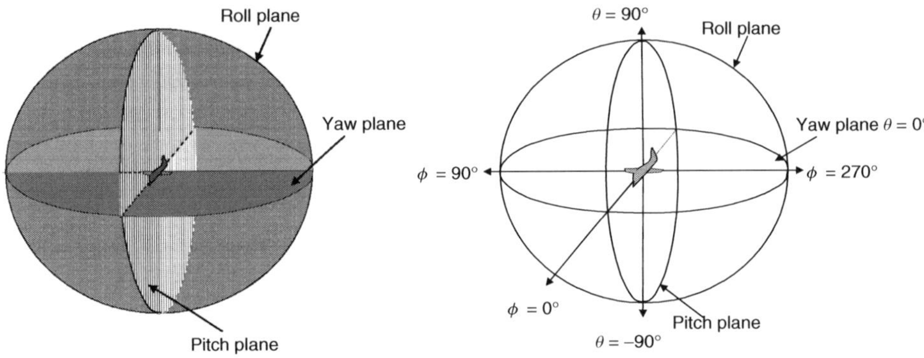

Figure 1.25 Radiation pattern cuts used for aircraft antennas.

In the roll plane the azimuth angle ϕ is 90° (or 270°) and the elevation angle θ varies over 360°, with angles between −90 and +90°.

In the pitch plane, the azimuth angle ϕ is 0 and the elevation angle θ varies over 360°, with angles between −90 and +90°.

For aircraft antennas, the angles are plotted as navigational angles also known as bearings and not as convention mathematical angles. Bearings go clockwise from 0° at the top/north, through to 90°, 180°, 270° and back to 0/360°, as shown in Figure 1.26a. Mathematical angles on the other hand start at 0° on the right and go anticlockwise to 90° at the top, then 180° to the left and then 270° at the bottom and back to 0/360° as shown in Figure 1.26b.

1.5.1 Radiation Patterns

The line radiation pattern can be plotted using either rectangular/Cartesian or polar coordinates. Rectangular plots can be read more accurately (since the angular scale can be

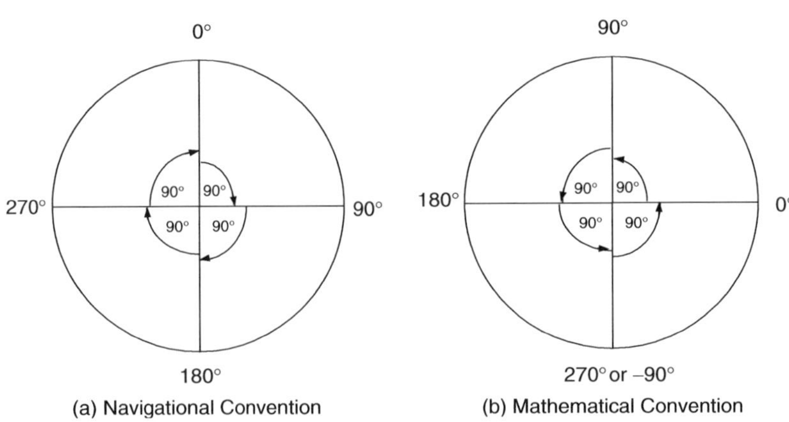

Figure 1.26 Navigational and mathematical convention for angles.

expanded), but polar plots give a more pictorial representation and are thus easier to visualize; rather like an analogue clock, or the plan position indicator (PPI) used in many radio assisted detection and ranging (radar) sets. Contour plots are another way of depicting the radiation from antennas.

1.5.1.1 Radiation Patterns of Different Types of Antennas

Antennas can be isotropic, omnidirectional or directional, depending on the directions in which they radiate.

An isotropic antenna radiates uniformly in all directions so that the radiated power at any point on a sphere (with the antenna at its centre) has the same magnitude as depicted in Figure 1.27. The darker colours indicate the higher powers found nearer the antenna. However, this cannot be realized in practice, and would require the antenna to be a point source. The nearest approximation to an isotropic antenna is a Hertzian dipole, which is a dipole that is very small in terms of wavelength.

Omnidirectional antennas such as monopoles, dipoles and biconicals radiate uniformly in one plane. Figure 1.28a shows the radiation from an ideal vertical dipole. In the vertical plane the cross-section of the radiation is in the shape of figure of eight on its side as shown in Figure 1.28b, and radiation is uniform in the horizontal plane as shown in Figure 1.28c.

A directional antenna is one that radiates most of its power in one particular direction. Examples of directional antennas are horns, reflector systems, log-periodics and Yagis. Figure 1.29 shows the radiation from a reflector antenna. For a circularly symmetrical reflector the uninstalled radiation pattern is also circularly symmetrical.

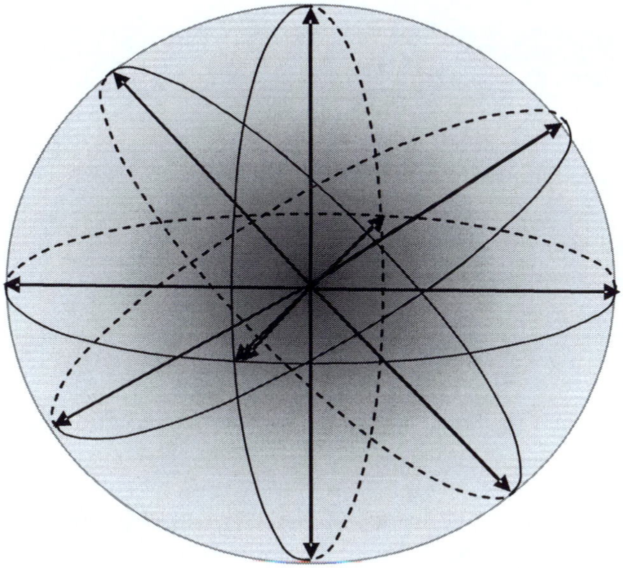

Figure 1.27 The radiation obtained from an isotropic antenna.

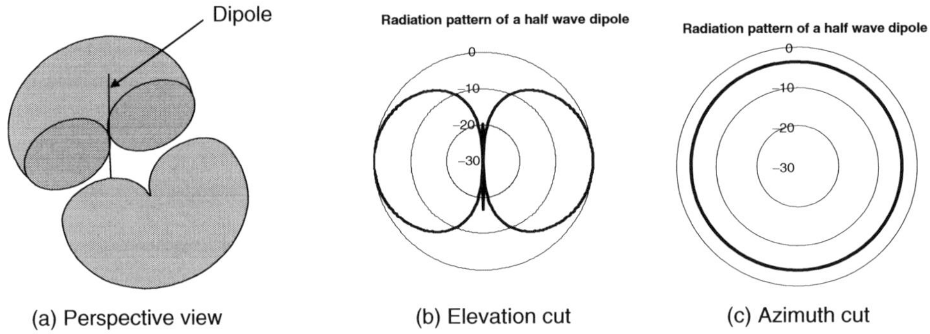

Figure 1.28 The radiation obtained from an idealized omnidirectional antenna.

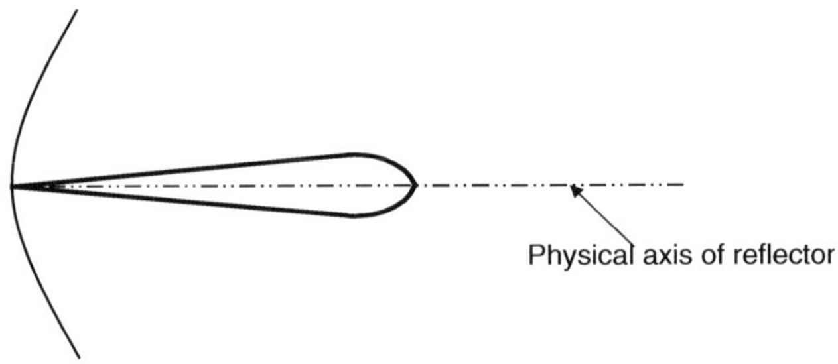

Figure 1.29 The radiation obtained from a directional antenna.

1.5.1.2 Representations of Radiation Patterns

The radiation pattern of an antenna can be presented as a 2D line graph or as a 3D representation, but shown in 2D. The line radiation patterns show the power in a specified plane, whereas the 3D representation could be shown as a perspective view, or as a contour plot projected onto a cylinder like Mercator's projection of a map of the world.

Line Radiation Patterns

The line radiation pattern could be taken at different cuts, but the most common ones are great circle and conical cuts.

Great Circle Cuts

If we imagine a sphere with the antenna at its centre, circles cut through the centre of the sphere will have the same diameter as the sphere and are known as great circle cuts as shown in Figure 1.30. If the great circle is horizontal (goes through the equator) then in the case of an aircraft antenna at the centre of the sphere, that would be known as a yaw plane cut, as shown in Figure 1.30. Similarly, if the great circle is vertical (goes through the north and south poles) then, in the case of aircraft antennas, that would be known as a pitch or roll plane cut, depending on the orientation of the aircraft – see Figure 1.25.

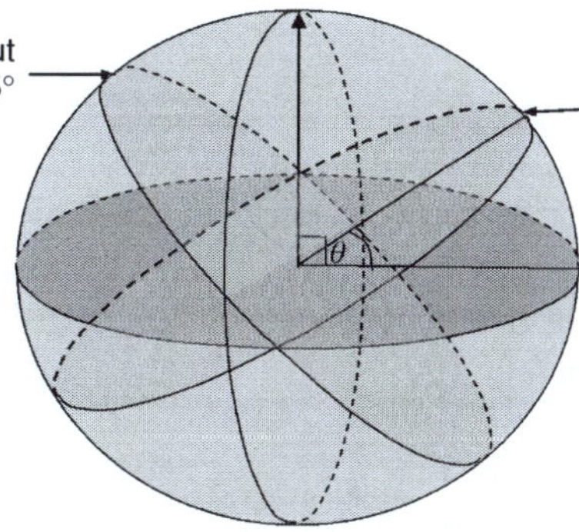

Figure 1.30 Great circle cuts.

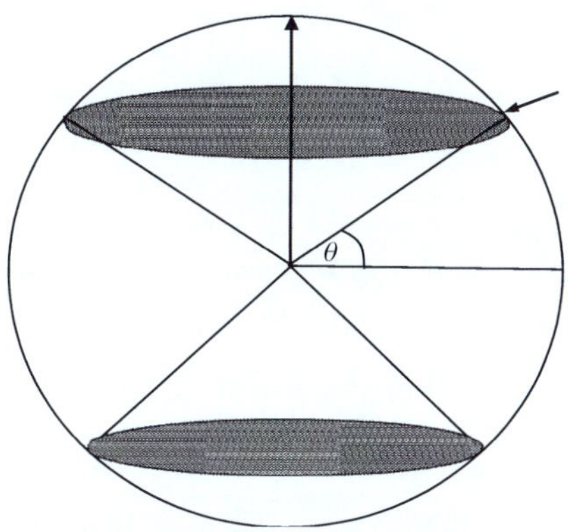

Figure 1.31 Conical cuts.

Conical Cuts

If we imagine cuts parallel to the azimuth plane of the antenna (the equator in the case of a globe), we would get circles of smaller diameter as we move away from the equator. The conical cuts are equivalent to the lines of latitude on the globe and the angle of elevation θ is constant for each cut as shown in Figure 1.31. Note that the cone semi-angle is in fact $90 - \theta$.

Rectangular/Cartesian Plots

Cartesian plots are named after the mathematician and philosopher, René Descartes. They are standard xy plots where the axes are plotted at right angles to each other. In a radiation plot the angle with respect to boresight (as defined in Section 1.5.3) is varied and the magnitude of the radiated power is measured. The x axis is used for the angle and the power radiated is plotted on the y axis. The x coordinate is known as the abscissa and the y coordinate is known as the ordinate. It is important to remember that the power radiated is measured in the far field. A typical rectangular plot of an antenna radiation pattern is shown in Figure 1.32. All values, whether negative or positive, are often shown without a sign. Thus the y axis in Figure 1.32, should be shown from 0 to −80 or −40 dB,

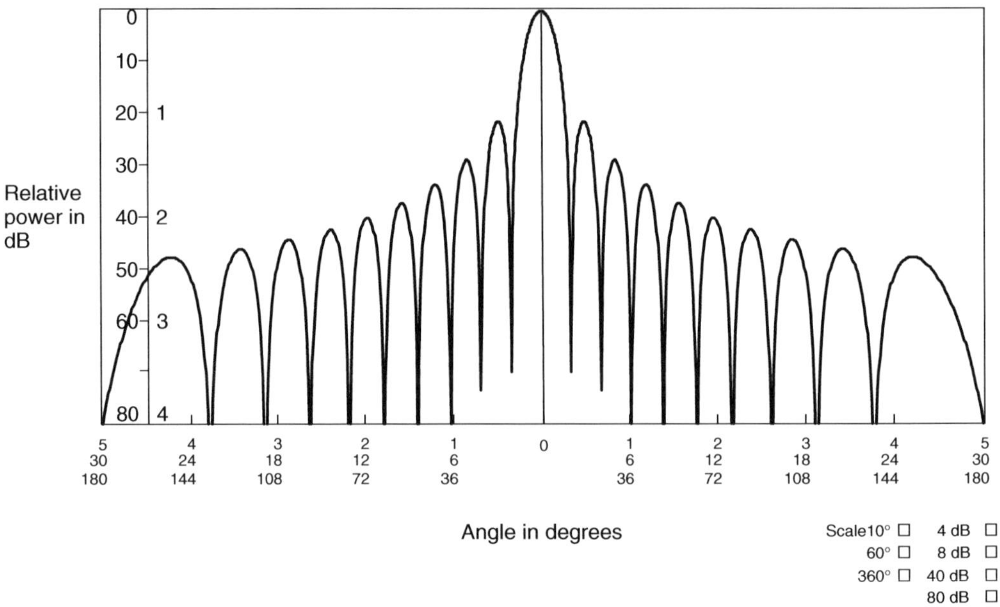

Figure 1.32 Rectangular plot of an antenna radiation pattern.

indicating that the maximum value is at boresight or 0°. The radiation patterns show the angles measured clockwise and anticlockwise from the boresight position, and in standard mathematical convention the x axis would be denoted by positive and negative signs, but on radiation patterns the signs of the angles are often omitted. The boxes on the lower right would be ticked/crossed to indicate the scale used for angles and relative power.

Polar Plots

In a polar plot the angles are plotted radially from boresight and the intensity or power is plotted along the radius, as shown in Figure 1.33. This gives a pictorial representation of the radiation pattern of the antenna and is easier to visualize than the rectangular plots. However, since the scale of the angular positions cannot be increased (i.e. they can only be plotted to scale from 0 to 360°), the accuracy cannot be increased as in the case of rectangular plots. The level of the intensity or power, however, can be varied as in the case of rectangular plots. On polar plots each circle represents a particular level, where the power has the same magnitude at all angles and is shown relative to the power at boresight. Since the power is in general a maximum value at boresight, this level is often shown as 0 dB and the levels elsewhere will always be less than the power at boresight. Thus these should be shown as negative values. However, they are usually written without a sign, and should be assumed to be negative, contrary to standard arithmetic convention. Figure 1.33 shows levels from 0 to -25 dB in 5 dB steps. Some polar plots may have a level of $+3$ or $+10$ dB as the maximum, go through the 0 dB, and then down to -30 dB. In almost all cases the signs are omitted.

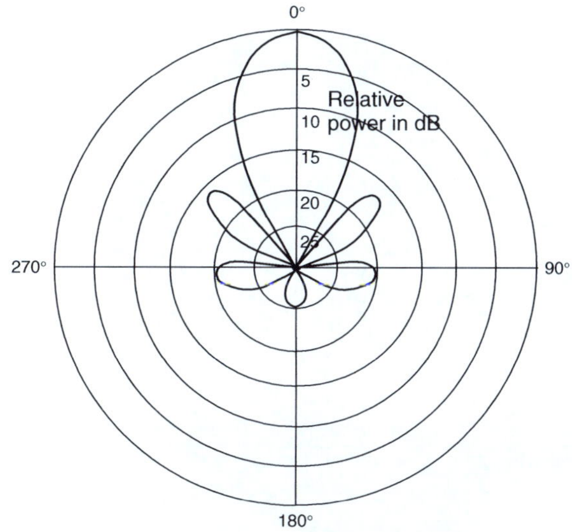

Figure 1.33 An idealized polar plot of a directional antenna.

Contour Plots

Contour plots can be plotted:

1. on a sphere and appear like a global map on a 2D surface
2. like Mercator's projection of a spherical surface onto a cylinder, like global maps
3. as line contour plots.

The contour plots are colour-coded, so that each colour represents a particular level of gain, and the levels go up in discrete steps. Thus, for instance, if the discrete steps are 2 dB, all levels between 2 and 4 dB are the same colour and the levels do not change continuously as in the case of line radiation patterns. Thus the gain at a particular angle cannot be read off accurately. However, they provide a good visual indication of the total spherical radiation pattern, and angular sectors of any 'dropouts' or nulls can be seen. These are particularly useful to systems engineers who have to evaluate the spatial coverage of avionics systems.

One version of Mercator's plots is the projection of the 3D spherical surface onto a cylinder which is then shown as a 2D rectangular surface, as if the cylinder has been opened out into a rectangle. These are the type of plots that we see in an atlas. They require some skill is deciphering the actual angles since, for instance, the whole of the top line represents just one angle of 90° elevation, as in some world maps where the north pole is represented by a line instead of a single point. A line through the centre/equator corresponds to the azimuth line radiation plot when plotted in polar or rectangular coordinates. The contour plot of Figure 1.34 is plotted as a Mercator's projection in Figure 1.35.

Instead of using solid colours for each discrete interval of power level, the contour plot can be presented as a contour line plot, where only the positions with the same power

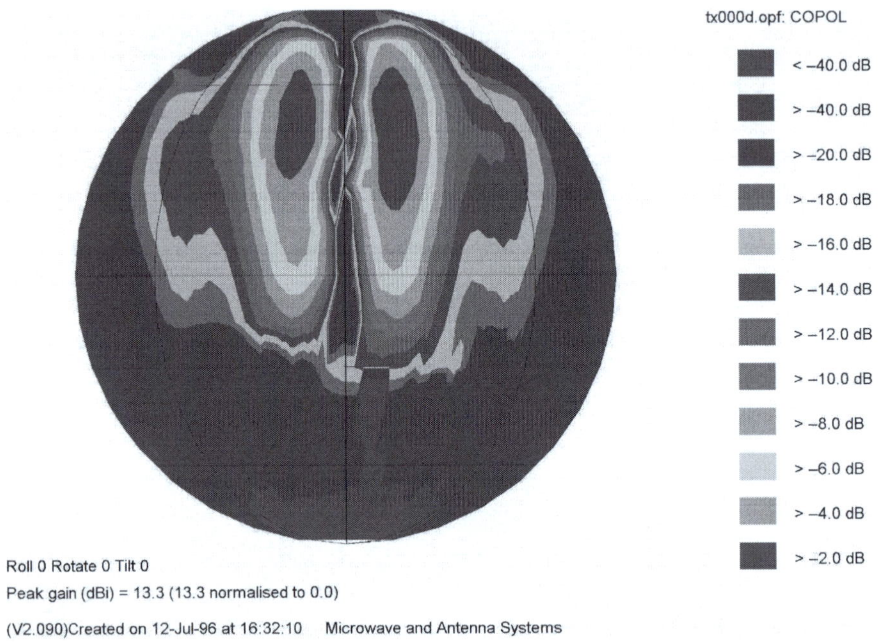

Figure 1.34 Contour plots on a spherical surface. Reproduced by kind permission of ASL [5]. See Plate 3 for the colour figure.

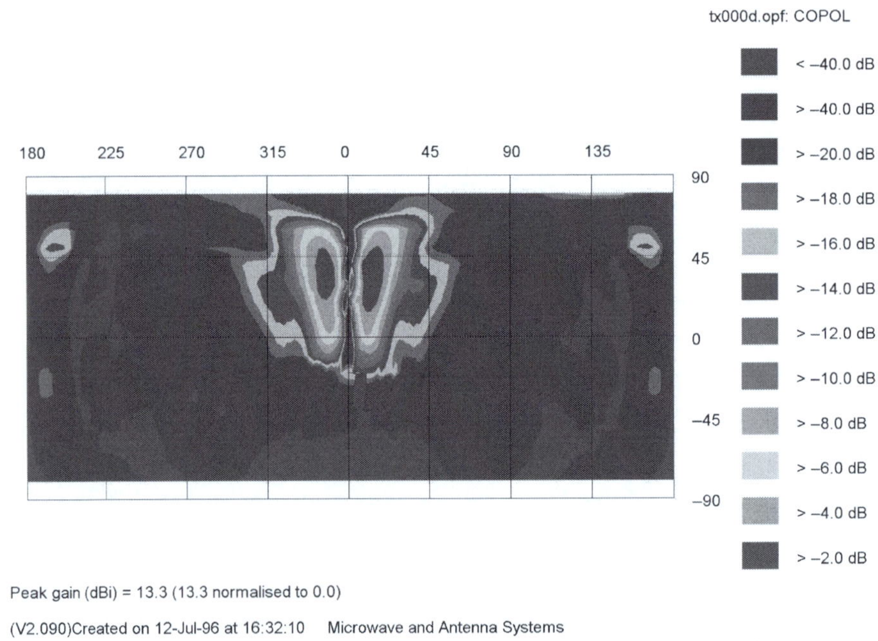

Figure 1.35 Mercator's projection of a contour plot. Reproduced by kind permission of ASL [5]. See Plate 4 for the colour figure.

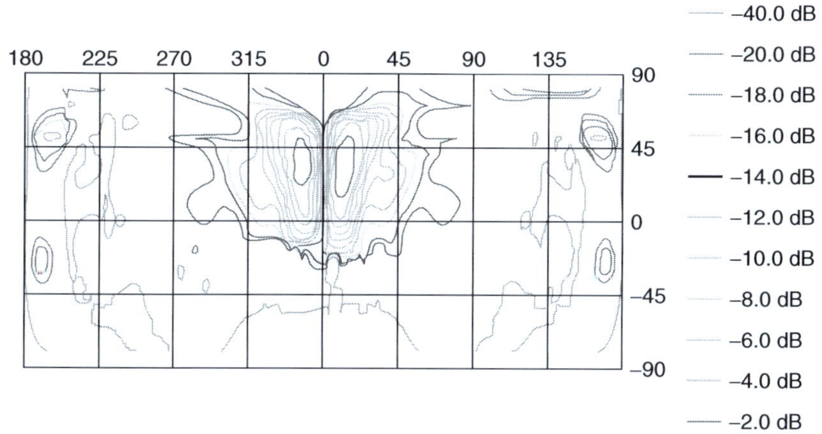

Figure 1.36 Mercator's projection of the radiation pattern, showing lines of equal gain/power levels. Reproduced by kind permission of ASL [5]. See Plate 5 for the colour figure.

level are plotted as lines (sometimes in different colours) each representing different power levels – see Figure 1.36. Thus each line represents a single different power level and we would not be able to correlate this plot with a contour plot where a number of power levels have the same colour. This looks similar to the contour plots on an Ordnance survey map or a weather map with isobars showing high and low pressures of the atmosphere.

1.5.2 Directivity, Gain and Efficiency

In the case of antennas the gain is not the same as the gain of an amplifier. We can think of the gain of an antenna as its ability to focus the total power in a particular direction or angular sector. An isotropic antenna is analogous to a candle with its light radiating in all directions, whereas a directional antenna is analogous to a torch with its light focused in one particular direction.

One definition of the gain of an antenna relates the power radiated by the antenna to that radiated by an isotropic antenna (which radiates equally in all directions) and is quoted as a linear ratio or in decibels isotropic (dBi). When we say that the gain of an antenna is, for instance, 20 dBi (100 in linear terms), we mean that an isotropic antenna would have to radiate 100 times more power to give the same intensity at the same distance as that of the directional antenna.

Directivity, or directive gain, is the power that would be radiated by an antenna if there were no losses. A tuned antenna such as a half-wave dipole has a resistance of $\sim$73 Ω. The RF cable feeding the antenna usually has a characteristic impedance (see Section 1.6.2.1) of 50 Ω. Thus there is a mismatch which results in some power being reflected, that is, not all the power is transmitted. Note that domestic RF cables usually have a characteristic impedance of 75 Ω and therefore match the impedance of dipole aerials more closely.

The efficiency η of an antenna is the ratio of the power delivered at the terminals of the antenna to the radiated power, expressed as a percentage:

$$\eta = \frac{G}{D} \times 100, \tag{1.12}$$

where

D is the directivity in linear terms
G is the gain in linear terms.

The efficiencies of antennas vary between about 50% and 100%.

If the gain and directivity are in dB then the efficiency would also be in dB and is not usually expressed as a percentage in this case:

$$\eta_{\text{dB}} = G_{\text{dB}} - D_{\text{dB}}. \tag{1.13}$$

Efficiencies of 50% and 100% are 0.5 and 1.0 and, since $10\log(0.5)$ is -3 and $10\log(1)$ is zero, the values in dBs would be -3 and 0 dB, respectively.

1.5.3 Electrical and Mechanical Boresight

For directional antennas that have a main lobe the position of the peak radiation is the electrical boresight of the antenna. In the case of an aperture antenna such as a horn or reflector, ideally the electrical boresight would be expected to be along the physical main axis through the centre of the aperture and coincident with the mechanical boresight. However, for the real antenna, as shown in Figure 1.37, this is not the case since the main lobe tends to be slightly squinted and at a small angle to the mechanical boresight. The radiation pattern is often positioned so that its electrical boresight is coincident with the zero angular position of the graph.

1.5.4 Beamwidth and Gain of the Main Lobe

The beamwidth is inversely proportional to the gain of a directional antenna. Thus the narrower the beamwidth the higher the gain on boresight.

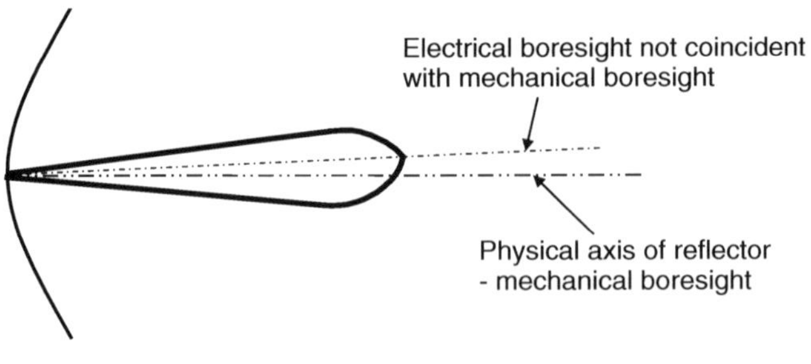

Figure 1.37 Difference between electrical and mechanical boresight.

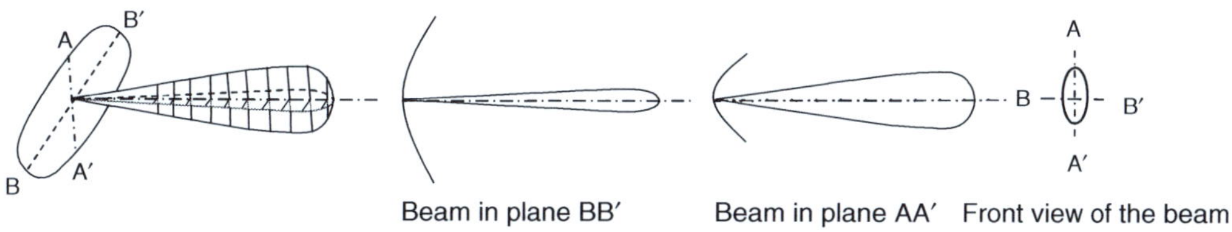

Figure 1.38 Beamwidths of an elliptical reflector antenna.

1.5.4.1 Beamwidth

When we refer to the beamwidth of an antenna we are referring only to the width of the main lobe/beam of the antenna and not the sidelobes. In general, the larger the antenna the smaller is its beamwidth for the same frequency, so that the beamwidth of an antenna is inversely proportional to its physical size. If the antenna does not have the same dimensions in all planes, the plane containing the largest dimension will have the narrowest beamwidth. In the case of a elliptical reflector, as shown in Figure 1.38, the beamwidth in the horizontal or azimuth plane BB′ is narrower than that in the vertical or elevation plane, AA′, since the major axis of the ellipse is horizontal.

The beamwidth of an antenna can be defined in three ways:

1. half power beamwidth
2. 10 dB beamwidth
3. first null beamwidth.

When the beamwidth is quoted, it is usually assumed that it is the half power or 3 dB beamwidth, that is, the width in degrees (or sometimes in radians) of the main beam across which the gain drops to 3 dB below that of its level at boresight.

In the case of highly directional antennas that have a narrow main lobe, the 10 dB beamwidth is often used, that is, the width at the points on either side of main beam where the radiated power is one-tenth of the maximum value. Large reflector antennas may have gains as high as 60 dBi, that is, linear gains 1 million times greater than that of an isotropic antenna.

For wider beamwidth antennas, the beamwidth is quoted to the first nulls, that is, the width across which the main beam drops to the first nulls. These are depicted in Figure 1.39, where the angle between the 3 dB points K and L, 10 dB points M and N, and first nulls are shown as 25°, 44° and 60°, respectively.

A tuned half wave dipole has a beamwidth of about 78°, whereas a Hertzian dipole has a beamwidth of about 90°.

1.5.4.2 Gain of Main Lobe

The radiation pattern of an antenna shows the power on boresight as 0 dB and the power in other directions as negative values. The gain in all directions is plotted relative to the gain on boresight. Thus we cannot find out the absolute power radiated in a particular direction or at a particular distance, unless we know the absolute gain of the antenna, as

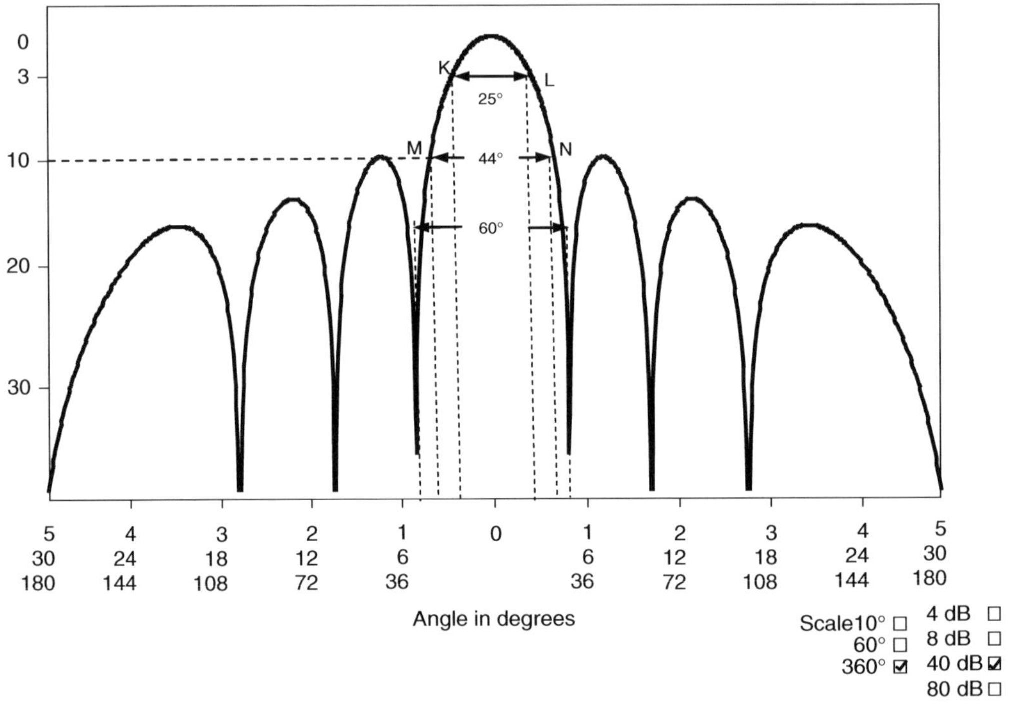

Figure 1.39 Definition of 3 dB and 10 dB beamwidths and the width to the first nulls of an antenna.

well as the power actually radiated by the antenna. In order to find the absolute gain in any particular direction the absolute gain on boresight must be known. If this gain is in decibels then this value can just be added to the gain at any point to give the absolute gain. The absolute gain on boresight is measured using a standard gain horn or other standard antenna. Often the plot is normalized. When a plot is normalized the peak value is set to the maximum (usually zero) shown on the radiation pattern and all other values are adjusted accordingly. Thus, for instance, if the absolute peak gain is 2.3 dBi this value is shown as 0 dB and the level of the near-in sidelobe (see Section 1.5.5) which is actually −14 dBi, would be shown as −16.3 dB (−14−2.3) on the plot.

1.5.5 Position and Magnitude of the Lobes

The lobes are the sidelobes as well as the backlobe that is 180° away from the boresight. When sidelobes are referred to, it is often just the near-in sidelobes, that is, the two lobes on either side of the main lobe marked A in Figure 1.40. It is very important to know where sidelobes are, as well as their magnitudes. In the case of a radar antenna, for instance, if a return is received on a sidelobe instead of on the main lobe, the return would be a lower level and thus it could be mistaken for a target that is further away than its true range.

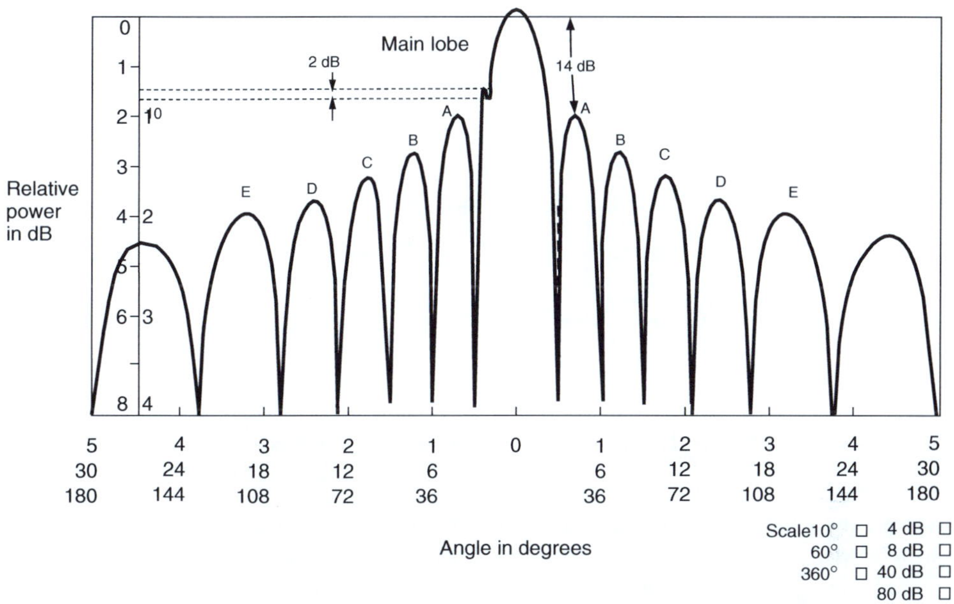

Figure 1.40 The sidelobes (near-in) and definition of sidelobe.

The sidelobe level is the level of the near-in sidelobe below the peak of the main lobe. For instance, in Figure 1.40 the near-in sidelobe marked A is 14 dB below the peak. Thus it is −14 dB compared with the peak. However, it is usually just stated that 'the sidelobe level is 14 dB' instead of stating that the near-in sidelobe is 14 dB below the peak.

Measured patterns of directional antennas are often not smooth or symmetrical and may have several dips and peaks. Thus it may be difficult to distinguish between a sidelobe and an irregularity of the radiation pattern. In some cases a sidelobe is defined as the case where the difference between a peak and a null is greater than 2 dB. For instance, in Figure 1.40 there is an irregularity shown where the difference is 2 dB, and therefore this would not be considered a sidelobe.

1.5.6 Bandwidth

When we refer to the bandwidth, we are usually referring to the frequency bandwidth. The IEEE defines the frequency band limits as exclusive of the lower limit and inclusive of the upper limit. Thus a frequency band of 3–30 MHz refers to frequencies above 3 and up to 30 MHz. An antenna has its maximum gain at its tuned frequency, but can still receive other frequencies, albeit with reduced gain. The response of an ideal tuned antenna over its frequency band is similar to that of the current in an inductance-capacitance-resistance (LCR) circuit, as shown in Figure 1.41. In the case of an antenna (as in LCR circuits) the resonant frequency is the point at which the impedance is purely resistive. The bandwidth is commonly defined as the frequency band over which the linear gain of the antenna is at least half of its gain at its resonant frequency. Expressed in decibels, the bandwidth would

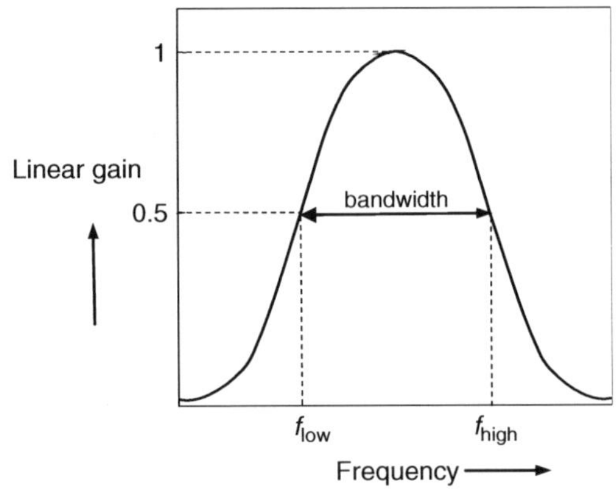

Figure 1.41 Frequency response of an ideal antenna.

be the frequency band over which the gain of the antenna is at least $-3\,\text{dB}$ compared with its resonant frequency gain. Thus this definition is similar to that of the beamwidth of an antenna, although we are dealing with frequency instead of angles. Note that since the gain has the dimensions of power, when the power is halved the electric field is $1/\sqrt{2}$ (0.707) times that of the electric field at resonance.

A wide bandwidth is usually achieved at the expense of gain. Thus if the antenna had the same size but a narrower bandwidth, its gain would be higher.

The bandwidth can be expressed:

1. by quoting the actual frequencies
2. as a percentage or as a fraction
3. as a multiple of an octave.

1.5.6.1 Bandwidth by Quoting the Absolute Frequencies

The absolute bandwidth is specified by quoting the upper and lower frequencies, or by quoting the midband frequency plus or minus ($\pm$) half the frequency band. Thus if the bandwidth of the antenna is from 300 to 1000 MHz, these frequencies are quoted, or the bandwidth is expressed as the centre frequency (arithmetic mean of the upper and lower frequencies) and the difference between the centre and band edge frequencies, that is, $650 \pm 350\,\text{MHz}$.

1.5.6.2 Bandwidth as a Percentage or Fraction

When it is expressed as a percentage bandwidth B_w, its centre frequency f should be quoted, and the percentage expressed relative to this centre frequency:

$$B_w = \left(\frac{\Delta f}{f}\right) \times 100, \tag{1.14}$$

where Δf is the difference between the upper and lower frequencies.

The bandwidth can be expressed as a percentage of its total frequency span or ± half the percentage of its frequency span, but we have to quote its centre frequency. Thus, for instance, if we have an antenna that is quoted as having a bandwidth from 100 to 300 MHz, its centre frequency f would be 200 MHz and its absolute bandwidth or frequency span is 200 MHz. Using Equation 1.14, B_w would be 100%, so we could either say that the antenna has a centre frequency of 200 MHz and its bandwidth is 100% or its bandwidth is ±50%.

Similarly, if an antenna has an operating frequency from 20 to 40 MHz, its centre frequency is 30 MHz and its bandwidth is 66.7% (20/30) or ±33.3%.

1.5.6.3 Bandwidth as a Fraction or Multiple of an Octave

When it is expressed in octaves, its lower and upper frequencies should be quoted. An octave is a band of frequencies between one frequency and another frequency that is double or half the first frequency. Thus, for instance, we have an octave between 100 and 200 MHz, and two octaves between 100 and 400 MHz. In this case we would not use the centre frequency since we would be unable to quote the bandwidth in terms of octaves.

The bandwidth as a fraction or multiple of an octave is defined as

$$B_w = \log_2\left(\frac{f_{high}}{f_{low}}\right), \tag{1.15}$$

where f_{high} is the highest frequency and f_{low} is the lowest frequency. Thus, for instance, if we have an antenna that is quoted to have a bandwidth from 100 to 300 MHz, its centre frequency f would be 200 MHz and its absolute bandwidth is 200 MHz. The ratio of the highest to lowest frequencies is 3, and $\log_2 3$ is 1.58.

Many calculators do not have logarithms to the base 2. In this case we can use logarithms to the base 10 of the ratio and divide the result by logarithms to the base 10 of 2. Thus $\log_{10} 3$ is 0.477 and if this is divided by $\log_{10} 2$ (=0.301) we get 1.58.

1.5.7 Polarization

The antenna has to be oriented so that it can receive the maximum electric field. When this occurs for a monopole, for instance, the antenna is said to be receiving the copolar field. If the monopole is then turned through 90° (at the same position) so that it receives the minimum electric field, it is said to be receiving the cross-polar field. In the case of a monopole antenna, when the antenna is vertical it would receive the maximum electric field of a vertically polarized incident wave and therefore this would be the copolar radiation. A horizontally polarized wave would be the cross-polar radiation. However, in the case of other linearly polarized antennas such as aperture antennas, the copolar orientation cannot always be determined by visual inspection and prior knowledge may be required.

All polarizations could be considered to be variations of elliptical polarization. In the case of elliptical polarization, the magnitude as well as the direction of the electric field will vary during a cycle.

Details of the different types of polarizations are given in Section 1.4.

1.5.8 Power Handling

The power that an antenna can handle depends mainly on the structure of the antenna, its feed or feeding network, and its frequency. Printed circuit board antennas, such as patches, cannot handle powers much above 5 W and are prone to delamination on aircraft at powers approaching this magnitude. In general, the higher the frequency the less power it can handle in the case of printed circuit antennas. This is because the substrate is thinner (in order to meet the characteristic impedance requirements) in the case of the higher frequency antennas, and hence the breakdown voltage is lower.

1.6 Propagation

EM waves can be guided along transmission lines or travel through free space. These waves, composed of electric and magnetic fields, change periodically in time and have clearly defined configurations that satisfy the boundary conditions of Maxwell's equations. The boundary conditions are those that would occur at the interface when a wave is travelling from one medium to a second medium. For instance, the electric field would reduce to zero at a perfect conductor. A qualitative explanation of the boundary conditions is given in Chapter 3 of [3].

In free space the electric and magnetic fields are at right angles to each other and to the direction of propagation of the wave. This type of wave is known as a transverse electromagnetic (TEM) wave, and it can also be propagated as a guided wave in any two-conductor transmission line, such as parallel wires, coaxial lines and striplines. However, a TEM wave cannot be supported by waveguides. The TEM wave is the most common mode and hence it is the main one discussed in this book.

1.6.1 Power Flux Density

In the case of the TEM wave, the electric and magnetic fields are perpendicular to each other and the plane containing them is perpendicular to the direction of propagation of the EM energy. The vector cross-product of the electric and magnetic fields is also a vector which is equal to the power flux density (power per square area) known as Poynting's vector P_d. For a simple explanation of the vector cross-product see Chapter 2 of [3].

We can see that if we multiply V/m by A/m, we would get VA/m^2, which is W/m^2. The relationship between the three vectors shown in Figure 1.42b is defined by the equation

$$P_d = E \times H, \qquad (1.16)$$

where

P_d is in W/m^2

E is the electric field vector in V/m

$\times$ represents the cross-product and

H is the magnetic field vector in A/m.

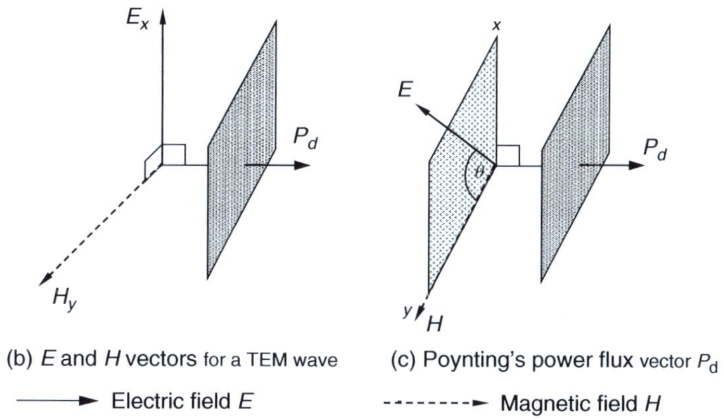

Figure 1.42 Wave propagation and Poynting's power flux density vector.

The magnitude of P_d is given by

$$|\mathrm{P}_d| = |E||H|\sin\theta, \tag{1.17}$$

where

$|E|$ is the magnitude of E
$|H|$ is the magnitude of H and
θ is the angle between the electric and magnetic field vectors.

Note that the magnitudes of the electric and magnetic fields are those in the plane perpendicular (i.e. the transverse plane) to the direction of propagation or power flow. The maximum values of these fields occur when the E and H vectors are in this transverse plane; any component parallel to the direction of propagation will not contribute to the power flux density. We can see from Equation 1.17 that algebraically the maximum value of P_d occurs when the sine of the angle between the electric and magnetic fields is 1, that is, angle θ is 90° or the fields are perpendicular to each other in the plane transverse to the direction of the propagation of the wave. Thus the TEM wave (which has these fields perpendicular to each other) has the maximum value of Poynting's vector. The power density through a surface S, as shown in Figure 1.42, has a maximum value when Poynting's vector is perpendicular to it.

1.6.1.1 Wave Impedance

The EM wave can be considered to have an impedance, depending on its configuration. If the wave is incident on a surface with the same impedance the surface can be said to be matched, in the same way as a load is matched to a transmission line. When this occurs, no energy will be reflected and the wave is totally absorbed/transmitted. This impedance is known as the intrinsic or characteristic impedance Z_w of the wave and in the case of a TEM wave in a medium it is given by

$$Z_w = \frac{E_x}{H_y} = \sqrt{\frac{\mu}{\varepsilon}\left(\frac{1}{1+\sigma/j\omega\varepsilon}\right)}, \tag{1.18}$$

where

Z_w is in ohms
E_x is the electric field along the x axis in V/m
H_y is the magnetic field along the y axis in A/m
μ is the permeability of the medium in H/m
ε is the permittivity of the medium in F/m and
σ is the conductivity of the medium in S-m.

If the TEM wave is in free space, $\mu = \mu_0$ and $\varepsilon = \varepsilon_0$ and the conductivity is zero. Thus the characteristic impedance Z_0 of TEM wave in free space is given by

$$Z_0 = \sqrt{\frac{\mu_0}{\varepsilon_0}} = 120\pi, \tag{1.19}$$

since $\mu_0 = 4\pi \times 10^{-7}$ H/m and $\varepsilon_0 = (36\pi \times 10^9)^{-1}$ F/m. The characteristic wave impedance is often quoted at its approximate value of 377 Ω.

1.6.2 Guided Waves

Transmission lines that can support TEM waves require two separate conductors. The electric (solid lines) and magnetic (dotted lines) fields for circular coaxial transmission lines are shown in Figure 1.43. Note that when the magnetic lines are evenly spaced, they will also represent the lines of equipotential (with the electric field lines perpendicular to them), and the closer the spacing of the electric field lines, the greater is the magnitude of the electric field. The electric field variation with distance in the direction of propagation (longitudinal direction) is sinusoidal. We should also note that there is a sinusoidal

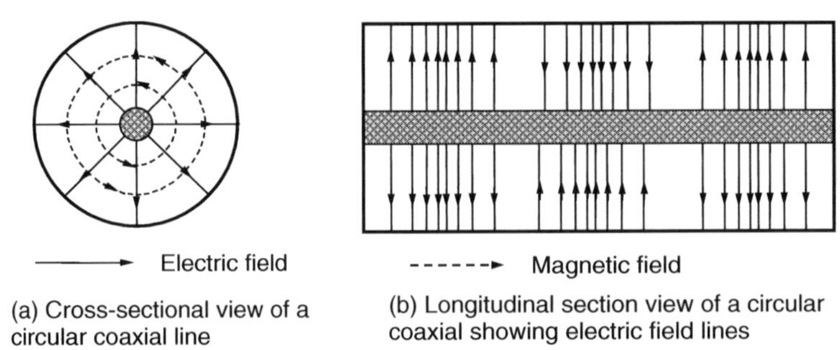

Figure 1.43 The electric and magnetic fields in a coaxial transmission line.

variation with time, which effectively moves this electric field pattern forward by half a sine variation every half period in time.

The coaxial line is one of the most efficient ways of containing the EM energy. However, the spacing between the inner and outer conductors in a coaxial line is maintained by dielectric (in the form of beads or a continuous hollow cylinder). As the frequency increases, the losses in the dielectric also increase and the energy is increasing attenuated. Another source of energy loss is the outer conductor, which is often of braided form. Energy can leak through holes in the braid, and the amount of the leakage is proportional to the electrical size of the hole, that is, the size of the hole as a fraction of the wavelength. Energy can escape through gaps that are as small as 0.01 times the wavelength. As the frequency increases the wavelength decreases and the gaps in the outer conductor become a larger fraction of this wavelength, causing the EM energy loss to increase.

1.6.2.1 Characteristic Impedance of Transmission Lines

When a line is greater than one-tenth of a wavelength at its highest operating frequency, we cannot ignore the properties of the line. The current distribution in it (due to the electromotive force at the input end) is not uniform. The line behaves as though it has resistive and reactive components distributed along its length. Each section of the 'go' and 'return' cables forms an unit loop, which can be represented by a shunt capacitance C and conductance G between the two cables; and a series resistance R and inductance L. If the line is a pair of parallel identical conductors, it is balanced and the resistance R and inductance L are equally divided between each cable, as shown in Figure 1.44. Normally an incident wave, travelling from left to right towards the termination, is partly reflected. The reflected and incident waves combine to form a standing wave. If the termination is 'matched' (made equal) to the impedance of the unit loop of the line, there is no reflected wave, and hence no standing wave.

The impedance of this unit loop is known as the characteristic impedance Z_0 and is given by

$$Z_0 = \sqrt{\frac{R + j\omega L}{G + j\omega C}}, \tag{1.20}$$

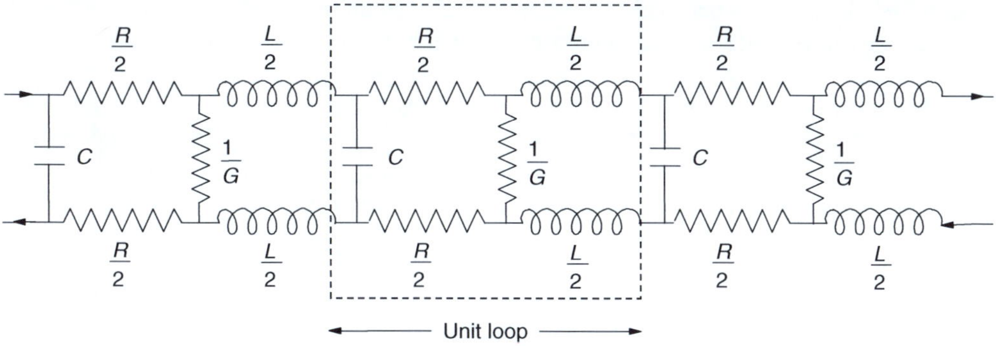

Figure 1.44 Equivalent circuit of a balanced transmission line.

where

R is the series resistance in Ω
G is the shunt resistance in S
L is the series inductance in H and
C is the shunt capacitance in F.

If the line is lossless the resistive parts are zero, so that $R = 0$ and $G = 0$.

Equation 1.20 for the characteristic impedance reduces to

$$Z_0 = \sqrt{\frac{L}{C}}. \tag{1.21}$$

This is the general formula for a lossless transmission line, and the magnitudes of L and C will depend on the shape and size of the particular transmission line and the dielectric material.

1.6.2.2 Matching Transmission Lines

When two transmission lines are connected together, they should have the same characteristic impedance. For domestic use the characteristic impedance of coaxial cable (used to connect a television to the antenna, for instance) is 75 Ω, but for professional/industrial use the characteristic impedance is 50 Ω. If the connected cables do not have the same characteristic impedance there is a mismatch and some of the power is reflected back towards the incident power. Thus in the case of an aircraft system, if a transmitter is connected using cables of the incorrect characteristic impedance the power reaching the antenna will be greatly reduced. Similarly, if the antenna is connected to a cable of the incorrect characteristic impedance, some of the radiated power reaching the antenna from outside the aircraft will be reflected back, thus reducing the sensitivity of the entire system.

When an antenna is electrically very small, that is, small in terms of wavelengths, its impedance changes with frequency, and at some frequencies it has a negative reactance (capacitance) and at other frequencies it has a positive reactance (inductance). In these cases there is a matching unit, called a tuner, to prevent a large mismatch at the antenna terminals (where it is connected to the RF cable). This tuner provides a conjugate match as well as an impedance transformation between the antenna and the RF cable.

The complex conjugate of a number has the same real part as the number but its imaginary part has sign opposite to that of the number. For instance, if we have a complex number $S = a + jb$, its complex conjugate is $S^* = a - jb$. If the antenna has an impedance of, say, $R + jX$ at a particular frequency, the tuner is adjusted to have an impedance of $R - jX$, where R is the resistance and X is the reactance. A positive reactance is obtained from an inductor and a negative reactance is obtained from a capacitor.

The same problem would occur if we have a load which has a susceptance (1/reactance) of $+jB$ in parallel with its conductance of $G_l = 0.2$ S, and we want to connect it to a line whose characteristic impedance is 50 Ω (i.e. a characteristic admittance of 0.2 S). Since the load does not have the same impedance as the transmission line, some of the transmitted wave would be reflected, and, together with the incident wave, this would result in a standing wave. This would result in a loss of power transfer, and in some cases could lead to voltage breakdown in high power systems. In order to match the load to the line, we would connect a stub, which is a short-circuited line, in parallel with the transmission line, as near to the load as possible. The conductance of the load can be varied by adjusting the position of the stub so that the conductance presented to the line equals 0.2 S. The susceptance of the stub can be varied by adjusting its length so that its susceptance is equal and opposite to that of the load, that is, $-j$B. Thus the resultant susceptance is zero, and the conductance is 0.2 S. The load now presents an impedance that is the same as that of the transmission line so that it is matched, and maximum power transfer can be effected.

1.6.2.3 Relationship between Power Density and Electric Field Strength

In the far field the electric and magnetic fields are perpendicular to each other and the plane containing them is normal to the direction of propagation. The following relationship between the power density and the power radiated by the antenna only applies to the far field. If the formula is used in the near field, we would get an infinite magnitude for the power density when the distance d is zero.

The power density P_d at a distance d in the far field is given by

$$P_d = \frac{P_t G_t}{4\pi d^2} \tag{1.22}$$

where

P_d is the power density in W m^{-2}
P_t is the power transmitted by the antenna in W
G_t is the linear or numeric gain and
d is the distance in m.

For omnidirectional antennas the gain G_t is constant in the azimuth plane, but for directive antennas the gain is a function of the angle relative to boresight. The magnitude of the power density is also equal to the real part of the cross-product of the E and H field vectors,

$$P_d = \mathrm{Re}(E \times H), \tag{1.23}$$

where

E is the electric field intensity in V/m
H is the magnetic field intensity in A/m and
P_d is the power density in W m^{-2}.

The intrinsic wave impedance ξ of a plane wave is the ratio of the transverse E and H fields. Substituting E/ξ for H gives

$$P_d = \frac{E^2}{\xi} \tag{1.24}$$

for the power density. Combining Equations 1.22 and 1.24, we get the following expression for the electric field:

$$E = \sqrt{\frac{P_t G_t \xi}{4\pi d^2}}. \tag{1.25}$$

For a plane wave the intrinsic wave impedance is equal to 120π, and thus Equation 1.25 can be rewritten as

$$E = \sqrt{\frac{30 P_t G_t}{d^2}}. \tag{1.26}$$

This expression gives us the electric field at a distance d from the transmit antenna. In the case of a directional antenna the gain will be a function of the angle relative to its boresight.

1.6.3 Free Space Waves

The propagation of the EM wave depends on the frequency, the polarization, the medium it traverses and, in the case of free space waves, whether the path is over land or sea.

The energy from a transmitting antenna reaches the receiving antenna by traversing several paths and by the mechanisms of reflection, refraction and diffraction ([5], p. 608). Waves reaching the receiving antenna can be one or all of the following types:

1. ionospheric or sky waves that are reflected or scattered by the ionosphere, which is 100–300 km above the earth's surface.
2. tropospheric waves reflected or scattered by the troposphere, which is 10 km above the earth's surface
3. ground waves.

These paths are shown in Figure 1.45.

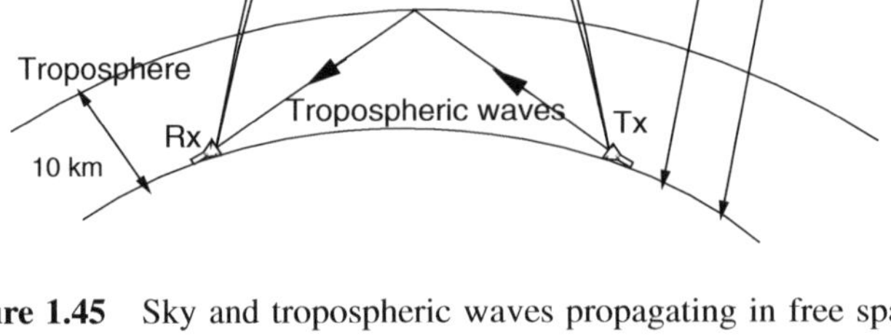

Figure 1.45 Sky and tropospheric waves propagating in free space.

1.6.3.1 Sky Waves

Sky waves are refracted by the same phenomenon of total internal reflection that occurs with light waves travelling from one medium to another with lower refractive index, as described in Section 1.2.2.1.

Sky wave propagation is used for high frequency communications, but above 30 MHz the refraction is insufficient for this type of propagation to be used. Horizontal polarization is used in the frequency range 1.5–30 MHz for 'sky wave' propagation. Sky waves are reflected by the ionosphere, which is 100–300 km above the earth's surface, and the round trip back to earth is 200–600 km.

In Figure 1.45 we can see that the receiver is not in line of sight (LOS) of the transmitter, so normally the signal received would be a result of diffraction alone and hence would be very low. However, because the waves are reflected off the upper layers of the atmosphere, the signals can effectively be 'seen' from a greater distance than would be the case for LOS. This increased distance is commonly called the radar horizon.

The additional path travelled by the wave can be considered as a direct path to the horizon of an earth with a modified radius of ka, where a is the true radius of the earth, and k is given by

$$k = \frac{1}{1 + a\dfrac{dN}{dh}10^{-6}}, \tag{1.27}$$

where

a is the true radius of the earth in km and
dN/dh is the rate of change of the refractive index with height in km^{-1}.

For a dN/dh value of $-40\,\text{km}^{-1}$ and taking the earth's radius as 6370 km, the value of k is 1.34 or approximately 4/3. Tropospheric propagation permits transmission beyond LOS between transmit and receive antennas. However, it entails the use of elaborate and expensive equipment, because the transmission efficiency is poor (see [6], p. 4). The extra loss is caused by the longer paths taken by the EM waves.

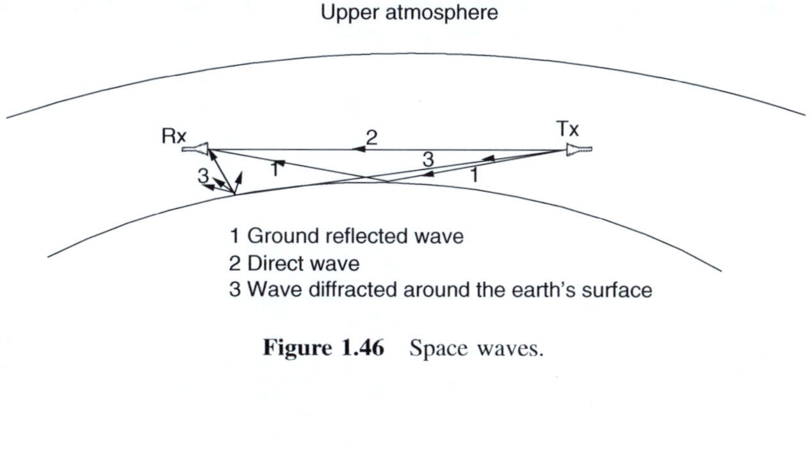

Figure 1.46 Space waves.

1.6.3.2 Ground Waves

Ground waves are made up of space and surface waves. Space waves predominate at high distances above the earth's surface, whereas surface waves are more important when an antenna is near the earth's surface.

1.6.3.3 Space Waves

A space wave is made up of three waves:

1. the ground reflected wave, by specular or diffuse reflection from the ground
2. the direct path between the transmit and receive antennas
3. the wave diffracted around the earth's surface.

Vertical polarization is used below 1.5 MHz for ground wave propagation. Ground wave propagation is also used for horizontally polarized waves, with the electric field perpendicular to the plane of incidence and parallel to the surface. The reflected wave is nearly 180° out of phase with the incident wave.

As the angle of incidence is increased the magnitude and phase of the reflection factor will change but not to any large extent. The change is greater for high frequencies and for lower ground conductivities.

When the angle of incidence is near to grazing (i.e. around 90°) the magnitude of the reflected wave is nearly equal to that of the incident wave for all frequencies and for all ground conductivities ([5], p. 612).

References

[1] Tooth, A.R. (2004) IPAS first year activity report, Deliverable D2C, IPAS/PAR/BAES/ART 041101, 1 November. Installed Performance of Antennas on AeroStructures (IPAS), an EU Specific Targeted Research Project, November 2003 to January 2007, Contract No AST3-CT-2003-503611.

[2] Thain, A. (2006) Computation and analysis of existing and advanced tools, Deliverable D17, IPAS/RP/EADS/AT 061219, 19 December. Installed Performance of Antennas on AeroStructures (IPAS), an EU Specific Targeted Research Project, November 2003 to January 2007, Contract No AST3-CT-2003-503611.

[3] Macnamara, T.M. (1995) *Handbook of Antennas for EMC*, Artech House, Boston, ISBN 0-89006-549-7.

[4] ASL (Antenna Software Ltd) P.R. Foster.

[5] Jordan, E.C. (1953) *Electromagnetic Waves and Radiating Systems*, Constable and Company Ltd.

[6] Powell, J. (1985) *Aircraft Radio Systems*, Pitman, London.

2

Aircraft Systems Using Antennas

2.1 Aircraft Systems

The systems described in this chapter are only those connected to antennas. The antennas provide the interface between the aircraft and the outside world. This is not an extensive description of all the systems used on airborne platforms and is intended as an overview of the most common systems using antennas.

In general, military aircraft have more systems than civil aircraft; however, both types of aircraft require the same basic types of systems for take-off, navigation and landing. Most large commercial and military aircraft duplicate essential systems as contingency measures. This is known as redundancy.

Navigation systems used by man on land have not been crucial where distances and speeds have been small and reliance could be placed on recognizable features and landmarks. At sea, where these landmarks are non-existent, the position on board a shipping vessel in terms of latitude could be determined relatively easily by measuring the sun's altitude using a sextant. However, the longitude was more difficult. The Greenwich Observatory was set up in 1675 by King Charles II, and in 1884 an international conference in Washington agreed that it should be considered to be at 0° longitude.

The system requirements with respect to the positioning of antennas relative to their system line replaceable units (LRUs) as well as with respect to the airframe are discussed in this chapter.

Some systems, such as VHF omnidirectional ranging (VOR) en-route navigation, are to be phased out and other systems, such as global positioning system (GPS), will be augmented by Galileo (the new European global positioning system) so that both systems will be used as a multi-constellation system.

Some systems such as instrument landing system (ILS) and microwave landing system (MLS) require ground stations, others such as GPS and SatCom require satellite infrastructure, and others such as Radio/radar Altimeter (RadAlt), weather and search radar are standalone systems and do not require any support from or interface with, other aircraft or ground/airborne/satellite stations.

Introduction to Antenna Placement & Installation Thereza M. Macnamara

2.2 Frequencies of the Most Common Aircraft Systems

Table 2.1 shows the most common avionics systems currently used on commercial aircraft. Not all these systems are used on all aircraft. For instance, GPS and the Global Navigational Satellite System (GLONASS) are not generally installed on the same aircraft. GLONASS is not fully operational and new satellites are to be launched.

2.3 Automatic Direction Finding

The automatic direction finding (ADF) system is used to find the heading of the aircraft relative to a ground station. The ground station provides signals between 190 and 1799 kHz with 0.5 kHz spacing. Radio transmitters used for air navigation are known as beacons. One type of transmitter that could be used is the non-directional beacon (NDB). Any terrestrial transmitter (e.g. broadcast) that transmits amplitude modulated (AM) signals can be used. However, because the location and identity of radio broadcast stations are not controlled like aviation NDBs, they are less reliable. Commercial AM radio stations broadcast on 540–1620 kHz. In the case of the NDB, the ADF receiver is tuned to the correct frequency, and the identity of the beacon is verified by listening to the Morse code signal transmitted by the NDB. The NDB frequency and identification information may be obtained from aeronautical charts and the airport facility directory (AFD). International Civil Aviation Organisation (ICAO) Annex 10 specifies the NDB frequency band as 190–1750 kHz, but in North America the frequency band is 190–535 kHz and the NDBs are categorized by their output power. NDBs with an output power of less than 50 W are rated as low power, 50–2000 W are rated as medium power, and above 2000 W are rated as high power. NDBs are identified by one-, two- or three-letter Morse code call signs which could include numbers in the case of some Canadian NDBs.

The position of the NDB can be verified by what is known as station passage. There is a null directly above a station. As an aircraft approaches an NDB station, the level of the signal increases so that the needle of the display oscillates erratically. When the aircraft passes directly over the station the needle swings from the 'ahead' to the 'behind' position. This provides an accurate fix. Since the null has a fixed angular width, the distance required to fly over and obtain a fix increases with altitude.

The antennas for this system consist of a loop and a sense antenna. Most antennas are now integrated into a single LRU. These antennas should be placed near the physical centre of the fuselage and are usually placed on top of the fuselage.

Some ADF receivers are capable of receiving the international maritime distress frequency of 2182 kHz, making them suitable for search and rescue and other offshore operations.

Operation of the on-board system is based on the received signal being detected by two perpendicular loops combined with the signal received by an omnidirectional sense antenna.

The two most common types of systems used are as follows:

1. A rotating directional loop antenna combined with an omnidirectional sense antenna to eliminate ambiguities in the loop antenna pattern. The sense antenna adds an additional 90° phase shift. The radiation pattern of the loop antenna alone is a figure of eight

Table 2.1 Frequencies of the most common aircraft systems.

System	Frequency range (MHz)		Wavelength (m)			
	Lowest	Highest	Longest	Shortest	Polarization	
ADF	0.19	1.8	1578.9	166.7	Vert	Automatic Direction Finding
Comms – Military	30	88	10	3.41	Vert	–
DF	30	407	10	0.74	Vert	Direction Finding
Distress – Civil	121.5	–	2.47	–	Vert	Also called 'Guard'/SAR
Distress – Civil	243	–	1.23	–	Vert	Also called 'Guard'/SAR
Distress – Maritime	2.182	–	137.5	–	Vert	Also called 'Guard'/SAR
Distress – Satellite	406.028	–	0.74	–	Vert	Also called 'Guard'/SAR
DME	960	1215	0.31	0.25	Vert	Distance Measuring System
ECM	500	40 000	0.60	0.01	All	Electronic Counter Measures
EL	121	243	2.48	1.23	Vert	Emergency Locator
ESM	500	40 000	0.60	0.01	All	Electronic Support Measures
GPS L1	1565	1586	0.19	0.19	RHCP	Global Positioning System
GPS L2	1217	1238	0.25	0.24	RHCP	Global Positioning System
GLONASS (Russian GPS)	1602	1615.5	0.19	0.19	RHCP	Global Navigation Satellite System
GLONASS (Russian GPS)	1240	1256	0.24	0.24	RHCP	Global Navigation Satellite System
HF	2	30	150	10	Horiz/vert	High Frequency
ILS – Glideslope	329	335	0.91	0.90	Horiz	Instrument Landing System
ILS – Localizer	108	112	2.78	2.68	Horiz	Instrument Landing System
ILS – Marker	74.75	75.25	4.01	3.99	Horiz	Instrument Landing System
In-flight telephony	824	895	0.36	0.34	Vert	–
MLS	5031	5091	0.06	0.06	Vert	Microwave Landing System
RadAlt	4200	4400	0.07	0.07	Horiz	Radio Altimeter

(continued overleaf)

Table 2.1 *(continued)*

System	Frequency range (MHz)		Wavelength (m)			
	Lowest	Highest	Longest	Shortest	Polarization	
Radar S band (search)	2000	4000	0.150	0.075	Vert	–
Radar X band (search)	8200	12 000	0.037	0.025	Vert/horiz	–
Radio Broadcast FM	88	108	3.41	2.78	Vert	–
SatCom Civilian downlink	1530	1559	0.20	0.19	All	–
SatCom Civilian uplink	1626.5	1660.5	0.18	0.18	All	–
SatCom	4000	6000	0.075	0.05		–
SatCom Ku band downlink	11 700	12 700	0.03	0.02	Vert	–
SatCom Ku band uplink	14 000	14 500	0.0214	0.0207	Vert	–
SSR Tx 1030 Rx 1090	1030	1090	0.29	0.28	Vert	Secondary Surveillance Radar
TACAN	960	1215	0.31	0.25	Vert	–
TCAS Tx 1030 Rx 1090	1030	1090	0.29	0.28	Vert	Traffic Collision Avoidance System
Telemetry	1900	2500	0.16	0.12	Vert	–
Telemetry ARIA	4160	4250	0.07	0.07	Vert	Advanced Range Instrumentation Aircraft
Telemetry Police	4900	6400	0.06	0.05	Vert	–
UHF	225	400	1.33	0.75	Vert	Ultra High Frequency
UMTS	1800	1900	0.17	0.16	Vert	Universal Mobile Telecommunications System
VHF Civil	108	137	2.78	2.19	Vert	Very High Frequency
VHF Maritime	150	174	2.00	1.72	Vert	–
VHF Military	108	152	2.78	1.97	Vert	–
VOR en-route navigation	112	118	2.68	2.54	Horiz	VHF Omnidirectional Range
VOR terminal	108	112	2.78	2.68	Horiz	Part of Instrument Landing System
Weather Radar C	5500		0.0545		Horiz	–
Weather Radar X	9330	9354	0.0322	0.0321	Horiz/vert	–

(with two nulls). Vectorial combination of the radiation patterns of the loop and sense antennas results in a cardioid pattern with only one null.

2. Fixed, crossed loops, with a motor-driven goniometer. The goniometer can detect the direction of maximum radiation. The antenna consists of two fixed loops that are placed at right angles to each other and their electrical outputs combined in a goniometer. The goniometer has two sets of fixed windings at right angles to each other, each set connected to one loop. Within the magnetic field of these windings is a rotor, with a winding connected to the receiver.

2.4 Distress/SOS

Aircraft usually have dedicated channels to detect distress signals, although they are often part of the VHF/UHF radios. The antennas are usually shared with the V/UHF system antennas as well, although in some cases dedicated whip antennas are used. These dedicated antennas are sometimes called ‘guard’ antennas.

The three most commonly used distress frequencies are 121.5, 243 and 406.028 MHz. For military applications 243 MHz is usually used, but military aircraft usually have systems that can detect all three frequencies. The maritime distress frequency for boats is 156.8 MHz.

An analogue 121.5 MHz distress beacon is a small electronic device that emits a radio signal to help rescue authorities locate its position in a life-threatening emergency. The signal is picked up by polar-orbiting satellites and aircraft monitoring the 121.5 MHz frequency.

When detected by satellite the position of the distress beacon can be narrowed down to a 20 km^2 region. Search and rescue (SAR) aircraft can home in on the signals of the beacon, leading them directly to the target.

The 406 MHz distress beacon emits both an analogue and a digital signal. The digital signal carries a code which identifies the beacon while the analogue signal is to enable aircraft to home in on the location. The analogue signal from the beacon is transmitted continuously, but the output power is only about 100 mW (+20 dBm). The digital message to the COSPAS-SARSAT satellites is a 300 ms pulse that is only transmitted every 50 seconds and its power is 5–7 W (+37 to +38.5 dBm). The COSPAS-SARSAT system consists of a constellation of four polar-orbiting satellites and 27 local user terminals (LUTs) linked to a mission control centre (MCC). The satellites store the data and transmit it to the LUTs, where it is processed and retransmitted to the MCC.

The digital code received can be cross-referenced with a database of registered 406 MHz beacon owners which identifies who is in trouble and what type of craft they are in. This enables the search and rescue authorities to tailor a response to the emergency situation.

A 406 MHz beacon narrows the position of the beacon down to 5 km^2. This can be reduced to just 120 m if the beacon includes a GPS. Detection of 406 MHz beacons can be instantaneous if a geostationary satellite is over the beacon.

2.5 Distance Measuring Equipment

In modern ILS installations distance measuring equipment (DME) is installed, co-located with the ILS, to augment or replace marker beacons. The DME continuously displays the aircraft's distance from the ground station. Paired pulses (~3.5 μs pulse width (PW)) at a specific spacing are sent out from the aircraft to interrogate the ground station. The ground station then transmits paired pulses back to the aircraft at the same pulse spacing but at a different frequency. The time taken for the round trip of this signal exchange is measured by the on-board DME transceiver and provides the direct distance (slant range) between the aircraft and the ground station.

Interrogation rates vary between 10 and 30 interrogations per seconds, with the newer systems requiring fewer pulses per second. Transmission and reception frequency spacing is 63 MHz, and 126 channels are available for transmit and receive. For reliability of operation, there is a 50 μs systematic delay between reception of the interrogation pulses from the aircraft and the ground station's response. There may be possible interference to the GPS L5 frequency of 1176.45 MHz, resulting in the removal of DME ground station frequencies around the L5 frequency band.

Since the system tracks for ground beacons ahead, this antenna is usually installed in the lower forward fuselage. Thus there could be obscuration or distortion of the radiation pattern when the landing gear is deployed. For this reason the radiation pattern of the antenna should be modelled and/or measured with the landing gear extended. Usually, two antennas are installed for redundancy. The airborne antenna may be a monopole or a vertical slot that behaves like a monopole. Since the wavelength at 1 GHz is 30 cm, a quarter-wave monopole would only be 7.5 cm high. Thus the airborne antenna is a tuned monopole enclosed in a 'blade' and is sometimes called a shark's fin because of its appearance. This antenna could also be integrated as a separate radiating element into a VHF or a combined V/UHF blade antenna.

2.6 Electronic Counter Measures

Electronic counter measures (ECM) are used to negate or confuse the electronic threat to the aircraft. This threat could be an electromagnetic or a physical threat such as a missile. The electromagnetic threat would most probably emanate from an enemy ground or airborne radar, whereas the physical threat would possibly be a missile fired at the aircraft.

The RF and microwave segment of active counter measures involves radiating a fixed frequency or a band of frequencies. The fixed frequency would be the particular frequency emitted by the threat such as a radar. The emitted signal would have the same characteristics of the incident radar signal. In its simplest form, it would thus provide a false return to the signal transmitted by the radar, and being of a higher level, would thus trick the enemy radar into thinking that the aircraft was nearer than it actually was.

The active counter measures could also include an off-board system used as a defensive aid, such as a towed decoy. A towed decoy is one that is deployed to about 200 m behind the aircraft to transmit the EM radiation in order to deliver a false indication of the aircraft's real position, to the threat.

Passive counter measures such as chaff and flare could also be used. In the case of chaff the aircraft would release metal foil pieces that would act as parasitic radiators and re-radiate the incident EM radiation. The chaff consists of foil/metal pieces of different lengths, each length being resonant at a particular frequency. The 'smart' chaff would be lengths of individual foil pieces of the correct length to resonate at the threat frequency.

2.7 Electronic Support Systems

Electronic support systems are usually passive systems that detect and obtain the signatures of threats and then 'hand over' to ECM systems.

A physical threat such as a missile would most probably be detected by an on-board optical surveillance system, an infrared detector or an ultraviolet detector. The optical surveillance system would usually be mounted on the lower fuselage to scan the lower hemisphere. The infrared detector would most probably consist of a number of fixed sensors installed at a number of positions on the airframe to detect the direction of the oncoming missile. The ultraviolet sensing system would detect the launch of the missile and could, for instance, consist of a number of individual detectors with each detector aligned to cover a particular angular sector. A tracking radar would be used to detect the position and velocity of the missile, but this is not considered part of the system of electronic support measures (ESMs).

Since the ESM system is a passive one, only threats that emit EM radiation can be detected and hence these threats are called emitters. The signatures of the emitters are the characteristics that can define the type and position of the threats and include the frequency, the position, the PW, the pulse repetition rate and the interval between pulses. A hostile radar is also detected by the radar warning receiver which then hands over to the ECM system.

2.7.1 Frequency

This is the base or fundamental frequency in the case of a continuous wave (CW) signal, the carrier frequency as well as the frequencies of the modulated signal, and pulsed frequencies. In addition, in the case of frequency hopping, it includes all the frequencies transmitted.

2.7.2 Positional Information

This is one of the most complex characteristics to measure from the antenna point of view. The angle of the emitter relative to the aircraft is determined from the direction of arrival (DOA). This is also sometimes known as the angle of arrival (AOA). However, the terms are not clearly defined and it is more common for the AOA to indicate the azimuth angle and the DOA to indicate both the azimuth and elevation angles of a threat.

In order to determine the DOA of an emitter, the most common passive methods are:

1. a movable antenna with a narrow beamwidth
2. a pair of antennas in an amplitude comparison system
3. a pair of antennas in a phase comparison system.

2.7.3 DOA from Antenna Position

The antenna is moved to obtain the maximum power from the target/emitter and thus the antenna pointing direction gives the position of the target. If the antenna can be moved in the elevation plane, then the angle of elevation/depression can also be determined. Since the beamwidth has to be small, in order to accurately determine the position of maximum radiation, this method can only be used at the higher frequencies, where the electrical size of the antenna is large, but its physical size is small. At low frequencies the antenna would be too large for airborne applications.

The type of antenna used is usually an aperture antenna such as a horn, phased array or reflector antenna. This antenna has to be mounted so that there is no obstruction between it and the target. Most radar antennas that are used for positional information are mounted in the nose of the fuselage. Thus only the position of targets in the forward sector can be obtained. However, where housing and other aerodynamic criteria can be met, a rotating slotted waveguide array has been used as in the case of the Boeing airborne early warning system (AWACS) that has a 'mushroom' mounted on two long vertical arms over the rear fuselage. This gives 360° azimuth coverage as well as limited vertical sector coverage – see Figure 2.9.

There is an ambiguity for intermittent emitters because of the sidelobes in the antenna radiation pattern. The signal detected could be a distant emitter on the main lobe or a less distant emitter on the sidelobes. Hence, the position cannot be determined. The sidelobes can be suppressed, but this also reduces the sensitivity of the system, since the boresight gain is reduced. However, several samples will be taken of the same emitter as the aircraft antenna rotates, with the particular emitter being identified by its other characteristics.

2.7.4 DOA Using Amplitude Comparison

In this case at least two antennas are required for each frequency band. The antennas are oriented so that the angles between their boresights are equal to their half power beamwidths (HPBWs). Thus in the case of the radiation patterns shown in Figure 2.1 the antenna has an HPBW of 60°, so that in the case of an aircraft, for instance, the antennas would be placed with their boresights 30° to port and starboard to detect targets in the 60° forward sector, as shown in Figure 2.2a. Note that these radiation patterns are for ideal uninstalled antennas. Antennas installed on aircraft do not have such smooth radiation patterns. If total 360° coverage is required then six antennas with equal angular spacing, and each with 60° HPBW would be used.

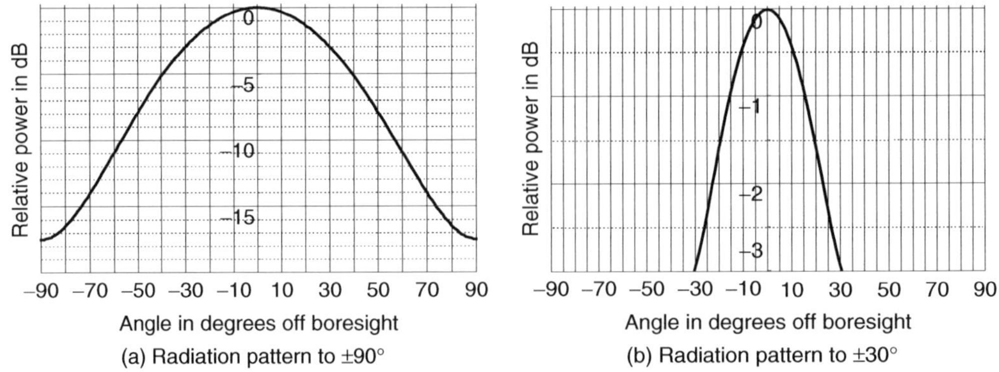

Figure 2.1 The radiation patterns for an antenna with an HPBW of 60°.

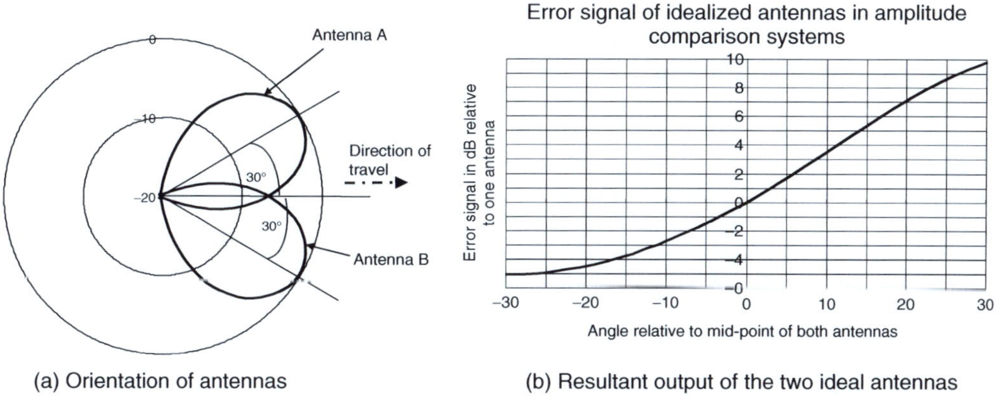

Figure 2.2 The orientation of antennas with 60° HPBW to cover the forward sector in an amplitude comparison system.

The outputs from the two antennas are subtracted so that if the resultant is zero then the target is exactly between the two antennas, indicating that the target is directly forward. If the resultant is obtained by subtracting the output of antenna A from that of antenna B, then when the maximum output is obtained, the target is on the boresight of antenna B, and when the minimum output is obtained, the target is on the boresight of antenna A. The angle of the target relative to boresight is shown in the error signal plot of Figure 2.2b. The angle of the target can be read off from this type of plot, which is sometimes known as the calibration curve. Note that this plot is obtained for antennas that are idealized.

Real uninstalled antennas would have radiation patterns that are not perfectly smooth and their error signal would also not be smooth. The error signal for two real antennas whose radiation patterns have been measured in an anechoic chamber is shown in Figure 2.3a, and that obtained by using the radiation patterns for the same antennas measured on an outdoor range is even worse, as shown in Figure 2.3b. It can be seen that

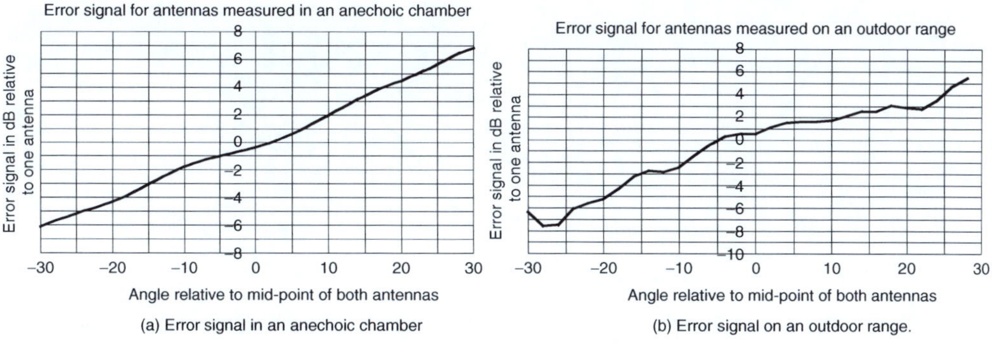

Figure 2.3 The error signals obtained for real antennas measured in an anechoic chamber and on an outdoor range.

the angle read off the error signal curve does not give a unique value in all cases. Thus if the value of $-6.4\,\text{dB}$ is obtained the angle could be either -30° or -24°.

In the case of antennas installed on an aircraft the radiation patterns would be more 'spiky', with many peaks and nulls, and the resultant error signal would be result in even greater ambiguity. However, since the aircraft is moving, the AOA of a particular target would change and, in addition, by identifying the characteristics of the emissions from the target, the ambiguity of the AOA can be resolved.

2.7.5 DOA Using Phase Comparison

In this case at least two antennas are required for each band of frequencies, and the antennas have to be placed half a wavelength apart. This separation is known as the baseline. In this case the phase difference between the incident waves on the antennas is measured and equated to the difference in path lengths from the emitter. Referring to Figure 2.4, if the two antennas are a distance d apart at Q and P, the difference in the path lengths travelled by parallel waves from a distant source is

$$\text{PS} = d \sin\theta. \tag{2.1}$$

This path length difference results in the wave reaching the antenna P at a later time and thus its phase will be different, unless the time was a full period and then the phase difference would be 360° which is the same as 0°. The phase difference measured equates to the path difference PS.

Since d is known and the phase difference is measured the angle θ can be calculated. For instance, if the distance d corresponds to $\lambda/2$ and a phase difference of 90° is measured, we know that a phase difference of 90° corresponds to $\lambda/4$. Equation 2.1 becomes

$$\frac{\lambda}{4} = \frac{\lambda}{2}\sin\theta,$$

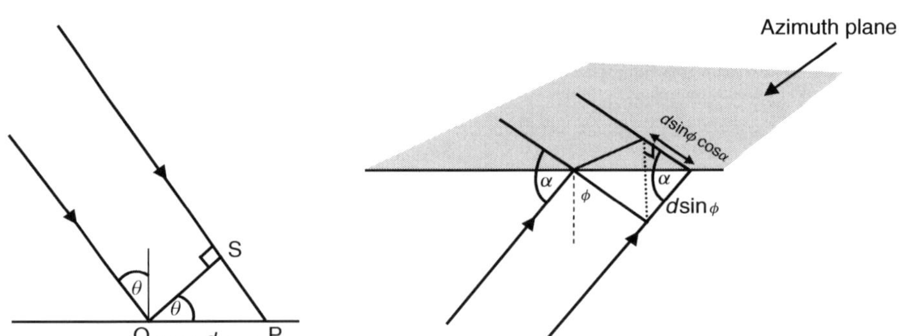

(a) Emitters in azimuth plane (b) Emitter at an angle of depression α

Figure 2.4 Phase difference between two antennas separated by a distance d for an emitter in the azimuth plane and at an angle of depression α.

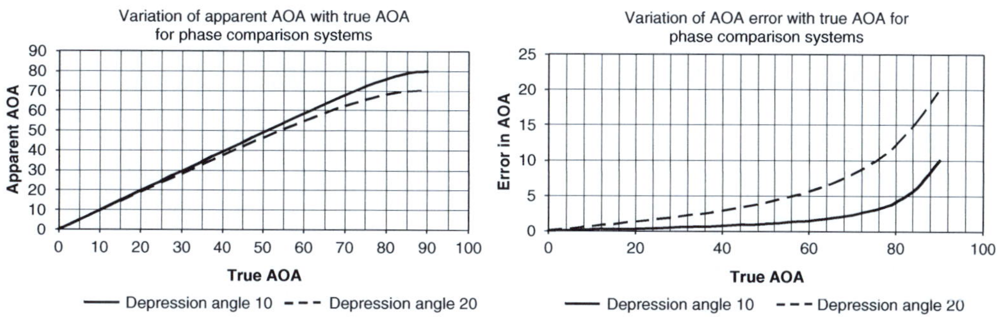

Figure 2.5 The variation of the apparent AOA and the AOA error with the true AOA for 10° and 20° angles of depression.

giving

$$\frac{1}{2} = \sin\theta$$

and thus angle $\theta = 30°$.

The longer the baseline the greater the resolution, but when the baseline is a whole wavelength or greater, the ambiguity is also increased since 360° gives the same phase difference as 0° and 361° is the same as 1°, and so on.

In the case of omnidirectional antennas such as monopoles there is a further ambiguity with regard to whether the emitter is in the forward or aft sector. In order to avoid ambiguities and improve the accuracy of the system several pairs of antennas are used with different baselines for the same emitter.

Another problem with phase comparison systems is what is known as the coning angle error. The DOA can be measured quite accurately if the emitter is in the same plane as the antennas, that is, if the emitter is in the azimuth plane for an aircraft in straight and level flight. However, if the emitter is in any other plane at angle of depression α, the path difference that should be measured is $d \sin\phi$, but the path difference actually measured is $d \sin\phi \cos\alpha$. This gives an apparent AOA which differs from the true AOA and the difference increases with increasing angle α. Since this same error occurs for all angles in a cone of semi-angle α, this is called the coning angle error. The variation of the apparent AOA with the true AOA is shown in Figure 2.5 together with the AOA error for two angles of depression, 10° and 20°.

2.8 Emergency Locator Transmitter/Emergency Position Indicating Radio Beacon

These are most commonly used in the marine world. This device may be a single monopole located on the top of fuselage, often located to the rear or in the tail fin, and is designed to float in water. It should also be carried by anyone planning a boat trip. Most emergency position indicating radio beacons (EPIRBs) automatically float to the surface if the vessel/aircraft sinks.

There are two types of marine EPIRBs. One type transmits an analogue signal on 121.5 MHz. The other type transmits a digital identification code on 406 MHz and a low-power 'homing' signal on 121.5 MHz.

All 121.5 MHz EPIRBs are manual activation units. Although these units do work with the low earth orbit (LEO) satellite system, they do not work as well as 406 MHz beacons, and they cannot be detected by the geostationary (GEO) satellites that provide instantaneous alerting for 85% of the globe. Furthermore, 121.5 MHz beacons are a large source of wasted effort by search and rescue forces. Most 406 MHz false alerts can be resolved easily with a phone call. In contrast, every 121.5 MHz false alert must be tracked to the source using direction finding (DF) equipment.

The air emergency frequency of 121.5 MHz is also used to warn civil aircraft if they have violated the combat air patrol (CAP) airspace. If the warning is not heeded, the aircraft risks being fired upon. Therefore, it is a good idea to be tuned into this frequency at all times.

2.9 Global Positioning System

Satellite navigation is based on the reception of the signals of at least four satellites of a constellation of satellites such as those of the US GPS. The GPS satellites are in approximately 12-hour orbits at an altitude of approximately 20 000 km. The orbits have an inclination of 55°. The minimum constellation consists of at least 24 satellites (6 orbital planes with 4 satellites in each plane). In 2006, 28 satellites were active. The signals are transmitted at 1575.42 MHz (L1) and 1227.60 MHz (L2).

Future satellites will also transmit on L5 (1176.45 MHz). Currently military signals are transmitted on L1 and L2, but civil signals only on L1. In the near future civil signals will also be transmitted at L2 and L5. This new L5 signal is protected worldwide for aeronautical radio navigation use, and will support aviation safety-of-life applications.

Each of the 24 GPS satellites emits signals to receivers that determine their location by computing the difference between the time that a signal is sent and the time it is received. GPS satellites carry atomic clocks that provide extremely accurate time. The time information is placed in the codes broadcast by the satellite so that a receiver can continuously determine the time that the signal was broadcast. The signal contains data that a receiver uses to compute the locations of the satellites and to make other adjustments needed for accurate positioning. The receiver uses the time difference between the time of signal reception and the broadcast time to compute the distance, or range, from the receiver to the satellite. The receiver must account for propagation delays, or decreases in the signal's speed caused by the ionosphere and the troposphere. With information about the ranges to three satellites and the location of the satellite when the signal was sent, the receiver can compute its own three-dimensional position. An atomic clock synchronized to GPS is required in order to compute ranges from these three signals. However, by taking a measurement from a fourth satellite, the receiver avoids the need for an atomic clock. Thus, the receiver uses four satellites to compute latitude, longitude, altitude and time. The ground receiver will have in its memory the precise details of the orbits of all the satellites.

The GPS antenna used on aircraft is normally a patch antenna type. The antenna should have coverage over the most of the upper hemisphere (from elevations of approximately +5°). The directivity of the antenna is low. GPS uses right-hand circular polarization (RHCP). The antenna may have a built-in low noise amplifier (LNA). The normal GPS antennas are called fixed radiation pattern antennas (FRPAs). Military aircraft may also use controlled radiation pattern antennas (CRPAs). These CRPAs have the capability of minimizing the interference-to-signal ratio by placing a null in the radiation pattern, in the direction of the interfering source. CRPAs are circular arrays with patch antenna elements similar to the FRPAs. Many antennas have an antenna electronics unit (AEU) that has to be placed within 1 m of the antenna. The AEU contains the RF mixer stage that has a local oscillator and effectively lowers the frequency so that the signal is less susceptible to loss, pick-up and electromagnetic interference (EMI) through the coaxial cable connected to the GPS receiver and processor.

To increase the accuracy, availability and integrity of satellite navigation, aircraft can use augmentation systems. The two types of augmentation systems used are space based (via additional satellites) for en-route navigation and approaches to airports, and ground based (via additional ground stations) called pseudolites.

Examples of space based augmentation systems (SBASs) are the European Geostationary Navigation Overlay Service (EGNOS) and the US wide area augmentation system (WAAS). These systems use geostationary satellites to broadcast differential global positioning system (DGPS) correction signals and integrity information. DGPS is based on the use of reference stations at known locations that receive the same GPS signal as the GPS receiver in the aircraft. This reference station processes the GPS signals and derives pseudorange, delta-pseudorange and pseudorange-rate errors with respect to its accurately known location and then transmits these corrections to the users in the area. EGNOS transmits at L1 (1575.42 MHz).

Ground based augmentation systems (GBASs) can be received with a normal GPS antenna. An example of a GBAS is the US local area augmentation system (LAAS). LAAS is an all-weather landing system based on real-time differential correction of the GPS signal. Local reference receivers send data to a central location at the airport. This data is used to derive a correction message, which is then transmitted to users via a VHF data link. A receiver on an aircraft uses this information to correct GPS signals, which then provides a standard ILS-style display to use while flying a precision approach. GBASs require an additional antenna on the aircraft for a VHF data link.

The Galileo system which is currently being developed in Europe by European Space Agency (ESA) and the European Union will be interoperable with the GPS system. The Galileo constellation will consist of 30 satellites in three planes. The inclination of these orbital planes is 56°. The altitude of the satellites will be 26 300 km. As in the case of GPS, the ground receiver will have in its memory the precise details of the orbits of all the satellites and when signals from at least four satellites are received the positional accuracy can be obtained down to 1 m. The frequency bands are 1164–1215 MHz (E5), 1260–1300 MHz (E6) and 1559–1590 MHz (E1/E2).

GLONASS is the Russian/Soviet global positioning system. It is sometimes used on military aircraft to supplement the GPS. The spaceborne segment consisted of 21 satellites (plus three spares) at 19, 100 km with orbital planes inclined at 64.8° and with an eight-day repeat track. The satellite signals are RHCP and their effective isotropically radiated power (EIRP) varies between 25 and 27 dBW. There are 25 channels at 0.525 MHz intervals in the frequency bands 1240–1260 MHz and 1602.5625–1615.5 MHz. The accuracy for the Coarse Acquisition (C/A) is 100 m, and in P-mode (precision), used mainly for the military, the accuracy can be as high as 10–20 m. However, GLONASS is no longer used because the system is not fully operational (since there are an insufficient number of satellites). Russia is currently (2009) launching new satellites.

2.10 HF

The operating HF frequency band for aircraft is 2–30 MHz. In order to obtain a quarter-wave resonant monopole at 2 MHz (wavelength 150 m), the monopole would have to be 37.5 m long, and at the midband frequency of 16 MHz, the length would be 18.75 m. In large aircraft the HF antenna is implemented as a long wire that stretches between the tail fin and the front or middle of the upper fuselage. However, since most aircraft are much smaller than the 18.75 m required, an electrically small antenna is used. This could be embedded in the front of the tail fin and/or the dorsal fin. Other implementations are surface mounted 'towel rails'. In the case of special aircraft such as the Apollo Range Instrumentation Aircraft (ARIA) EC-135E a trailing-wire antenna is used. This wire is wound on a capstan/drum and the length of the wire that is deployed corresponds to approximately a quarter wavelength of the transmitting frequency.

The propagation of HF through the atmosphere is very dependent on atmospheric conditions. These conditions vary with the time of day (TOD), sunspot activity and other atmospheric variations. In Chapter 1, we have seen how total internal reflection results in sky hop. The signals can sometimes reach distances further away, but cannot be picked up at shorter ranges. The TOD and the angle of launch have a large impact on the path and distance travelled. The frequency of the signal can influence the height at which the waves are reflected and hence the hop distance. Thus if communication is required between two stations (be it between two aircraft or between an aircraft and the ground) the choice of frequency is very important. HF radios have a number of frequencies spaced over the frequency band, each with a specified frequency bandwidth known as a channel.

Modern HF systems employ what is known as automatic link establishment (ALE). A link is automatically established depending on the quality of the signal between two stations. The quality is determined by analysing the signal based on the bit error rate (BER) and the signal to noise and distortion (SINAD). This link quality analysis (LQA) automatically memorizes the short-term channel quality on designated frequencies using active or passive measurements. The active measurements consist of special transmissions called soundings that enable the listening stations to measure the quality of the frequency/channel being transmitted at scheduled time intervals. All stations in a network agree in advance on the frequencies in a particular scan group that will be used for ALE attempts. Most units can scan up to 100 channels at the rate of 2 to 5 channels per second. Whilst the receiver is scanning it continuously monitors for incoming ALE signals which are evaluated and stored for future reference. When the station wants to contact another station it checks its LQA memory for the best channel to use. It then transmits a digital call signal that follows a specific protocol. If the called station receives and decodes the call properly, it will respond with a specific signal acknowledgement. The calling station will then send out a confirmation signal and the link is established on one of the channels in the LQA memory.

As mentioned at the start of Section 5.6.3, an electrically small antenna has a reactive impedance at most frequencies. Thus it is conjugatively matched to the transmission line by a tuner. If the antenna has a negative impedance (capacitance), the tuner provides a positive impedance (inductance) to cancel out the capacitance. For this reason the HF antenna is always tested with its tuner and requires a full-scale mock-up of all or part of the aircraft, or the real aircraft, during its measurement programme. The output of HF antennas is around 400 W on average and 1 kW peak and towel rails have gain of the order of −20 dBi.

2.11 Instrument Landing System

The first landing of a scheduled US passenger airliner using an ILS was on 26 January 1938, when a Pennsylvania-Central Airlines Boeing 247-D flew from Washington, DC, to Pittsburgh and landed in a snowstorm using only the ILS system.

The ILS ground segment guides the aircraft to the touchdown point on the runway. This consists of three associated systems, namely:

- ILS marker
- ILS glideslope
- ILS localizer.

The marker system is used in conjunction with the glideslope and localizer. However, the ILS marker beacons consists of three transmitters, two of which are available on approach to the runway, and the outermost marker is the point at which the glidepath signal is first acquired by the aircraft. The markers inform the aircraft of its distance from the touchdown point on the runway. All the systems operate using horizontal polarization.

2.11.1 ILS Marker

An equipped runaway generally has outer, middle and inner marker beacons. These beacons are low-power transmitters that operate at a frequency of 75 MHz with 3 W or less rated power output. Each beacon radiates a fan-shaped beam pattern of elliptical cross-section upward from the ground, the axis of the fan being at right angles to the airway. At an altitude of 1000 ft, the beam dimensions are 2400 ft long and 4200 ft wide. At higher altitudes, the dimensions increase significantly. The beacons have different modulation tones of repeated Morse-style dashes and the tone of the beacon is audible to the pilot.

The airborne marker antenna is used to sense the beacons installed on or near the extended axis of the runaway. A cockpit indicator is also activated on the pilot's instrument panel and flashes in unison with the received audio code. In the case of the outer marker a blue light flashes, whereas in the case of middle and inner markers the colours are amber and white, respectively.

1. The outer marker beacon has a modulation of 400 Hz and is ideally installed 7.2 km (3.9 nmi) from the runway. However, where this distance is impractical, the outer marker may be located between 6.5 and 11.1 km (3.5 and 6 nmi) from the runway threshold. It is located within 250 ft (broadside) of the extended runway centreline. The outer marker beacon intersects the glide slope vertically at between 1100 ft and 1900 ft above runway elevation, depending on its distance from the runway and assuming that the glidepath angle is 3°. It also marks the approximate point at which aircraft normally

intercept the glideslope, and designates the beginning of the final approach segment. The signal is modulated at 400 Hz, which is an audible low tone with continuous Morse code dashes at a rate of two dashes per second. Where geographic conditions preclude the positioning of an outer marker, a DME unit may be included as part of the ILS system to enable the pilot to make a positive position fix on the localizer. In the United States, an NDB is often combined with the outer marker beacon in the ILS approach (called a locator outer marker, or LOM). The NDB is a low-frequency non-directional beacon with a transmitting power of less than 25 W and a frequency range of 200–415 kHz. The reception range of the radio beacon is at least 15 nmi. In Canada, low-powered NDBs have replaced marker beacons entirely.

2. The middle marker is located between approximately 3250 and 3750 ft from the threshold on the extended runway centreline. The beacon is modulated at 1300 Hz and crosses the glideslope at approximately 170–200 ft above the runway elevation. It is near the missed approach point for the ILS Category I approach. Middle markers have been removed from all ILS facilities in Canada but are still used in other countries.
3. An inner marker may be installed 100 ft from the runway threshold, modulated at 3000 Hz. The inner marker is not used very often – only with Category II ILS equipment. This is used for a runway visual range (RVR) of 350 m or more. These beacons are used by the pilot to obtain an aural indication of the runway proximity and by autopilot computers that have inherent rules for adjusting the gain of the loop antennas.

The airborne antenna is installed in the bottom part of the fuselage, parallel to the axis of flight and as far aft as possible. It sometimes comprises a quarter-wave element; however, in many cases it is an electrically small antenna that is foreshortened by capacitive loading. In the case of high-speed aircraft it is a suppressed antenna that is recessed into the lower fuselage and covered by a dielectric sheet/radome.

2.11.2 ILS Glideslope and Localizer

The glideslope and localizer ground stations are located on the airfield. Aircraft approach the runway of the airfield along the centreline of the runway and at an angle defined by the glidepath. The glidepath is the line along which the aircraft should travel on approaching the airfield to land. The angle of the glidepath that the aircraft adopts is ideally 3° to the horizontal.

The glideslope provides guidance to the aircraft in the vertical plane, while the localizer provides lateral guidance, that is, it guides the aircraft in the horizontal plane.

Two signals are transmitted on one of 40 ILS channels transmitted from separate but co-located antennas. One is modulated at 90 Hz, the other at 150 Hz, each with a 20% depth of modulation. The difference between the two signals varies depending on the position of the approaching aircraft from the centreline. If there is a predominance of either 90 or 150 Hz modulation, the aircraft is off the glidepath. When the two signals are equal the aircraft is on the centreline.

Some aircraft possess the ability to route signals into the autopilot, allowing the approach to be flown automatically by the autopilot.

Localizer signals are transmitted on a carrier frequency between 108.10 and 111.95 MHz, with 90 Hz modulation to the left and 150 Hz to the right of the runway

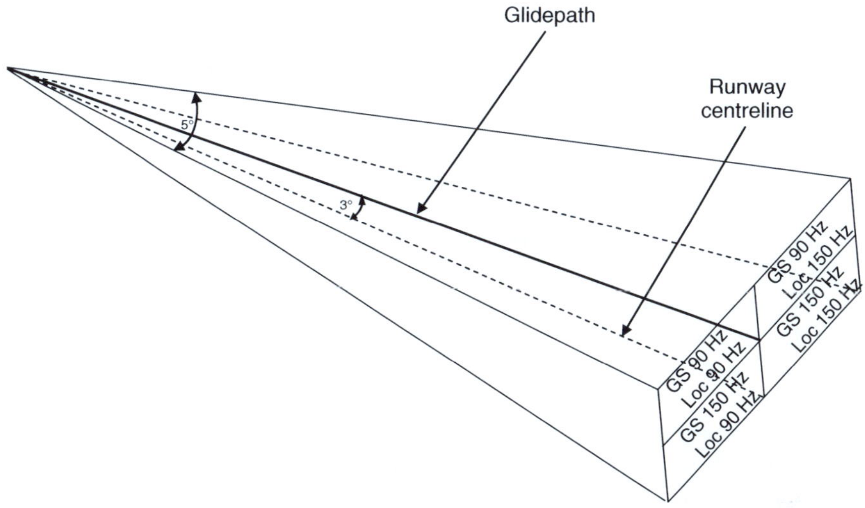

Figure 2.6 ILS glideslope and localizer modulation areas from the pilot's viewpoint.

centreline, the left and right being defined from the pilot's viewpoint as shown in Figure 2.6. The localizer also provides an ILS facility identification by periodically transmitting a 1020 Hz Morse code identification signal.

Glideslope signals are transmitted on a carrier frequency between 329.15 and 335 MHz, with 90 Hz modulation above the glidepath and 150 Hz below.

Each of the 40 localizer carrier channels is coupled with the glideslope channels to provide a complete ILS system. Thus only one selection is required to tune both receivers. An aircraft that has intercepted both the localizer and the glideslope signal is said to be established on the approach. Typically, an aircraft will be established by 6 nmi (11.1 km) from the runway.

In the case of the ground station, the localizer transmit antenna is installed on the runway axis at the end of the runway. The glideslope antenna is installed next to the runway near the threshold. In modern ILS installations a DME is installed, co-located with the ILS, to augment or replace marker beacons. A DME continuously displays the aircraft's distance to the runway.

The signal transmitted by the ground station localizer antenna consists of two fan-shaped patterns that overlap at the centre. The total width of the radiated beam may be varied from approximately 3° to 6°, with 5° being normal. It is adjusted to provide a track signal approximately 700 ft wide at the runway threshold. Runways differ in length, but a typical length is 1.5 statute miles or 8000 ft. The width of the beam increases with distance, so that at 10 nmi from the transmitter, the beam is approximately 1 mile wide.

The ILS ground station antenna is susceptible to distortion in its radiation pattern caused by multipath due to other aircraft. This can result in false guidance beams, and thus aircraft are not permitted to enter this 'sensitive area'. No aircraft or vehicles may enter this area without air traffic control (ATC) permission during low-visibility landing operations, since their presence may reflect the ILS signals and create false guidance to the landing aircraft. Taxiways to Category II and III runways therefore bear signs showing

Table 2.2 ILS landing categories.

ILS category	DH decision height[a]	RVR runway visual range
I	≥60 m	≥550 m or visibility >800 m
II	30–60 m	≥350 m
IIIA	<30 m	≥200 m
IIIB	<15 m	30–50 m
IIIC	0 m	Zero

[a]The decision height is the height at which the pilot must have visual contact with the runway or abort the landing. Thus the most onerous is the lowest decision height, that is, the aircraft with the best instrumentation are cleared to Cat. IIIC and can land with zero (visual) visibility.

'Cat II Hold' and 'Cat III Hold' points well clear of the active runway. The categories are shown in Table 2.2.

Similarly, the ILS guidance signals received by an aircraft following another aircraft on final approach can reflect off the aircraft ahead, resulting in false guidance. This is usually not critical if the pilot of the following aircraft can see the runway ahead, but in low visibility this can create a potentially hazardous condition. Because of this effect, the normal 2- to 3-mile separation between following aircraft is effectively tripled during UK Cat III ILS operations, resulting in a reduction of runway throughput to about 10 landings per hour, compared to about 30 in normal visibility.

The LAAS Category II/III are signal correction systems that will make GPS landings possible with a greater degree of accuracy than current ILS systems. Until LAAS systems are available, the Federal Aviation Administration (FAA) plans to meet Category II/III requirements with ILS.

In good weather the pilot will be able to see either the runway or the approach lights and able to execute a safe visual landing. The FAA defines good weather as a visibility of at least 3 statute miles and the ability to maintain at least 500 ft clearance from the clouds above. Under these conditions flying by visual flight rules (VFRs) is permitted. In practice, because the minimum altitude one can fly in an uncongested area is 500 ft above the surface, the cloud ceiling for good weather is effectively 1000 ft. Aircraft with Category III landing must have two ILS marker antennas.

RVR readings usually are expressed in hundreds of feet. For example, 'RVR 24' means that the visual range along the runway is 2400 ft. In weather reports, RVR is reported in a code: R36/4000 FT/D means that the RVR for Runway 36 is 4000 ft and decreasing.

To ensure the utmost reliability in any conditions, the ILS equipment has to be tested to a 10-million-to-one failure factor. An ILS is required to shut down upon internal detection of a fault condition. With the increasing categories, ILS equipment is required to shut down faster since higher categories require shorter response times. For example, a Cat I localizer must shutdown within 10 seconds of detecting a fault, but a Cat III localizer must shut down in less than 2 seconds.

The ICAO ([1], Section 3.1.5.2.2) states that the glidepath equipment has to be capable of adjustment to produce a radiated glidepath between 2° and 4° to the horizontal. For Categories I and II the glidepath angle has to be maintained within 0.075θ (where θ is the glidepath angle). Thus for the maximum angle of 4° the glidepath angle has to be maintained within ±0.3° (i.e. between 3.7° and 4.3°), and for Category III this angle has to be maintained within 0.04θ (i.e. ±0.16° for the maximum angle of 4°).

For an automatically monitored system, a warning is issued to designated control points and radiation ceases if the mean glidepath angle shifts by more than -0.075 to $+1.10\theta$ ([1], Section 3.1.5.7). Thus for a glidepath of 4°, if the angle decreases by 0.225° or increases by 4.4°, the radiation will cease. This sets the maximum angle that the glidepath is ever likely to be at 8.7° (4.4° + 4.3°). Radiation will cease for higher angles.

Because of the horizontal polarization, the dipoles are horizontally oriented and the main lobes strike the ground. These are specularly reflected and result in several sidelobes to the main beams, giving integral multiples of the glidepath angles. Thus 'false' glidepaths occur at 6°, 9°, 12°, and so on, for a true glidepath of 3°. At the higher angles the level of the signals is progressively reduced. Because of these false glidepaths, the aircraft usually captures the glidepath from below. The transmitting ground antenna is located 750–1250 ft down the runway from the threshold, offset 400–600 ft from the runway centreline, and monitored to a tolerance of ±½°.

The thickness of the overlap area is 1.4° or 0.7° above and 0.7° below the optimum glidepath. With 1.4° of beam overlap, the area is approximately 1500 ft thick at 10 nmi, 150 ft at 1 nmi, and less than 1 ft at touchdown.

The positioning of the airborne localizer and glideslope antennas depends on the clear field of view, as well as the proximity of obstacles (including other antennas) that affect their radiation patterns. The field of view for the antennas are the forward sector of the flight path. The glideslope antenna is often installed in nose cone above the radar.

Since it operates in the same frequency band as VOR (112–118 MHz band) and uses the same receiver, the localizer and the VOR system may share the same antenna. VOR is due to be discontinued and its frequency band is to be used by the Localizer system. The localizer antenna is often mounted either in the radar nose cone at the front of the aircraft, or on either side of the tail fin, as long as a clear line of sight to the ground segment is available to it. In some cases the VOR antenna can be combined with the VHF antenna and looks like a horizontal boomerang on top of the VHF blade. In other cases the glideslope and localizer antennas share part of the same radiating element, as discussed in Section 5.7.

2.12 In-Flight Telephony

The situation is continuously changing as improvements in the mobile phone industry implement changes in terrestrial cell phones. Unlike the case of terrestrial networks, most passengers will only make one or two phone calls on the flight and a 'pay-as-you-go' service is required. On-board calls have met with limited acceptance by passengers, and Verizon Airfone's president was quoted in a *New York Times* article as stating that only two or three people per flight make a call.

AT&T abandoned its 800 MHz air-to-ground offering in 2005, and Verizon Airfone (formerly GTE Airfone) was scheduled for decommissioning in late 2008. In 2008 the main player was Aircell, using a 3 MHz band licence. The signal from the aircraft is transmitted on a 850 MHz carrier to either a receiving ground station or a communications satellite, depending on the design of the particular system. The call is then forwarded to a verification centre to process the credit/calling card information, before being routed to the public switched telephone network (PSTN). The return signal frequency is in the 894–895 MHz band. Three antennas were required on the aircraft. One was a Personal Communications Services/Global Positioning System antenna on the upper surface of the fuselage, and two blade antennas were used on the lower fuselage. The passengers accessed the wireless access points evenly distributed across the cabin ceiling. The ground segment consisted of 92 cellular towers in the continental USA.

2.13 Microwave Landing System

The MLS provides precision guidance for horizontal and vertical alignment and distance. It allows curved and steep approaches, as opposed to ILS which requires straight line approach. It provides signals of up to 40° in azimuth on either side of the runway centreline and up to 20° in elevation. The signal from the ground station can be picked up at an altitude of 20 000 ft and at a slant range of 20 nmi.

It has been developed to overcome the intrinsic limitations of ILS, which include single descent path, capture issues, large ground station antennas, sensitivity to reflections, monitoring, and spectrum saturation. In addition, the horizontal polarization of the ILS system necessitates an elevated position for the ground station antenna, whereas the MLS ground station antenna is vertically polarized (hence does not require an elevated position) and has horizontal and vertical scanning beams operating in the 5 GHz band.

Azimuth and elevation angle measurements of an aircraft may be derived by 'time-reference' processing. In this type of processing a reference signal pulse is radiated from a ground station to cover an area to be scanned. A scanning directional beam of radiation originating at the ground station is started to sweep from a known direction when the reference pulse is radiated. The beam is swept at a predetermined scan rate. The variations in radiation received at any point in space through which the scanning directional beam is swept represents the radiation pattern of the antenna, assuming that the pattern is not affected by obstacles that could affect the radiation pattern.

The scan rate and the time when the sweep was started are known, and hence by noting the time between reception of the reference signal pulse and that of the beam peak of the radiation pattern, an airborne receiver and processor can derive the angle between the initial direction of the scanning directional beam and the line of sight between the aircraft and the ground station. This gives the direction relative to a ground station. If the sweep is in elevation, the angle of elevation can be derived. Except for highly directional antennas, the shape of the scanning directional beam is relatively flat around its peak. It therefore difficult to determine the exact position of the peak. Instead the position of the peak is estimated by determining the times at which the 3 dB points on each side of the peak are received, and interpolating between those two positions.

A marker at either end of each scan is received by the aircraft, and the time between two markers allows determination of the angular position of the aircraft relative to the ground station. By using a precision DME (P-DME) 3D positioning can be obtained, on approach courses covering a sector up to 40° on either side of centreline and up to 15° elevation above the runway.

MLS ground station antennas are smaller than ILS ones, and less subject to multipath reflections. However, the directivity must be very high for accurate positioning and the financial cost of the ground station infrastructure makes this system uneconomic for most small and medium sized airfields.

MLS pays off handsomely for large airports with a high density of aircraft (landing and taking off), since the MLS scanning beam ignores reflected signals. A preamble is transmitted from a low-gain fixed pattern antenna. The azimuth beam width is approximately 2° and the elevation beam width may be as small as 1°. The MLS ground station transmits both angle and data functions on one of 200 frequencies between 5031.0 and 5190.7 MHz. It allows MLS-equipped aircraft to maintain normal operational spacing on their final approach, and therefore enables airports to recover the capacity levels lost during low-visibility periods.

On the aircraft, the forward fuselage antennas must provide unobstructed coverage in the forward sector. It is common to have one antenna on the lower fuselage as well as one on the upper fuselage at the front or on the nose of the aircraft. The antenna on the upper fuselage of the aircraft is often a directional one, such as a Yagi or pyramid horn. An additional omnidirectional antenna is required at the tail of the aircraft for missed approaches and take-offs. This is usually installed on the lower surface to allow unobstructed view of the airport from the rear, as well as broadside coverage when the aircraft is banking. This rear antenna may require a built-in or through-line amplifier, since the RF cable (to a receiver at the front of the aircraft) is subject to considerable loss at these frequencies (~5 GHz).

2.14 Radar

All radar-based systems are covered in this section. These include the RadAlt, weather radar, search radar (used in military aircraft) and synthetic aperture radar (SAR) used for surveillance and mapping.

Although the RadAlt is not usually classed as a radar, it is based on the principle of radio detection and ranging. The electromagnetic wave detects that there is an object because part or all of the wave is scattered back to the source, and the range (distance of the object from the source) is established because the time for the return (backscattered wave) is measured and the speed of EM wave is known – approximately 300 000 000 m/s (3×10^8 m/s).

The search and weather radar antennas are mounted in the nose cone and are protected by a radome that is transparent to the radar frequencies as well as protecting the antenna and all the devices inside the radome from the external environment.

2.14.1 Doppler Shift

Doppler radar is based on the same principle as applied to sound waves, although the latter are longitudinal rather than transverse waves. We can tell when a train emits a whistle, whether the train is approaching us or moving away from us by the frequency of the sound wave. The pitch/frequency is higher as it moves towards us and lower as it moves away.

When the train is stationary as shown in Figure 2.7a and the whistle is operated, the sound waves are emitted spherically. If the circles represent the peaks of the waves then the distance between the circles gives us the wavelength λ_0, corresponding to a frequency f_0.

Now consider the situation when the train is moving towards the observer at X as shown in Figure 2.7b. The waves are squashed together by an amount that is determined

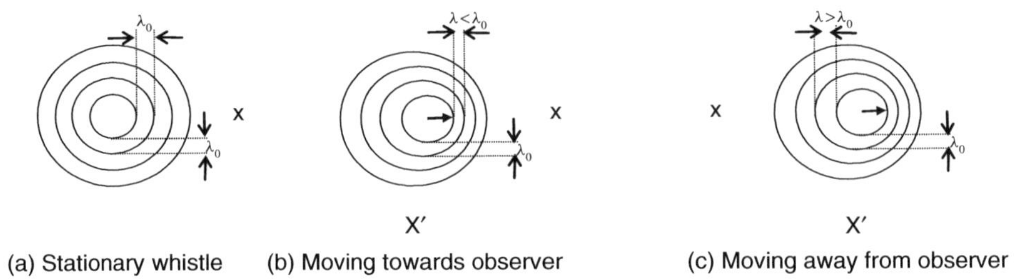

Figure 2.7 The change in wavelength and hence frequency of the whistle on a train due to the Doppler shift.

by the speed of the train, making the distance between the circles smaller in the direction of the observer and hence the wavelength is smaller, resulting in a higher frequency f_h (since frequency = speed/wavelength). When the train is moving away from the observer the distance between the wavefronts is greater as shown in Figure 2.7c, resulting in the longer wavelength and thus a lower frequency f_l. Note, however, that if the observer is at position X' at right angles to the direction of motion of the train, the wavelength towards X' is the same throughout and hence the frequency is also constant.

The difference between the frequency f_0 of the stationary whistle and the frequency when it is moving is $f_h - f_0$ or $f_0 - f_l$ and is known as the Doppler shift.

If we consider a stationary observer at X and a train going past at a velocity represented by the vector, there is a component of the velocity in the direction of the line joining the train to the observer. This is known as the resolute. We can see from Figure 2.8a that the wavelength is shorter and hence the frequency is higher when is approaching the observer. For the spilt second when the train is broadside on to the observer, that is, neither moving towards or away from the observer, there is no component of the velocity along the dotted line as shown in Figure 2.8b, and at that point its true frequency is heard. The train must have a component of its velocity (resolute) in the direction of the observer for a frequency shift to occur. In the last position shown in Figure 2.8 the component of the velocity is in the direction away from the observer (i.e. negative), resulting in a frequency that is less than the true frequency of the whistle.

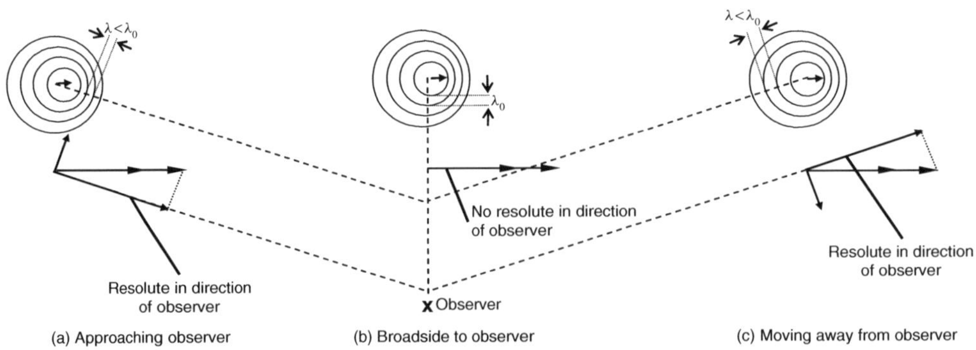

Figure 2.8 The change in wavelength of a train driving past a stationary observer.

In the case of radar, the frequency emitted by the radar is f_0 and the Doppler frequency is f_h if the object is moving towards the observer and f_l if the object is moving away from the observer. In radar systems the Doppler principle is used to detect moving targets. The received signal is mixed with the transmitted one to determine the Doppler shift by heterodyne detection.

2.14.2 RadAlt

RadAlt stands for radar/radio altimeter and operates around the 4.3 GHz frequency band, and is used to measure the distance above the ground of the landing gear, as opposed to the barometric altimeter that measures the height above sea level. Horizontal polarization is used in the frequency range 4200–4400 MHz.

The pair of antennas used for this system have to be installed on the lower surface of the aircraft and may be on the fuselage or wing. Older versions of these antennas are rectangular pyramid horns, but since the power transmitted is only about 20–500 mW printed circuit patch antennas are used on more modern systems.

The aircraft installation delay (AID) is defined as the distance from the ground to the landing gear, and allowances made for the RF cable runs. The landing gear is used as the zero height reading of the aircraft. To ensure independence of aircraft attitude and dynamics, especially for fighter aircraft, these antennas have a wide beam pattern. The HPBWs of the antennas vary between about 60° and 80°.

In some systems a pulse of around 200 ns duration is transmitted by one antenna, and the other one receives the signal reflected from the ground. The distance to the ground is deduced from the time between transmission and reception of the pulsed signal and taking into account the on-board coaxial cable length.

Other systems transmit a frequency modulated (FM) continuous wave and these achieve greater accuracy at low altitudes. The transmitted signal is FM, and the greater the frequency difference between received and transmitted frequencies the further the distance travelled – the higher the altitude. The RadAlt is not normally used at altitudes/heights below 500 ft or above about 5000 ft. Most large aircraft carry two independent RadAlt systems for redundancy. If the pair of antennas are on the lower surfaces of each of the wings, then when the aircraft is in roll (one wing lower than the other) the altitude of the fuselage would be the average of the two heights recorded.

2.14.3 Search Radar

Search radar is used for military and surveillance applications. The antenna is normally mounted in the nose and the metal nose is replaced by a radome made of material with a low dielectric constant such as fibreglass. The lightning protection for the radome is implemented by conducting strips that could be in the form of discontinuous discs of metal, called button strips. CW radar is used for detecting moving targets using the Doppler principle. The range cannot be obtained from the CW signal alone, but by using FM-CW the target range can be obtained from the beat frequency, which is the frequency difference between the instantaneous transmitted frequency and the received 'echo' frequency.

The main frequency band for search radar is the X band (NATO I band) and covers the frequencies from 8.2 to about 12 GHz. Other frequencies are used for specialized search

Figure 2.9 The Boeing 707 airframe adapted to E3 Sentry with the radar in the rotodome.

radar applications such as the Airborne Warning and Control System (AWACS), that uses S band (NATO E–F band, 2–4 GHz). The Westinghouse AN/APY-1 and AN/APY-2 radars are used on the E3 Sentry as part of the AWACS and the antenna consists of a slotted waveguide array canted 2.5° downwards (see [2], pp. 74–75). The E3 Sentry is a modified Boeing 707 that has received the C-137 designation. The radar antenna is supported (by two 3.4 m struts) above the rear fuselage in the 9 m diameter rotating enclosure known as a rotodome – see Figure 2.9. Azimuth coverage is obtained by rotating the antenna (at 6 rpm) about a vertical axis and elevation coverage is provided by electronic scanning using phase shifters.

Details of all the current radar systems are given in Chapter 3 of [3].

2.14.4 Weather Radar

The usual frequency band for weather radar is X band 9.33 and 9.5 GHz, but other frequency bands for weather radar are 5350–5470 MHz, 8750–8850 MHz and 13.25–13.4 GHz. Weather radar detects the presence of clouds and water vapour as well as turbulence.

The X band (NATO I band) is more commonly used since the higher frequency of the X band can provide a narrower beam for the same physical size since it is electrically larger than the 5350–5470 MHz band radar. In addition, the shorter wavelength of around 1 cm provides better resolution and enables the same antenna to be used as a search radar on military aircraft. Flat plate antennas are used in modern systems, eliminating the requirement for a feed used in reflector antennas. Slotted waveguide flat plate antennas can handle higher powers up to around 12 kW from vacuum oscillators such as klystrons and magnetrons, but the printed circuit arrays can only handle up to about 50 W and solid state transmitters are used in these cases. PWs of 1–20 μs are used and HPBWs of the pencil beam antennas vary between 2.5° and 8° depending on the physical dimensions of the antenna. In order to produce these pencil beams the antennas are circularly symmetrical. The boresight gains of the antennas vary between 26 and 36 dBi. Although the antennas scan in the horizontal direction up to 120° the movement in the vertical direction is usually limited to about ±15°. By varying the illumination across the antenna, the sidelobe level can be reduced to as low as 31 dB below peak.

Heavy rain results in large raindrops that flatten as they fall and the returns from horizontally polarized electrical fields are larger than those obtained from vertically polarized electric fields. Some weather radars transmit both polarizations and, depending on the relative returns, the nature of the precipitation can be established. Solid ice has a much smaller dielectric constant (~3 at −12 °C and 10 GHz) than water (77 at 25 °C) and hence the returns from water are much greater than those from hail. The maximum range for high specification radars is about 180 nmi; however, the normal range is more likely to be 30–80 nmi.

2.14.5 Synthetic Aperture Radar (SAR)

SAR is usually a sideways looking radar (SLR) and is either mounted in a pod below the aircraft or on one or both sides of the fuselage, so that its boresight is normal to the fuselage axis, that is, it looks to port and/or starboard. If it is mounted on one side of the fuselage, then it can only be used for the other side if the aircraft turns around. It is used to detect targets as well as for surveillance and mapping.

The SAR is effectively an antenna with a larger aperture than the physical size of the antenna. The larger aperture is generated in the time domain by the motion of the aircraft. The resolution obtained is equal to the aperture dimensions of the real antenna in the direction of generation of the synthetic aperture. Thus if the aperture of the antenna is 2 m, the resolution obtained for any ground feature is also 2 m.

When used for surveillance, the selected area is irradiated by the EM signal, and the physical characteristics of the area are built up by analysing the characteristics of the return signal. An optical image is then produced from these EM returns, which is colour-coded to match the contours of the terrain. The SAR data is processed in real time and available on board to the crew. Doppler signal processing is also available to distinguish between moving and fixed targets, with the usual proviso that there must be a component of the target velocity perpendicular to the direction of motion of the SAR antenna. The data can be processed on board and since the output is largely in graphical or visual format it requires a wider bandwidth to transmit the data to the nearest ground node. It might therefore be more secure, convenient and more economically viable to transfer the data at the end of each sortie rather than in mid-flight.

2.14.6 Secondary Surveillance Radar

The secondary surveillance radar (SSR) system is part of the control/management of air traffic, and is used in conjunction with the primary radar for providing aircraft identification (mode A) or altitude (mode C). The ground system interrogates the transponder on the aircraft through the aircraft's ATC antenna. All aircraft entering this airspace must therefore have a transponder. In this function, it is similar to the military identification friend and foe (IFF) system, and all IFF systems are compatible with SSR to enable military aircraft to enter civilian airspace. Currently, a mode S transponder is also used as a data link for the on-board traffic collision avoidance system (TCAS) (collision avoidance). Since the operating frequency is around 1 GHz band as in the case of the DME, a similar type of blade antenna is used.

The ground station uses a directional array of dipoles that is mechanically rotated and a fixed omnidirectional antenna both transmitting specific pulses. The gain of the omnidirectional antenna is larger than the gain of the sidelobes of the directional antenna. If the aircraft receives a pulse through the sidelobe of the directional antenna, it is weaker than the pulse from the omnidirectional antenna and the pulse is discarded. This is effectively sidelobe blanking.

The ground segment antenna is mounted above the primary airfield radar and has a field of view of up to 45° in elevation and 360° in azimuth. The range is up to 200 nmi for aircraft altitudes up to 100 000 ft.

An improved version of SSR is monopulse secondary surveillance radar (MSSR), which eliminates the problem of distinguishing between two aircraft whose DOAs are very close to each other. The MSSR uses a larger ground station antenna with a series of independent dipoles that can distinguish between aircraft by acquiring the phase delay from the replies. This allows an angular resolution of as little as 0.5° instead of about 3° with the standard SSR. The minimum spatial separation of 10 nmi (18.5 km) between aircraft, obtained in the case of SSR, is reduced to 5 nmi using MSSR.

2.15 SatCom Civilian

In the case of communication to a satellite, because of the distances that the EM wave has to travel, the attenuation to which it is subjected is high. Bearing in mind that a 6 dB reduction in the signal halves the range, it is important that the satellite is kept within the electrical boresight of the antenna. This is achieved by an antenna with either a gimballed positioner or by using an electronically steered phased array. In both cases an input will be required from the aircraft inertial navigation system to track the satellite. There are several players in this field using different satellite systems.

2.15.1 INMARSAT

The International Marine Satellite Organization (INMARSAT) initially launched a constellation of four geostationary satellites (GSSs) for maritime use. The Inmarsat 3 fleet of four GSSs have a number of spot beams (up to a maximum of seven) for coverage of the more populated areas. Although the coverage is good over most of the globe, at latitudes above about 85° N and 55° S the coverage is poor for aircraft using the conventional flight paths. On the aircraft communication with the satellite is in the NATO D band, with uplink in the 1626.5–1660.5 MHz frequency range, and the downlink in the 1530–1559 MHz frequency range. The ground station communicates with the satellite using the NATO G band (4–6 GHz). Stratos Global Corp. launched Inmarsat's SwiftBroadband service in 2008. There were five main types of configurations using SwiftBroadband currently available on an aircraft in 2009:

1. Aero-H, which provides a near global capability and requires a high-gain aircraft antenna. It is used on long-range routes and provides multi-channel cockpit data, cockpit voice and passenger telephone services at 10.5 kbps for fax and data.
2. Aero-H+, which provides a similar service to Aero-H but at lower cost, since it only accesses the Inmarsat 3 satellite spot beams and the Inmarsat 4 satellite.
3. Aero-I, which provides a multi-channel spot beam service at 4.8 kbps circuit-mode data and fax, requiring an intermediate-gain aircraft antenna. It is used on medium-

and short-range routes and also provides cockpit data, cockpit voice and passenger telephone services.
4. Mini M Aero, which provides a single-channel voice, fax or 2.4 kbps data, for general aviation and smaller corporate aircraft.
5. Aero-C, which provides passenger data services allowing data to be sent to and from a personal computer.

2.15.2 *Globalstar*

The Globalstar constellation consists of 48 LEO satellites in eight planes inclined at 52° to the equator. It is designed for coverage by a single satellite between the latitudes of ±70°, and coverage using two or more satellites between the latitudes of 25° to 50°. The antennas on the satellite provide an elliptical beam, with the major axis of the ellipse being in the direction of motion of the satellite.

2.15.3 *Iridium*

The Iridium fleet of 66 LEO (Low Earth Orbit) satellites, provide global coverage through spot beams. Although there are 3168 beams, only around 2150 will be active at any one time, as some beams will be switched off around the earth's poles where beam overlap occurs. Each satellite is connected to its four neighbouring satellites through inter-satellite links (ISLs). Connections between the Iridium network and the PSTN occur via earth station (ES) gateway installations. In cases where the ES cannot be located, the ISLs are used.

2.15.4 *SKYLink*

ARINC provides SKYLink using Ku band GSS. The downlink frequencies are 11.7–12.7 GHz, and the uplink frequencies are 14–14.5 GHz. The higher frequency allows the airborne aperture antenna to be much smaller (0.3 m diameter) whilst still achieving a 6° HPBW.

At the upper end of the aviation communications market, Arinc's SKYLink service allegedly delivers the highest throughput of any commercially available satellite communications service to aircraft, with inbound speeds between 1 and 3.5 Mbps and a return channel with speeds of 128 to 512 kbps. This is predicted to be the best option for delivering broadband services to passengers on board aircraft.

2.15.5 *Teledesic*

The planned Teledesic satellite system was the most ambitious and consisted of 840 active satellites (924 satellites, including in-orbit spares) in 21 planes in sun-synchronous, inclined, circular LEOs. However, this system now appears to have been abandoned.

2.15.6 *SatCom Airborne Antennas*

The airborne system uses a high-gain antenna steered to a satellite. In the past, steering was generally mechanical, but electronically steered antennas are now becoming available.

Table 2.3 Typical characteristics of some communication satellites/services and associated aircraft antennas.

Satellite/service	Frequencies	Typical type of aircraft antennas	Aircraft antennas gains in dBi
Connexion by Boeing	Uplink 14–14.5 GHz Downlink 11.2–12.7 GHz	Phased array or mechanically steered reflector antenna	High gain – approximately 32 dBi
Globalstar	Uplink 1610–1626.5 MHz Downlink 2483.5–2500 MHz	Various	4 dBi approximately
INMARSAT	Uplink 1626.5–1660.5 MHz Downlink 1525.0–1559.0 MHz	Phased arrays, mechanically steered helix and reflector antennas	High gain 12–16 dBi, medium gain 7–9 dBi and low gain below 7 dBi
Iridium	1616–1626.5 MHz	Printed circuit array of patch antennas	Low gain
SKYLink Ku band	Uplink 14–14.5 GHz Downlink 11.7–12.7 GHz	Phased array or mechanically steered reflector antenna	High gain – approximately 32 dBi

The main disadvantages of a mechanical installation on the upper fuselage are:

1. the length (at least 80 cm)
2. a lack of real estate that would permit a clear LOS view rearwards, because of the tail fin and
3. the requirement to provide spatial isolation from the GPS frequencies which operate around the same frequency band.

By using a number of conformal antennas, one on the top fuselage and one on each side of the fuselage, the use of a large steerable antenna on the top fuselage is avoided.

The characteristics are shown in Table 2.3.

2.16 Signals Intelligence

This system is used in military applications. Signals intelligence (SIGINT) requires the acquisition of the signals as well as the decryption of these signals to extract the information. The SIGINT system is a passive one and thus will only detect emitters of EM radiation, (hereafter referred to as 'emitters') in the 20 MHz to 40 GHz frequency range. The SIGINT system consists of a communications intelligence (COMINT) and an electronic intelligence (ELINT) system.

The decryption of the signals could be done by the ground segment equipment, since the SIGINT would normally have a millimetre wave data link, in order to preserve its integrity. The short-range capability is achieved by atmospheric attenuation, which occurs naturally for millimetre waves. This will only be operated very near a ground node, thereby reducing the risk of detection. The use of a short-range link could also obviate the need for encryption equipment (usually used to transmit the information about the emitters) to be installed on board the platform. The omission of on-board encryption equipment reduces the cost, power budget and total mass, with the subsequent reduction in the stress, environmental control and aerodynamic problems associated with large power budgets and masses. Additionally encrypted data requires transmission in a broader frequency bandwidth than unencrypted data, and hence is more prone to detection.

Aircraft with SIGINT usually only operate at long stand-off ranges, causing a subsequent loss in sensitivity. They tend to be restricted to operation in allocated time frames, which results in restricted information update. The mission programme is essentially the same as for peacetime surveillance, except that the moving target indicator will most probably be switched on more frequently, depending on the treat environment and the operational requirements. The use of longer stand-off ranges in a theatre of war (than in peacetime) necessitates smaller look down angles. This in turn results in greater obscuration of areas on the leeward side of tall structures. There is an overlap of the frequency bands covered by the ELINT and COMINT systems. However, the COMINT system only processes the data and voice of the received signals, whereas in the case of the ELINT system all the characteristics of the intercepted signals are analysed and an attempt is made to identify the emitter and its 3D location.

2.16.1 Communications Intelligence

The COMINT signals cover the lower part of the frequency range, namely 20 MHz to 2 GHz. The voice and data signals are extracted from the intercepted EM signal and direction-finding equipment is used to establish the DOA of the EM signal. Decryption of the signals is also undertaken as part of the COMINT system.

2.16.2 Electronic Intelligence

The ELINT covers the upper part of the EM spectrum, from 500 MHz to 18 GHz. The ELINT antennas acquire the EM signals, and the receiving and processing equipment measures the following parameters:

(a) frequency
(b) PW
(c) pulse repetition interval (PRI)
(d) time of arrival (TOA)
(e) amplitude
(f) antenna rotation period
(g) polarization characteristics
(h) DOA.

The characteristics (a)–(g) are measured by one set of antennas, whereas the DOA is measured by a different set of antennas. The characteristics are compared with those of emitters in an emitter library, which usually forms part of the pre-flight message. A match is established with a degree of confidence, depending on the number of the characteristics of the acquired signal that correspond to the characteristics of the emitters in the library. If no match is found, the emitter is regarded as new or unknown, and its characteristics may be recorded for more detailed analysis at the master ground station. Some ELINT systems also analyse the other characteristics of the emitters such as pulse stagger and jitter.

The AOA in the azimuth plane is obtained if the antenna(s) are only in this plane, but the DOA can be obtained if antennas (and the processing equipment) can separate targets in the elevation plane. Ideally the sensor should be capable of distinguishing between land and airborne threats, and should therefore be capable of performing DF in the azimuth as well as the elevation planes. However if the sensor has a long stand-off range compared with the altitude, then the angles subtended at the sensor by land and air threats will be very close to each other and the angular resolution required in the elevation plane will necessitate the use of very accurate and hence expensive DF equipment in the elevation plane. Thus DF in the elevation plane is more likely to be undertaken at short stand-off ranges than at long stand-off ranges. At an altitude of 38 000 ft (11.4 miles) and a slant range of 50 miles, the angle of depression is of the order of 13°. If the resolution in the vertical plane is 1°, a system with elevation plane DF equipment will only be able to distinguish between targets that are separated by about 1 mile (4600 ft) in altitude at this range. The antennas and processing required are similar to those described in Section 2.7.

2.17 Tactical Air Navigation

Tactical air navigation (TACAN) is used by military aircraft. This system uses the 960–1215 MHz frequency band. It provides the range and bearing relative to the ground station, but can also provide these parameters relative to other points and hence it can, for instance, provide the range and bearing to a theatre of war or military formation. The carrier signal contains squitter (randomly variable pulses), interrogation pulses at 15 Hz and 135 Hz modulation. The PW and spacing of the pulse pairs are 3.5–12 μs but the spacing between the pulse pairs is differently encoded for the squitter, DME, identity, and north reference and auxiliary reference signals. The aircraft receiver decodes the signals. Modern TACAN ground station antennas rotate the beam by electronic means, but the older ones used mechanical rotation. In order to understand the modulated beam produced it is easier to consider the older TACAN ground station antennas. The antenna is a circular array with a number of elements (usually dipoles) in the vertical dimension to provide a shaped directional beam canted upwards in the vertical plane. The circular array would normally result in an omnidirectional pattern in the azimuth plane, with the diameter at each horizontal cut varying depending on the vertical beam shape.

However, by using a small parasitic element a cardioid radiation pattern is obtained as shown in Figure 2.10a. The rectangular plot would be sinusoidal as shown in Figure 2.10b.

If the linear array is rotated at 15 Hz we can see that the cardioid pattern would rotate at this rate. If we now have an outer cylinder with nine linear arrays we would get a pattern with nine lobes instead of the one lobe that we got in the cardioid pattern with the single linear array. In the absence of the inner cylinder we would get a radiation pattern like the rectangular plot shown in Figure 2.11.

If we now consider the parasitic elements on both cylinders we would get the horizontal radiation pattern shown in Figure 2.12a since the nine parasitic element pattern would be

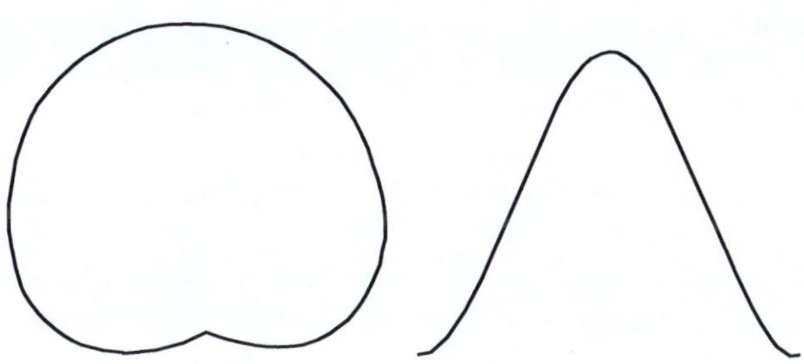

(a) The polar plot (b) The rectangular plot

Figure 2.10 Radiation pattern using a single parasitic element on the inner cylinder.

Figure 2.11 Radiation pattern using nine parasitic elements on the outer cylinder.

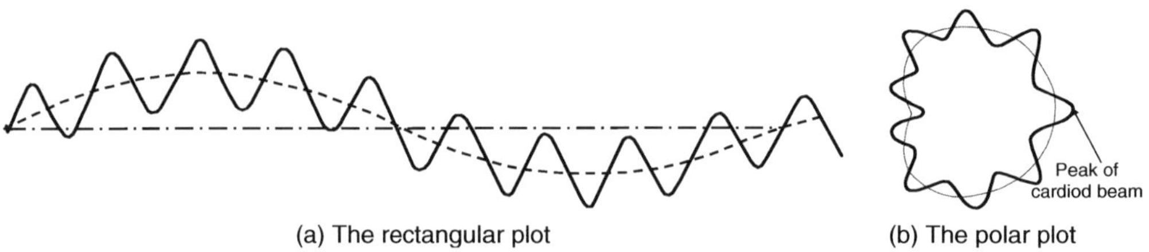

Figure 2.12 Azimuth radiation pattern using one parasitic element on the inner cylinder and nine parasitic elements on the outer cylinder.

superimposed on the single parasitic element pattern. In polar form the cardioid pattern would be altered as shown in Figure 2.12b.

If both cylinders are now rotated at 15 Hz the pattern would rotate at this rate. The ground station sends out a reference pulse each time the peak of the pattern points in a particular direction. The aircraft can then derive its bearing from the received signals.

2.18 Traffic Collision Avoidance System

An aircraft collision avoidance system (ACAS), more commonly known as a traffic collision avoidance system (TCAS), requires two antennas, one on the upper fuselage and one on the lower fuselage. Both antennas are ideally passive four-element phased arrays, but in some cases the lower fuselage antenna could be an omnidirectional one. It is used to receive and provide directional information for 1090 MHz ATC transponder interrogations, mode S, and air traffic control radar beacon system (ATCRBS) replies. Proper phasing of the four antenna elements provides for directional as well as omnidirectional transmission of 1030 MHz broadcast or coordination messages and ATCRBS or mode S interrogations. Mode S only responds when interrogated rather than every time the ATC radar sweeps past. Maximum forward operational range is about 30 nmi.

If the antenna installed on the bottom of the fuselage is an omnidirectional one, and nearby traffic is masked from the antenna mounted on the top, TCAS continues to compute avoidance manoeuvres from range and altitude information but may not be able to display the relative bearing of the traffic.

The TCAS-94 system (from Rockwell Collins) interrogates nearby aircraft transponders in the same way as ATC radar but with an architecture that increases efficiency so that overloading the transponders is avoided. TCAS-94 typically tracks transponder equipped aircraft traffic within 35 nmi. Traffic is displayed if the reported altitude is within ±2700 ft of the aircraft and ±9900 ft if the 'above/below' feature is available and selected. TCAS can display up to 30 aircraft and tracks all aircraft within range. It selects for display those that pose the greatest collision threat to the aircraft.

The ground SSR antenna rotates together with the primary radar antenna and provides additional information to the ground control operator.

Currently (in 2008) TCAS works in three different modes: A, C and S. In the case of mode A, a four-digit octal code attributed to the ground control is set in the aircraft and allows identification of a primary radar plot (of the other aircraft or the ground station radar) among all the displayed radar echoes. In the case of mode C, the pressure altitude can be transmitted from the aircraft, and gives a 3D position to the ground control operator. In the case of mode S, the main feature of the transponder is to respond to selective interrogations. It responds to TCAS only when addressed but always responds to the ATC's radar beam sweep. It allows automatic identification of each aircraft by sending an unique 24-bit aircraft address. It may also be interrogated in mode A or C. In addition, mode S SSR has a data link capability which allows transmission of data necessary to improve air safety. The timing of the mode S interrogation/response protocol is measured to ascertain the distance of an aircraft from the TCAS aircraft.

TCAS and ATC mode S work together in providing protection for a predefined volume of airspace around the TCAS-equipped aircraft with a surveillance area range of approximately 14 nmi minimum to 40 nmi typical. Interrogation of ATC mode C or mode S transponders of nearby aircraft can also take place to determine their bearing, altitude and range.

The directional antenna (array) determines bearing to the other aircraft, the mode C data replies from the other aircrafts' transponders (to provide altitude information) and the TCAS system determines how fast the threat aircraft altitude is changing. The target aircraft's time to respond allows the TCAS computer to measure range and rate of change of range. Avoidance information is provided to the crew via vertical manoeuvre resolution advisories. Nearby traffic is also indicated to the crew on a display for visual acquisition of targets.

TCAS I allows the pilot to see the relative position and velocity of all aircraft within a range of 10–20 miles, and has a traffic advisory capacity that provides a warning when an aircraft in the vicinity gets too close. TCAS I does not provide instructions on how to manoeuvre in order to avoid the aircraft, but does supply the pilot with important data for him to make the manoeuvres.

TCAS II provides pilots with airspace surveillance, intruder tracking, threat detection and avoidance manoeuvre generations and is able to determine whether each aircraft is climbing, descending, or flying straight and level. It suggests an evasive manoeuvre necessary to avoid the other aircraft. If both planes in conflict are equipped with TCAS II, then the evasive manoeuvres will be well coordinated via air-to-air transmissions over the mode S data link. The mode S squitter, which normally transmits altitude information, has now been enhanced as mode S extended squitter that transmits positional information that comes from a source of global navigation, such as a GPS receiver. Under automatic dependent surveillance-broadcast (ADS-B) each aircraft periodically broadcasts its identification, position, and altitude. These broadcasts can be received by ground sensors and other aircraft for surveillance [4].

2.19 Telemetry

In the case of aircraft, telemetry can be defined as the transmission of the data and results of measurements of the aircraft characteristics to the ground station for display, recording, processing and analysis. Telemetry antennas are only installed on the prototypes of commercial or military aircraft and therefore do not appear on the entire fleet. However, in the case of police helicopters, for instance, they would be fitted to the entire fleet, and be relayed to the ground station in real time.

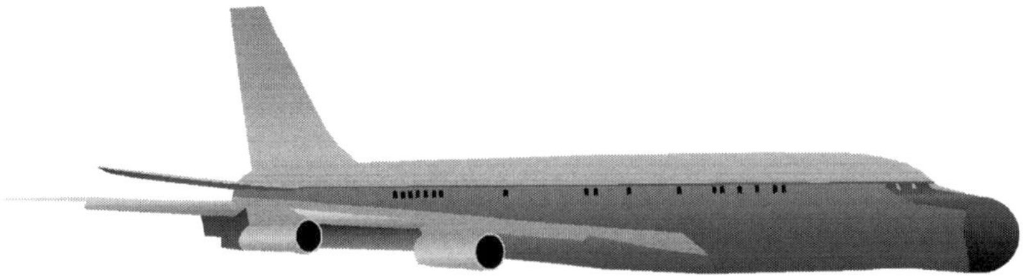

Figure 2.13 ARIA aircraft with large nose radome.

The frequency band used is usually 1.9–2.5 GHz, although other frequency bands (such as 4.15–4.25 GHz) are used for certain applications as in the case of the Advanced Range Instrumentation Aircraft (ARIA), and 4.9 GHz as well as 6.4 GHz are used for police helicopters.

In the case of the prototype or first production aircraft, the telemetry antennas would be installed on the lower surface and consist of one or two blade antennas and used for the flight trials in certification.

In the case of the police helicopters higher gain antennas such as Yagis would be used and they would be steerable so that continuous contact with the ground station could be maintained.

The ARIA which is a modified Boeing 707 has a bulbous, 'droop snoot' nose, consisting of 2.1 m reflector antenna housed in a 3 m diameter radome (Figure 2.13).

2.20 UHF

The aircraft frequency band for UHF comms is 225–400 MHz. This frequency band is used mainly for military communications. Instead of using encryption for security the UHF radios rely on frequency hopping, using HAVE QUICK. The security is maintained by providing authorized users with a word of day (WOD) key and a net number. The radios are initialized with accurate time of day (TOD) usually from a GPS receiver. The WOD, TOD and net number are input into a cryptographic pseudorandom number generator that controls the frequency changes. Although HAVE QUICK can be used with encryption equipment, most of the transmitted data is not encrypted. Newer radios support the Single Channel Ground and Airborne Radio System (SINCGARS) – the frequency hopping system used by ground forces.

2.21 VHF Comms

This very basic system is fitted on all aircraft since all primary communications use VHF comms. The operating frequency is between 118 and 136.975 MHz with 8.33 kHz frequency spacing.

Civil ATC uses the 118–136 MHz frequency band in the VHF band and UHF for military aircraft. Recent expansion of the VHF aeronautical band has taken the upper limit from 135.975 to 136.975 MHz. ATC facilities are equipped to transmit on both the VHF and UHF frequency bands simultaneously.

Control towers use the 118.000–121.400 MHz frequency band to communicate with the aircraft, and 121.600–122.900 MHz is used for ground and apron control. The air-to-air universal communications (UNICOM) frequency is 123.450 MHz, and 124.000–128.800 MHz is used for arrivals and departures. En-route area control centre communication uses the 132.000–135.975 MHz frequency band. The frequency band 136.000–136.975 MHz is used for shared ATC/company operations and data link.

2.22 VHF Omnidirectional Ranging

The VOR system operates in the frequency range 108–117.95 MHz but is due to be phased out and the frequency band used for ILS localizer channels. The phasing out was scheduled to commence in 2005 and to be completed by 2010. However, it was still in use in 2009. There are two different sectors in the VOR system, namely en-route and terminal navigation.

En-route navigation can provide coverage up to 130 nmi or more, whereas terminal navigation provided by 'terminal' or T-VORs have a relatively small geographic area up to about 25 nmi. In the case of en-route VOR, because of the earth's curvature and the ground station radiation pattern, the range is much longer at high altitudes (above 18 000 ft). Thus, for instance, at low altitudes the range might only be 40 nmi, whereas at high altitudes this could extend up to 200 nmi.

The aircraft can fly from one VOR station to the other in straight lines between these stations or it can use the VOR station to guide it along its pre-determined flight path.

A VOR ground station broadcasts two signals, a primary signal and a secondary signal. Early vacuum tube transmitters had mechanically-rotated antennas and were widely used in the 1950s. However, with the advent of solid state transmitters, the mechanical rotation was also replaced by phase rotation of the beam in the early 1960s.

The primary signal provides a phase reference via an omnidirectional antenna and radiates in a circular pattern. This phase is constant throughout the entire 360° of azimuth. The carrier signal is modulated at 30 Hz and the modulated radiation pattern is rotated at 30 revolutions per second or 1800 rpm using the same principle as scanning phased arrays. This results in a phase that varies at a constant rate as the field is rotated.

The VOR uses the phase relationship between a reference-phase and a rotating-phase signal to encode direction. The carrier signal is omnidirectional and contains an AM station Morse code or voice identifier. The reference 30 Hz signal is FM on a 9960 Hz sub-carrier. A second, AM 30 Hz signal is derived from the rotation of a directional antenna array 30 times per second. When the signal is received in the aircraft, the two 30 Hz signals are detected and then compared to determine the phase angle between them. The phase angle is equal to the direction from the station to the aircraft, in degrees from local magnetic north, and if the aircraft flies towards a VOR station it is said to be flying on a radial. The bearing of the VOR is obtained regardless of the aircraft heading, as long as the aircraft remains in the same track envelope. Then the pilot tunes to the second station and the intersection of the two bearings gives the aircraft's position.

Generally, VOR and DME transmitters are collocated for providing a rho/theta-type position (distance and bearing of an aircraft to a beacon) and VOR and DME frequencies are coupled so that both systems operate in unison.

At high altitudes the distances at which the VOR signals can be received is large, and this results in reception from more than one ground station. If the two signals are at the same frequency erroneous indications could result. In order to avoid this, stations on the same frequency are spaced as far apart as possible. Since there are only 160 frequencies available, the separation in frequency between collocated stations is not always possible. The solution has been to design and classify VORs according to the usable cylindrical service volume. This is the system by which VOR frequencies are assigned to stations far enough apart to prevent overlapping, confusing signals. As long as pilots use the proper chart, they are protected from interference between two VORs. The pilot uses low-altitude charts below 18 000 ft and the high-altitude charts at and above 18 000 ft.

The accuracy of the VOR system is theoretically $\pm 1.4^\circ$. However, errors of less than $\pm 0.35^\circ$ have been found 99.94% of the time. Internal monitoring of a VOR station ensures that it will shut down if the station error exceeds 1.0°. The on-board equipment must be strictly maintained and tested for errors. The maximum bearing error that can be tolerated is $\pm 4^\circ$. The pilot can test the accuracy of the on-board VOR equipment in four different ways.

1. A VOR test facility signal can be used; these are available at many airports.
2. VOR checkpoint signs which are located beside taxiways can be used. In these areas, there is sufficient signal strength from a VOR to check aircraft VOR equipment against the radial designated on the sign.
3. Dual VOR checks can be used if the aircraft has dual independent VOR systems. In this case both the systems are tuned to the same VOR station and the difference in the readings noted.
4. Airborne checks can be obtained by flying over a landmark located on a published radial from a VOR station, and noting the indicated radial.

2.23 Equipment Designation

Equipment is given an unique designation under the Joint Electronics Type Designation System (JETDS). Under this system the equipment is denoted with the letters AN followed by three letters and then numerals to denote the model number.

AN used to stand for 'Army-Navy' initially, since this designation was used during World War II by the joint US Army and Navy. When the Air Force separated from the Army in 1947, this system continued to be used and was later extended to encompass the new types of equipment. The letters AN are still used to preface the three other letters, and are simply an indicator for the JETDS. The first JETDS letter (after AN) is used to denote the type of location on which the equipment is installed, the second letter denotes the type of equipment, the third letter denotes the purpose for which the equipment is used. The details of the designations for the first three characters are shown in Table 2.4.

The fourth character is used to denote the model number. Each installation–type–purpose letter combination uses its own model number sequence, starting at 1. Model numbers 501–599 and 2500–2599 are reserved for use by Canada, Australia uses 2000–2099, New Zealand uses 2100–2199, and the UK uses 2200–2299.

The fifth character is an optional suffix letter and denotes a specific version of the equipment. The first version uses no suffix, the first modification uses 'A', and so on.

Table 2.4 JETDS designation.

First letter – installation	Second letter – type of equipment	Third letter – purpose
A – Piloted aircraft	A – Heat radiation (i.e. infrared)	A – Auxiliary assembly
B – Underwater, mobile (submarine)	B – Cosmic (secure communications). Until 1957, B was used to denote pigeons	B – Bombing
C – Cryptographic equipment (is not an 'installation' letter in the proper sense)	C – Carrier (electronic wave or signal)	C – Communications
D – Airborne carrier unpiloted, for example missile, drone, uninhabited/unmanned air vehicle	D – Radiac (radioactivity detection, identification and computation)	D – Direction finding, reconnaissance and surveillance
F – Fixed ground	E – Laser. In the past E was used to denote NUPAC (nuclear protection and control)	E – Ejection and/or release
G – Ground, general (multiple types of ground installation). For instance, the equipment could be used as a fixed or mobile ground). G is used when two or more of 'F', 'M', 'P', 'T' or 'V' apply	F – Fibre optics. In the past F was used to denote photographic	G – Fire control or searchlight directing
K – Amphibious	G – Telegraph or teletype	H – Recording and/or reproducing
M – Ground, mobile. The letter 'M' is only used if the vehicle housing the equipment is solely designed for this purpose, that is, the vehicle is a part of the equipment	I – Interphone and public address	K – Computing
P – Man portable	J – Electromechanical	M – Maintenance or test
S – Surface ship	K – Telemetering	N – Navigation aid. In the past N was used to denote reproducing
T – Ground, transportable. It is intended for equipment which can be moved, but cannot be operated while in motion	L – Countermeasures	Q – Special or combination

(*continued overleaf*)

Table 2.4 *(continued)*

First letter – installation	Second letter – type of equipment	Third letter – purpose
U – General utility or combination	M – Meteorological	R – Receiving or passive detecting
1. Equipment items (e.g. a radio which can be used in different installations – aircraft, ship or ground)		
2. Systems, which consist of several components installed in different locations (e.g. system has an airborne component and a ground-based component)		
V – Ground, vehicle	N – Sound in air	S – Detecting, range and bearing search
W – Water surface/underwater combination	P – Radar	T – Transmitting
Z – Piloted/pilotless airborne combination	Q – Sonar and underwater sound	W – Automatic flight or remote control
	R – Radio	X – Identification or recognition
	S – Special or combination	Y – Surveillance (target detecting and tracking) and control (fire control and/or air control)
	T – Telephone (wire)	Z – Secure
	V – Visual, visible light	
	W – Armament (only used if no other letter applies)	
	X – Fax or television	
	Y – Data processing	
	Z – Communications	

However, the letters 'I', 'O', 'Q', 'S', 'T', 'X', 'Y' and 'Z' are not used as version suffices. The letters 'I', 'O', 'Q', 'S', are not used as version suffices, since they could be confused with numerals 1, 0, 0 and 5 respectively. The letter 'T' is used for training equipment and a number added after the letter to denote the training set. For instance, the number 2 after the letter T (i.e. T2) would indicate that it is the second training set. The letters 'X', 'Y' and 'Z' are used for modifications of an equipment which only change the electrical power input requirements, such as a change to the voltage and/or frequency.

If the 'V' symbol is used as the sixth character, it indicates equipment with variable components (sets, groups or units). A number following the 'V' is used to designate a *specific* version of the equipment, that is, with a specific component configuration. If a component of a set or system is of variable configuration, that is, carries a 'V' symbol, the set or system itself must also use the '(V)' symbol.

In the case of AN/APG-5A, the first letter A stands for Aircraft, the second letter P stands for radar, the third letter G stands for fire control or searchlight directing, the fifth character is the model number 5 and the letter A indicates that it is the first modification.

References

[1] International Civil Aviation Organization (1996) *International Standards and Recommended Practices Aeronautical Telecommunications*, 5th edn, vol. 1, (Radio Navigation Aids), ICAO, Montreal.

[2] Donald, D. and Lake, J. (eds) (1994) *Encyclopedia of World Military Aircraft*, vol. 1, pp. 74–76, Aerospace Publishing, London, ISBN UK 1 874023 51 4, US 1 880588 14 5.

[3] Moir, I. and Seabridge, A. (2006) *Military Avionics Systems*, John Wiley & Sons, Inc., Hoboken, NJ, ISBN-13 978-0-470-01632-9.

[4] www.rockwellcollins.com/ecat/br/TCAS-94_PrintFriendly.html.

3

The Antenna Siting Process

3.1 Introduction

The siting process is as equally applicable to entirely new layouts as it is to single retrofits. This is because the installation of even a single antenna could have a knock-on effect on all the other antennas.

Since the radiation from an antenna depends on its type and location, the selection of these sites is of paramount importance to a successful antenna layout.

The antennas also have to operate simultaneously in most cases, so the interoperability between the different systems on an aircraft has to be achieved, as far as is practically possible, within the constraints of the available real estate as well as the other constraints of an air vehicle.

Whether antennas are selected for new aircraft as entirely new layouts, or as retrofits for existing aircraft, RF interoperability issues have to be addressed, apart from all the other trade-offs that have to be applied to meet the limitations of aerodynamic, mechanical and physical constraints associated with the installation of antennas on airframes.

Most antennas are installed outside the airframe and are therefore classed as excrescences. Under the IPAS project, a code of practice was developed for the antenna siting process on aircraft [1].

3.2 New Antenna Layouts

The systems are usually selected first, and then antenna locations are selected that are near the systems to which the antennas are connected. Short distances between the antenna and its associated system are very important especially at the high frequencies where the losses in RF cables can be very high. For the radar antenna, for instance, which is usually in the nose cone, the transmitter and RF mixer/receiver are located directly behind the antenna. Waveguides are also be used in preference to coaxial cables in order to minimize losses and to handle the high RF powers transmitted. The positions of the displays relative to the processing LRUs also have to be established. Displays for commercial aircraft and fighter jets would be in the cockpit at the front, but large surveillance aircraft and bomber jets could have the displays distributed throughout the cabin, with various mission

Introduction to Antenna Placement & Installation Thereza M. Macnamara

stations allocated to specified locations and with the mission commander supervising all the different activities such as SAR surveillance, tracking radar, ESM, ECM, magnetic anomaly detector, Sonobuoy system, and so on.

Antennas with high priority systems and/or physically large dimensions would have their locations fixed first. These include antennas for systems listed below.

1. ATC/TCAS antennas have to provide substantially omnidirectional coverage of the aircraft's upper and lower airspace and are dedicated to both air traffic control and collision avoidance, as well as being within line of sight (LOS) of the ATC ground station. Thus they are usually mounted near the front of the fuselage.
2. DME systems determine the distance to a selected way point. These are installed on the lower fuselage. For redundancy, two DME antennas are usually installed on most military aircraft.
3. Doppler navigation radar has to be installed on the lower fuselage. This system is usually installed on older military aircraft.
4. GPS antennas have to be on the upper fuselage near the inertial navigation system, which in turn have to be near the centre of gravity on the upper fuselage. For redundancy, two GPS antennas are usually installed on most commercial airliners and large military aircraft.
5. ILS glideslope and localizer antennas have to get an unobstructed view forward when landing and are therefore usually installed in the nose cone above the radar, on the tail fin or on the forward fuselage.
6. The ILS marker antenna has to be installed on the lower fuselage. Aircraft with Category III landing must have two ILS marker antennas.
7. HF built-in shunt/notches have to be installed in areas such as the dorsal or tail fin.
8. HF towel rails have to be installed on the rear fuselage or tail.
9. HF wire antennas (not widely used now) formerly installed between the upper fuselage and tail fin, or the wings and tail fin. Because of their longer lengths they have higher gain than their shunt or towel rail counterparts.
10. MLS antennas have to provide an unobstructed view forward (when the aircraft is landing in nose-up attitude) as well as coverage to the side when banking and coverage to the rear during take-off or missed approaches. Three MLS antennas may have to be installed.
11. RadAlt antenna pairs have to be installed on the lower fuselage. Two antenna pairs may be required for redundancy, especially on military aircraft or large commercial airliners with a long wingspan.
12. The radar (weather or search) antenna has to be installed in the nose cone.
13. SAR antennas have to be located perpendicular to the direction of motion of the aircraft and are therefore either installed on the port and/or starboard side of the fuselage, or in a separate pod underneath the fuselage.
14. SatCom antennas must be mounted on the upper surface or dual phased arrays on sides of the fuselage and are shaped to conform to the shape of the fuselage.
15. VHF/UHF communication antennas and VHF navigation antennas must be installed to give substantially omnidirectional coverage, as well as even coverage in the roll planes within the constraints of physical phenomena. They are usually installed along the longitudinal/vertical plane of the symmetry of the fuselage but not above the wings – see Section 4.7.

The antennas of each system are required to provide the optimum spatial coverage at the correct frequency, as well as to provide adequate gain to attain the correct range.

The spatial separation between antennas is calculated to provide the optimal isolation between systems in the spatial domain. Where the RF interoperability cannot be achieved by spatial separation, separation in the time and frequency domains is used as well as the reduction in the power of transmitters and/or reduction of the sensitivity of the receivers.

The total antenna layout is achieved as a result of trade-offs due to the constraints of the mechanical and other characteristics of the structure.

The antennas are usually subject to mechanical, stability and aerodynamic considerations. The electromagnetic requirements of antenna locations are low in the pecking order of the various disciplines. This is to be expected, since there is no point in having prime locations for the antennas, if aircraft cannot fly.

Antennas installed on aircraft and other structures behave very differently from those measured on standard ground planes by antenna manufacturers. The spatial power that is dispersed by the antenna in all directions is very different when the antenna is located on a structure. Since antennas provide the interface with the world outside the airframe it is essential that the spatial distribution of this power is known and suitable measures are taken for any deficiency in a particular direction. For details of the effects of wings, tail fin, and so on, see Chapter 4.

3.3 Optimum Positions for Blades

Blades are basically monopoles encapsulated in a covering to provide better aerodynamic properties. The higher frequency narrow band blades such as MLS are tuned quarter-wave monopoles that could be simple rods/wires approximately a quarter of a wavelength long.

The broader band antennas such as V/UHF blades that span several octaves are often on printed circuit boards and are shaped to give broad band performance. In general broad band antennas will have reduced gains over the entire frequency band and the antenna will only act as a quarter-wave monopole over a very narrow band of its frequency range. They may also have more than one element and thus have more than one RF connector. Some broadband antennas are top loaded (i.e. there is a bar or disc at the free end) to give an increased effective length and hence higher gain at the lower frequencies. However, these horizontal sections introduce a component of horizontally polarized radiation and usually result in a reduction in the cross polar discrimination (XPD). In the case of a vertical monopole, the XPD is the ability of the antenna to reject the horizontal or cross polar radiation. For details of the radiation patterns of blades at different positions on the fuselage see Chapter 4.

3.4 Design Phase

The process of siting antennas on the aircraft is undertaken in the initial stages of the installation of antennas when all the details of the system(s) may have not been finalized or are unknown.

Currently the typical design phase of placing antennas on an airframe consists of:

1. an initial paper design stage

2. an investigative and computational modelling stage, leading to a design freeze
3. a verification and implementation stage.

This is an iterative process and may be looped through several times before the first airframe is cut.

3.4.1 *Initial Paper Design Stage*

The process adopted in the initial paper design stage of the antenna layout is likely to be as follows:

1. collation of data
2. selection of antenna sites
3. positioning of neighbouring antennas – spatial separation
4. spreadsheet analyses
5. computer modelling.

3.4.1.1 Collation of Data

Collation of data is required in order to define the requirements and generate the preliminary antenna specifications after consultations with systems and antenna suppliers. The antenna may also require a tuning unit, as in the case of the HF antenna, and this unit has to be placed as near to the antenna as possible. It is usually a fairly large unit and also may require shielding, so its positioning is a major consideration. Other details of the system such as RF connectors may not be available but their details would not be essential in the early stages of the design.

Data affecting antenna characteristics need to be obtained, such as:

(a) frequency band
(b) the polarization
(c) the range and the power output for transmitting systems
(d) the range and sensitivity for receiving systems
(e) the spatial coverage of the system.

Frequency Band

The operating frequency band of the antenna must match that of the system. The frequency band of the antenna could be wider than that of the system, which would still result in satisfactory operation of the system. However, this is not desirable from the interoperability point of view, because it would mean that frequencies outside the band of the system could affect its performance. In the case of voice transmission it is desirable to have as wide a band as possible, since the fidelity of the speech is affected if harmonics are not received. It is the harmonic content of speech that characterizes the sound of a person's voice. Thus, for instance, the old telephones made it difficult to recognize the voice at the other end of the line. This also accounts for the fact that the same note on two different instruments sounds so different, and the instrument can be recognized by its characteristic sound. Tuning forks are used to tune pianos because they only resonate at the base or carrier frequency – see Section 5.12.1 for an explanation of the Fourier transform. The frequency band also impacts the spatial isolation since the latter depends on the electrical distance between the antennas.

Polarization

The polarization of the electric field will obviously determine the orientation of the antenna. Thus, for instance, the glideslope and localizer systems used for the ILS both operate on horizontal polarization, so the antennas are mounted horizontally. The VHF and UHF communications systems, on the other hand, have to transmit and receive vertical polarization and are therefore mounted vertically. This will affect the relative positioning of antennas since ideally there would be no coupling between a horizontally and a vertically polarized antenna and they could be installed close to each other as long as physical LOS obscuration is avoided. In practice there will be coupling at the XPD level as described in Section 1.4.

Range and Power Output of Transmitting Systems

It is often difficult to obtain any data on the output power of equipment from suppliers in the initial stages of the supplier selection process, since they would seek not to divulge any details until they are selected. Additionally, so-called commercial off the shelf (COTS) equipment is a misnomer, since most equipment is adapted for use in a particular fleet of aircraft. Thus, suppliers themselves do not often know the power output that their equipment will transmit. However, from the specification of the range for the system the power output can be calculated by working out the space loss at the highest frequency and then making allowances for the losses in the RF cables, antenna gain, and so on. Details of the formula for calculating space loss are given in Section 6.5.1.

Range and Sensitivity of Receiving Systems

As in the case of the power output of equipment, details of the sensitivity are rarely available in the initial design stages. However, the range required of the system will be specified and the space loss can be worked out for the highest frequency of the system. When working out the loss experienced by the RF cables it is always wise to err on the side of caution and use a factor of 1.5–1.75 times the shortest distance between the antenna and its receiver, for the cable lengths. This because the RF cables are usually tied in looms and may therefore follow tortuous paths. It is also wise to allow for one or two cables breaks. Low-loss cable can be used, but this type of cable usually has less dielectric between the centre and outer conductor and is therefore less robust as well as being much more expensive. Realistic sensitivities should also be assumed since the highly sensitive receivers are also more expensive than those with average sensitivity.

Spatial Coverage

The spatial coverage required by each antenna will depend on the requirements of the associated system.

3.4.1.2 Selection of Antenna Sites

Initially the antenna sites are selected to provide the required LOS spatial coverage with respect to the aircraft dynamics, within the limitations of other constraints. This will require inputs from other disciplines, such as aerodynamics, weights, stress, and stability. Simple LOS obstruction and first-order obvious specular reflection off other parts of the airframe and excrescences can also be undertaken at this point. As a first pass activity, 3D obscuration plots can be obtained using a computer-aided design (CAD) package such as CATIA (Computer Aided Three-Dimensional Interactive Application). A 3D external profile of the airframe is required and the obscuration plots are obtained by assuming that the antenna is a point source. Graphs are produced like Mercator projections on to a cylinder (opened out onto a rectangle) giving the obscuration in elevation and azimuth presented to the antenna by the rest of the fuselage and any excrescences. For a totally new antenna layout on an airframe the locations of the antennas will be prioritized as listed in Section 3.1. Any placement requirements such as the positioning of the system equipment would also be considered and if there is a restriction on the loss that can be tolerated, as in the case of the GPS antenna, for instance, then a compromise has to be reached between the optimum location for the spatial coverage of the antenna and the system losses.

3.4.1.3 Positioning of Neighbouring Antennas

After investigating the LOS obscuration between antennas, simple spatial separation (using Friis' formula – see Chapter 6) is used to provide the required isolation, when the details of systems and the antennas are not known. Often a separation such as two or three wavelengths (at the lowest operating frequency of the two systems) is chosen. If two antennas are one wavelength apart the spatial isolation is 22 dB. If this is doubled to two wavelengths the spatial isolation is 28 dB. Doubling this distance again to four wavelengths results in a spatial isolation of 34 dB. Each time the electrical distance is doubled the spatial isolation increases by 6 dB. The electrical distance is the distance divided by the wavelength. For further details of spatial isolation see Section 6.5.1. If the separation is three wavelengths this corresponds a spatial isolation of 31.5 dB. Note that this only gives the isolation due to the LOS electrical distance between the two antennas. It does not take into account any atmospheric attenuation, gain of the antennas, RF cable losses, transmitter power or receiver sensitivity.

Although standards give indications on recommended space isolation they tend just to state a distance from other antennas – for example, SSR antennas must be placed 1 m from other antennas. They do not state the type or frequency of the 'other antennas', nor does this allow for the transmitter power or receiver sensitivity, and thus it does not give the isolation between the systems. As mentioned in Section 3.4.1.1, antennas that are horizontally polarized can be placed nearer those that are vertically polarized since the XPD provides additional isolation. Thus if a vertically polarized antenna has an XPD of 15 dB that means that there will be an isolation of 15 dB (when it is placed near a horizontally polarised antenna) in addition to the space loss provided by the separation of the antennas.

3.4.1.4 Spreadsheet Analyses

Spreadsheet analyses are undertaken if details of the system and antenna are better defined or can be deduced. Since the repositioning of one or more antennas can have a 'knock-on' effect on the entire antenna layout, several iterations may have to take place before the first general assembly (GA) of antenna layout is established.

The isolation between pairs of antennas (that are within LOS) is calculated using the formulas given in Sections 6.5 and 6.6. on RF interoperability. These formulas provide values that are as near to the measured values as computer modelling predictions, and in some cases the calculated values correlate better with the measurements.

The spreadsheet is set up for the lowest frequency common to the two systems. Thus, for instance, if the two antennas being considered are a VHF antenna operating between 30 and 225 MHz and an UHF antenna operating between 225 and 400 MHz, the electrical distance we have to consider is the distance divided by the wavelength at 225 MHz. The spreadsheet would be used to calculate the coupling between each pair of antennas and thus if there are six antennas being considered the minimum number of calculations would be ${}^6C_2 = 6!/2!4! = 15$. For further details see Section 6.5. Since the exact figures for the gains of antennas, RF losses, transmitter power and receiver sensitivity may not be known, several values for these parameters would be used to determine bounds for the coupling calculations. More than one frequency may also be used, although the lowest frequency would provide the worst case for the coupling. Thus a fairly large spreadsheet would result and it could be populated with the correct data as this becomes available.

3.4.1.5 Computer Modelling

Computer modelling of the most critical antennas on the structure may also be undertaken at this stage. For instance, in the case of a radome some modelling may be undertaken to determine the type of radome to be used and whether a sandwich radome is required. The lead times for construction of the radome may also be a governing factor in this decision.

The quicker and less expensive forms of computer modelling may be undertaken such as geometrical theory of diffraction (GTD) using canonical models of the airframe surface as described in Section 7.5.

A flow chart outlining the different stages and iterations in the initial paper design phase is shown in Figure 3.1.

3.4.2 Investigative and Computational Modelling Phase

The more detailed investigation of all the aspects of antenna selection would take place at this stage, as summarized in the flow chart of Figure 3.2. This would involve the various disciplines (airframe, mechanical, integration, aerodynamics, stress, stability systems, electromagnetic health (EMH) and would result in decisions being taken on the following:

1. The exact positioning of antennas with respect to ribs, frames, doublers and other excrescences. The ribs and frames (see Figure 6.1) that form the skeletal framework to which the plates are riveted, should not be punctured if possible. However in some cases such as TCAS it is essential that the antenna array is installed on the spine of the fuselage.
2. Sharing of antennas between systems. For instance, the GPS antenna might be shared with systems such as the radar that require positional information.
3. Multiple antennas for systems. VHF and HF radios often have an antenna on the upper as well as the lower fuselage for each system. Note that the term 'radio' denotes a radio that transmits as well as receives, unlike domestic radios that only receive.

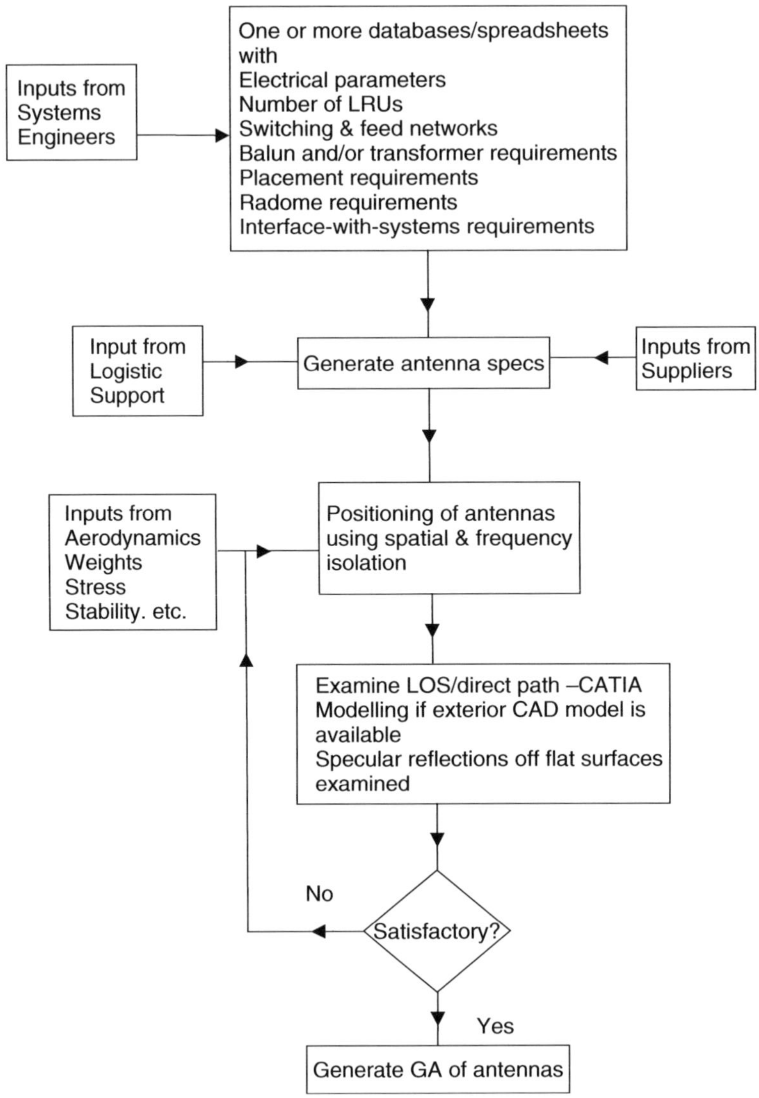

Figure 3.1 Simplified flow chart of the initial design phase.

4. Suppressed and conformal antennas. Suppressed antennas are those that are installed so that they do not protrude from the aircraft skin. They are usually covered with a fibreglass cover that acts as a radome. Examples of suppressed antennas are the older type RadAlt antennas that are rectangular pyramid waveguide horns. Most modern aircraft now have patch antennas instead of horns. The ILS marker antenna could also be a suppressed one.
5. Frangible antennas. These antennas are usually installed on the lower fuselage. They break off in the event of a 'belly' landing, when the aircraft landing gear fails to

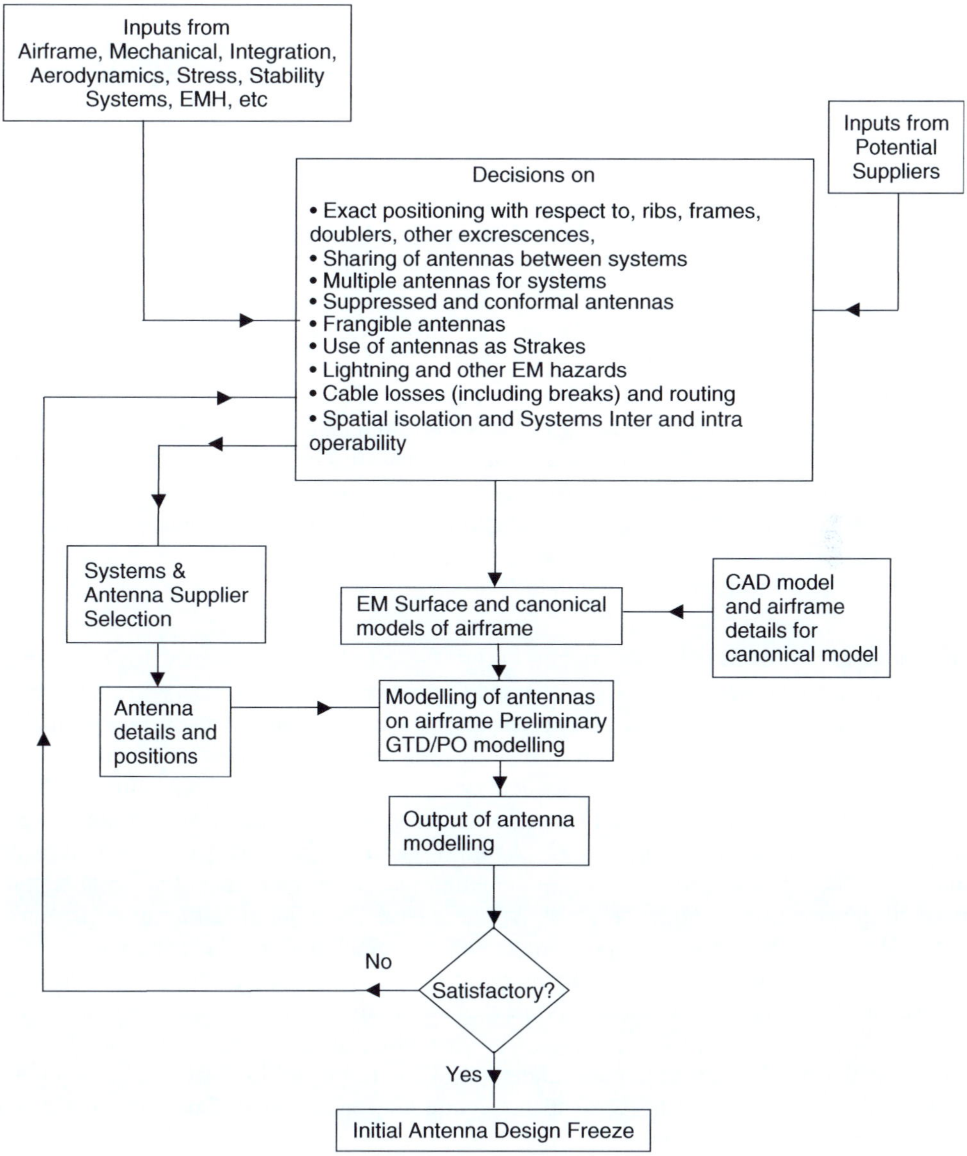

Figure 3.2 Simplified flow chart of the computational modelling phase.

deploy. This prevents damage to the airframe, which would occur if the antennas were not frangible.

6. Use of antennas as strakes. For aerodynamic reasons the turbulence caused by some protrusions requires the use of 'fins' behind them to provide lateral stability. These fins are similar to blade antennas, so in some cases, especially in the case of small fighter aircraft, a blade can serve a dual role and act as a strake. The Nimrod MR2 had 'finlets'

(vertical stabilizers) installed on the tail plane that provided the additional stability required when a refuelling probe was installed on some aircraft for the Falklands war.
7. Lightning and other EM hazards. Particular areas of the airframe are usually identified by the EMH specialist and these have to be avoided as antenna locations since they serve as catchment points for lightning. The radomes usually have lighting strips to protect the radar equipment inside the radome. Some blade antennas also have an embedded lightning strip at the leading edge.
8. Cable losses (including breaks) and routing. The losses of RF cables increase with frequency. In some cases, such as the MLS rear antenna that operates at 5.2 GHz, the losses are so large that an amplifier has to be used at the antenna terminal to boost the signal and thus compensate for the losses. Low-loss cable may also be required.
9. Spatial isolation and systems interoperability and intraoperability. The system intraoperability is the responsibility of the supplier of the equipment. However, although individual system LRUs may pass the electromagnetic compatibility and electromagnetic interference test, when several systems are connected the inter-modulation cross-products (see also Chapter 6) may result in a failure for the connected systems. The resultant of waves of different amplitudes and phases is shown in Figure 4.14. The greatest permissible spatial separation is used as a contingency, especially when all the system details are not available.

The above measures may be undertaken before the suppliers are selected for the systems and/or the antennas, but input from potential suppliers would still be required. This is because although the electrical characteristics of the antennas from different suppliers may be very similar, the external profiles of the antennas may be very different. Thus, for instance, some antennas may be top loaded, whereas others may be longer instead. Some so-called active antennas may require a logic unit to tune the antenna. These antennas give better gain at certain frequency bands, but the logic units have to driven by compatible equipment. It is prudent to purchase the antenna and equipment from the same supplier to avoid demarcation disputes between suppliers, with each supplier blaming the other.

The problems of logistic support may play a crucial part in the supplier selection. Logistic support includes the provision of spares. It is practical and cost-effective to choose the same model of antenna where possible in order to reduce the number of spares. Thus for an aircraft with two or more radios, the same antennas for all radios used over the whole fleet of a particular type of aircraft would be the preferred choice. This would reduce the number of spares required, especially if these are required at geographically distant locations.

In some cases the system supplier will also be responsible for the antenna(s) connected to the system. However, the performance of the antenna on the aircraft will still be the responsibility of the aircraft integration team, since in general the supplier would only accept responsibility for the antenna performance if the antenna is mounted on a standard ground plane.

The creation of the geometric model of the surface of the airframe would involve using the CAD surface model to provide an EM surface for the computer simulation. Details of the airframe would also be required to produce the canonical models of the airframe that are used for higher-frequency GTD/UTD (uniform/unified theory of diffraction) modelling, as described in Section 7.5.

The modelling of antennas on the airframe can be undertaken at this stage and the antenna position and other details may have to be altered to give a better radiation pattern. This will be an iterative process, since the disciplines outlined in point 1 above would have to be consulted over any planned changes. Several iterations may take place before the initial antenna design freeze is reached.

A parallel activity of supplier selection may also be undertaken whilst the geometric modelling of the airframe surface is under way, since antenna details and installation characteristics will be required for the computational modelling phase.

3.4.3 Verification and Implementation Phase

The next stage in the design phase is summarized in the flow chart of Figure 3.3. Whilst the initial computer modelling is being performed, a simultaneous activity of designing and building a scaled model would most probably take place. A scaled model is used to:

1. provide verification of computed data

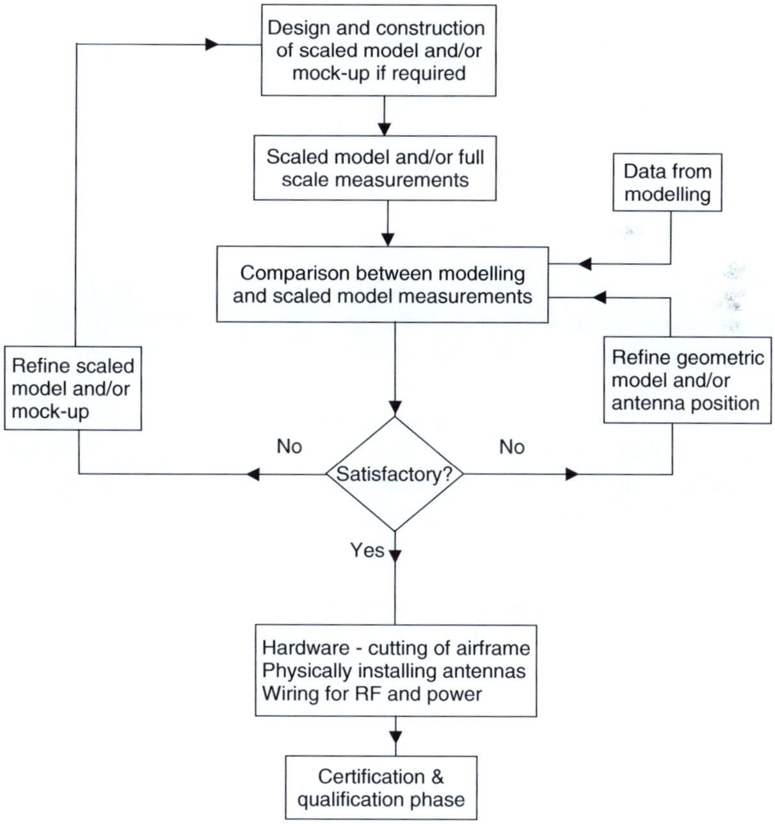

Figure 3.3 Simplified flow chart of the verification and implementation phase.

2. bridge the frequency gap between low- and high-frequency codes
3. replicate features that cannot be included in the CAD model of the airframe, such as the nose radome.

The EM modelling can be used to cover many frequencies and many cuts, but the measurements on the scaled model are usually restricted to a few frequencies and perhaps just one or two cuts, such as the azimuth and roll plane cuts. Once correlation is obtained with the measured data, it is assumed that all other predicted cuts and frequencies are also verified.

A full-scale mock-up of part or all of the airframe may also be required. The activities in this phase would typically be:

1. design and construction of a scaled model
2. scaled model measurements
3. refinement of the computer-generated geometric models.

3.4.3.1 Design and Construction of a Scaled Model

The choice of scaled model is discussed in Section 8.3. The scaling factor dependencies would include the:

(a) frequencies of the antennas
(b) frequencies of the test site
(c) handling capabilities of the test facilities
(d) physical size of the scaled model
(e) features required to be scaled down.

The scaled VHF and UHF antennas are usually just monopoles. If the real full-scaled antennas are top loaded then the radiation pattern of the monopoles will be quite accurate for the copolar but the cross-polar response may not be replicated accurately.

Most test sites have equipment that operates up to 26 GHz, although some do go up to 40 GHz and above. The higher-frequency equipment is more expensive and the power levels of the transmitters are usually lower, thus the dynamic range of the measurement facility is much lower at the higher frequencies.

Design and construction of a full-scale mock-up of part or all of the airframe may also be required. This would be the case for performing measurements especially in the case of the low-frequency antennas such as the HF that tend to be embedded in the tail, ventral or dorsal fins, or externally mounted as a 'towel rail' or long wire(s). The radar may also be measured separately using a mock-up of forward fuselage or just the nose cone.

3.4.3.2 Scaled Model Measurements

Scaled model measurements may be undertaken at an open test site or in an internal anechoic chamber, if the scaled model is small enough. Increasingly near field test sites are used and these are particularly useful since 3D spherical data can be obtained and the required great circle or conical cut processed as required. However, they cannot usually accommodate large scaled models. Open test ranges, on the other hand, can usually only be used to give line radiation patterns for each measurement run and they are subject to the vagaries of the weather. Pitch plane cuts are also difficult to obtain with booms used on standard positioners. A brief description of measurement facilities is given in Chapter 8.

3.4.3.3 Refinement of the Computer-Generated Geometric Models

In addition, the computer-generated geometric models will be refined to obtain good correlation between the predicted and scaled model measured data. This iterative process increases further the confidence in the selected antenna sites and may obviate the need for extensive in-flight trials. However, flight trials are still necessary to provide an overall evaluation of the whole system, including the antennas, in the final configuration. Flight trials are also useful in establishing the level of RF interoperability between the systems, especially between systems that have antennas on opposite (upper and lower) sides of the fuselage.

3.5 Certification and Qualification Phase

Once the design phase as been undertaken and the design freeze implemented, the more detailed modelling can be undertaken, before the metal is actually cut, to install the antennas. The actual cutting of the metal can be part of the certification/qualification phase, although there is no clear division between the design and certification/qualification phases. There is also no clear distinction between the certification and qualification phases. It is usually the qualification phase that is undertaken in order to obtain certification.

Only one aircraft of the total fleet is required to be qualified in order to obtain certification for the whole fleet. However, in many cases there are many variants for the same airframe, since different countries and organizations often require different systems for their particular fleet. This is particularly the case for military aircraft. In these cases qualification may be required for the variations where it cannot be shown that qualification is transferable to the variants.

Usually the whole system connected to the antenna has to be tested, not just the antennas. The systems may often be tested in labs using test rigs, but the actual signal to and from the antenna could be inputted/outputted from a simulator instead of the antenna.

3.6 Typical Antenna Layouts

The actual antenna layout will depend on the type of aircraft. The small aircraft tend to be either private/corporate jets or fighter/combat aircraft. The large aircraft are either commercial air freight or passenger airliners, or military transport, surveillance or bombers.

3.6.1 Small Aircraft

The number of antennas on small aircraft is restricted because of the real estate available. The spatial coverage required by antennas on small fighter jets is far more stringent than those on small commercial aircraft, since the coverage has to be maintained even during fast manoeuvres and dog fights. Links have to be maintained during the rapid vectorial changes to the aircraft's attitude and position.

In the case of the small corporate jets the changes to the aircraft's attitude and position are relatively slow and do not occur very often. Thus, for instance, on small commercial aircraft in the case of the UHF comms, although the spatial coverage in the azimuth plane has to be 360°, the coverage in elevation only has to cover a few degrees of depression and elevation, to allow for the attitude of the aircraft for communication with ground stations. Furthermore, we can see from the roll plane patterns of the Fokker 100 scaled model shown in Figure 4.31 that in the case of aircraft with small fuselage diameters (3.2 m) the antenna coverage in the lower hemisphere at 120 and 240 MHz is very good for an antenna placed on the upper fuselage.

In the case of the fighter aircraft, links may have to be maintained during the continuously changing dynamics, with other fighter aircraft and with the mission command on the ground and/or the air. Thus the angular coverage in the elevation plane has to be much larger to accommodate these aerobatics. Ideally full spherical coverage is required, which cannot be achieved in practice. Trade-offs and risk management studies have to be applied to select the optimum layouts. Although the small fuselage diameters help with spatial coverage at the lower frequencies, the lack of real estate at the middle and higher frequencies presents real problems. The antenna layout for a small passenger aircraft is shown in Figure 3.4. That for a small military aircraft is shown in Figure 3.5.

3.6.2 *Large Aircraft*

Large aircraft have the real estate to allow more than one system for redundancy. Commercial aircraft have to fly within the designated air corridors, whereas military aircraft do not have the same restrictions and theoretically do not even have to maintain any

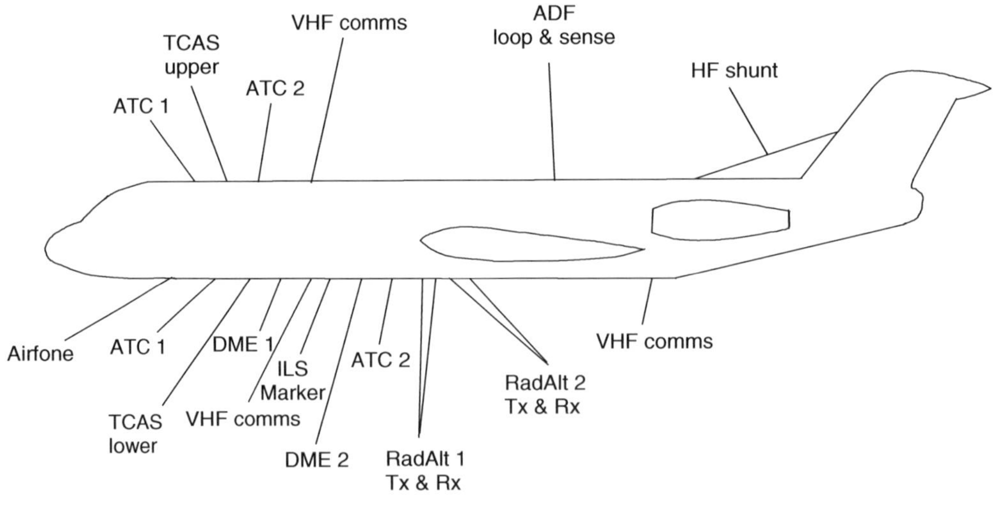

Figure 3.4 The layout for a small passenger aircraft. Reproduced by kind permission of NLR.

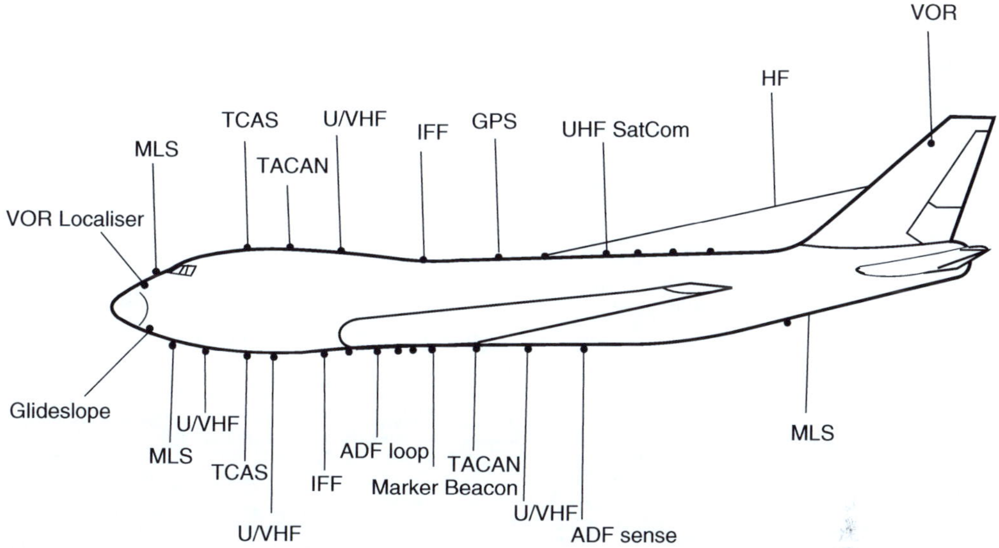

Figure 3.5 The layout for a small military aircraft.

specified altitude. They often fly at very low altitudes to avoid enemy radar or when operating in search and rescue missions. The layout for a single-aisle aircraft that has a fuselage diameter of 4.3 m, length 30.4 m and wingspan 32.54 m is shown in Figure 3.6. We can see the dual antennas used for GPS, RadAlt, ATC, DME and ADF, as well as the three antennas used for MLS as discussed in Chapter 2.

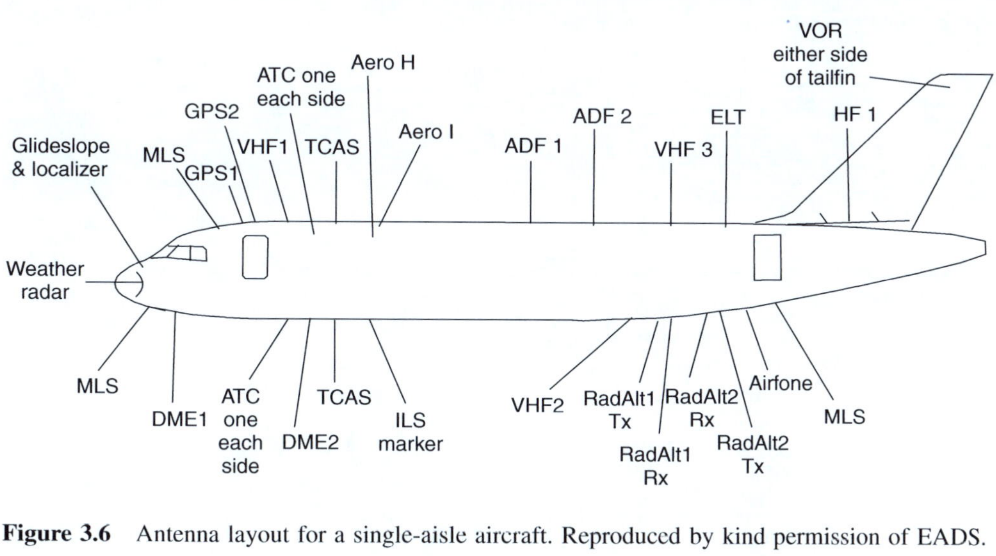

Figure 3.6 Antenna layout for a single-aisle aircraft. Reproduced by kind permission of EADS.

The antenna layout for the Boeing 777 shown in Figure 3.7 only has dual antennas used for GPS and DME, but it has four antennas for SatCom.

The commercial jet shown in Figure 3.8 has dual antenna systems for RadAlt, ATC, DME and ADF and uses two long wire antennas for HF.

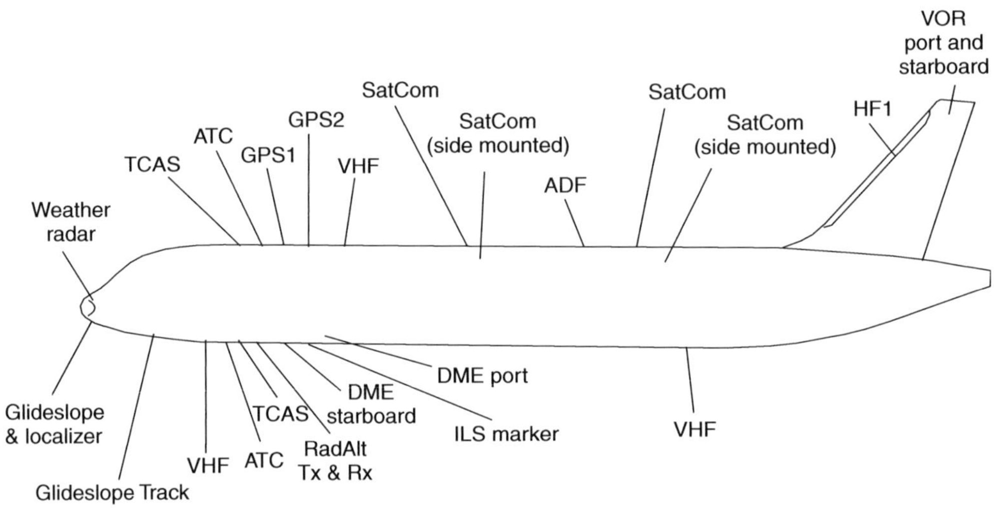

Figure 3.7 Antenna layout for the Boeing 777. Adapted from [2], p. 132.

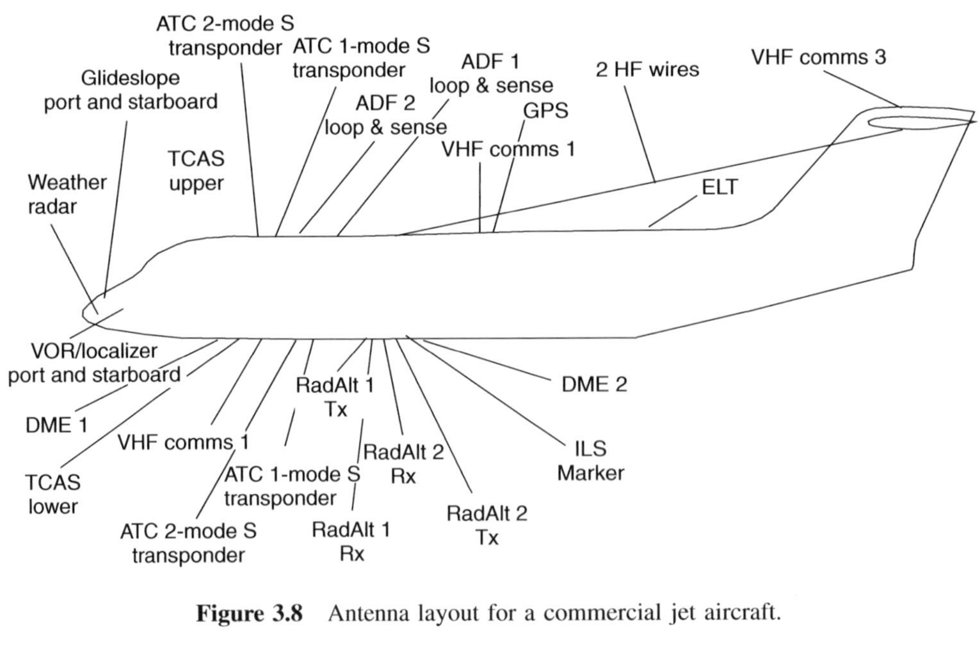

Figure 3.8 Antenna layout for a commercial jet aircraft.

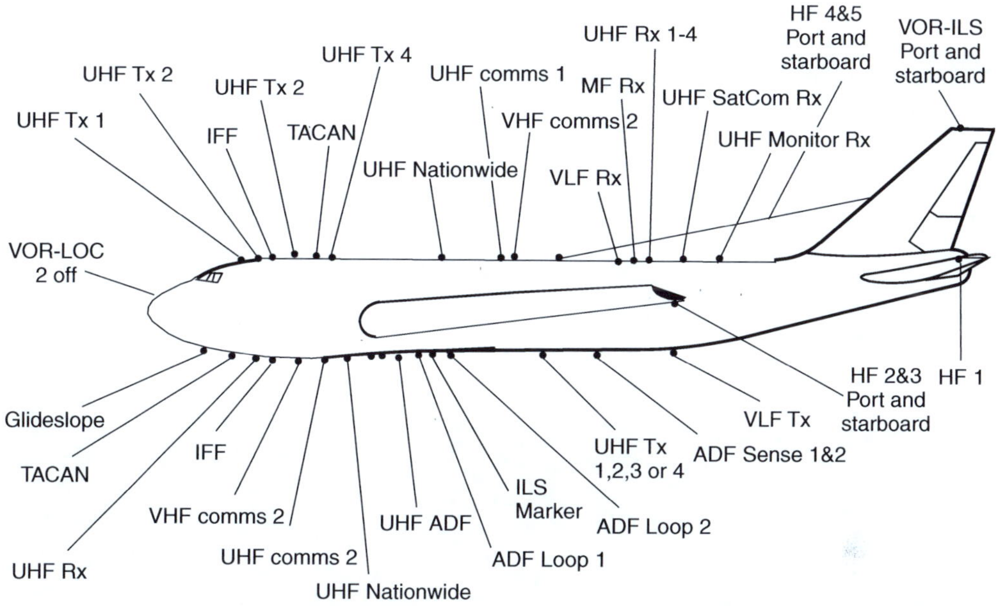

Figure 3.9 Antenna layout for a large aircraft. Reproduced by kind permission of NLR.

The antenna layout of the large aircraft shown in Figure 3.9 has five HF antennas, and uses separate antennas for transmitting and receiving UHF. It also uses UHF as well as ADF for direction finding. Details of the aircraft are not known but it is possibly not a commercial aircraft since it has TACAN and IFF which are usually only used by military aircraft.

References

[1] Macnamara, T.M. (2007) Code of practice for the design phase, Deliverable D20, IPAS/D_20/BAES/TM070110, 10 January. Installed Performance of Antennas on AeroStructures (IPAS), an EU Specific Targeted Research Project, November 2003 to January 2007, Contract No AST3-CT-2003-503611.

[2] Moir, I. and Seabridge, A. (2006) *Military Avionics Systems*, John Wiley & Sons, Inc., Hoboken, NJ, ISBN-13 978-0-470-01632-9.

4

Frequency and Spatial Coverage Considerations

4.1 Introduction

The antenna uninstalled radiation pattern is measured on a standard ground plane, in the case of monopoles. On the aircraft the ground plane is very different. For antennas installed on the upper fuselage, the ground plane is a cylinder, so it is no longer flat or symmetrical around the antenna. Furthermore, the shape and size of this aircraft ground plane does not vary with frequency. There are also many obstructions, such as the tail fin, wings, and other antennas, that result in reflection and diffraction affecting the resultant radiation pattern. Protrusions from the airframe are usually collectively called excrescences. The ground plane affects mainly the EM conducted waves, whereas the excrescences affect the radiated wave. Of course the conducted wave is also radiated into free space and the radiated wave is conducted by the metallic surface of the airframe.

The distortion of the radiation pattern by reflections at one or more surfaces is commonly referred to as multipath.

4.1.1 Standard Ground Planes

The radiation pattern is measured on a standard ground plane. For a monopole this is usually circular, with the diameter being inversely proportional to the frequency. Thus at 3 GHz the diameter could be about 3 m, but at 100 MHz the diameter is likely to be 10 m. This ground plane is flat, that is, in the shape of a disc. The antenna is measured at the centre of this disc, and thus the ground plane is symmetrical around the antenna. At the lower frequencies the ground plane could be made of mesh, instead of being solid. The mesh is more suitable for elevated ranges, because of the lower wind resistance and weight. Additionally, solid metal is prone to warping, whereas a mesh can be stretched onto a metal frame. The mesh spacing is selected depending on the frequency of the antenna. The spacing interval between adjacent wires should be about one-tenth of the wavelength at the highest operating frequency. We can think of the spacing requirement as the gaps between cobblestones. If adults walk on them, they can walk over the tops of the cobblestones, whereas a child's foot would fall into the crevices. The foot size is

Introduction to Antenna Placement & Installation Thereza M. Macnamara

analogous to the wavelength, so that low frequencies that have long wavelengths can be transmitted over the mesh, whereas high frequencies that have short wavelengths cannot bridge the gaps that are greater than $\lambda/10$.

4.2 Effect of the Structure on the Spatial Characteristics of the Antenna

If the spatial radiation pattern is not constant over the whole sector, this results in:

1. the range (the distance that the power from the antenna is transmitted) varying with the angle
2. the DOA being ambiguous or incorrect
3. the coupling between antennas varying with the angle between the antennas.

In order to understand the above effects we must compare the uninstalled and installed radiation patterns of the antenna(s).

4.2.1 Uninstalled and Installed Patterns

When we refer to uninstalled patterns we could be referring to the measured or computed patterns of

1. ideal antennas or
2. a sample of real antennas – suppliers often refer to the 'typical radiation pattern' of the antennas.

The computed patterns are usually modelled in free space or on an infinite ground plane that is completely isolated.

The measured patterns would be tested in an anechoic chamber or on an open test site/range, but the 'free space' would just be electrically far from obstacles and the ground plane would be a 'standard' one.

Whilst the computed patterns may be provided at a number of frequencies and cuts, the measured patterns would only be at a few spot frequencies, such as the low, middle and upper end of the operating frequency band, and usually only the azimuth cut.

Three of the most common types of antennas used on aircraft are monopoles, patches and planar spiral antennas, with the monopole being the most common. The radar antenna is a reflector or phased array, and although it is usually installed to have a clear field of view, the evaluation of the installation on the aircraft is a major task and requires the skill of highly specialized engineers.

The monopoles have uninstalled patterns that are omnidirectional in the azimuth plane, whereas the patch and spiral antennas have a main lobe along their axes of circular symmetry, as shown in Figure 4.1. Spiral antennas are broadband antennas and usually consist of a combination of Archimedean and equiangular/logarithmic spirals. These antennas operate by using a part of the spiral that has a circumference of approximately λ, which

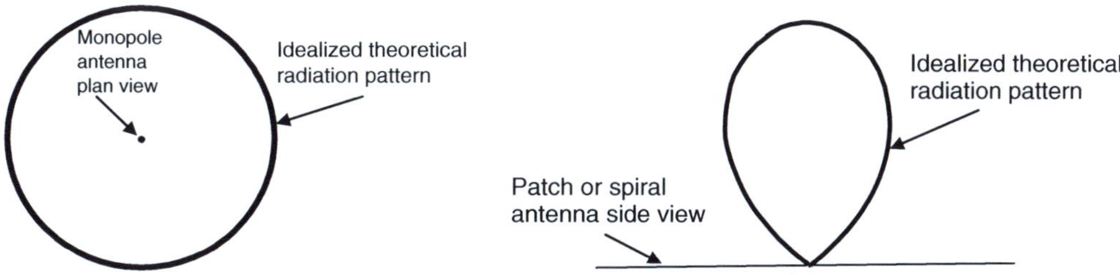

Figure 4.1 Uninstalled theoretical idealized patterns for a monopole on an infinite ground plane and a cavity-backed patch or planar spiral antenna.

corresponds to a diameter of λ/π or $\lambda/3.142$. Thus at the low frequencies the whole of the spiral radiates, whereas at the high frequencies only the central part of the spiral radiates. A qualitative explanation of the way a spiral operates is given in Chapter 6 of [1]. In theory the spiral has the same beamwidth and gain over the whole of the frequency band of operation, but in practice this is not the case. The boresight pointing angle also varies with frequency.

The ground plane greatly influences the radiation pattern of a monopole antenna. The main differences between a standard ground plane and the ground plane on the fuselage of an aircraft are shown in Table 4.1 on page 106.

The main effects on the radiation pattern for a monopole on an aircraft are summarized in Table 4.2 on page 107. However, it should be noted that some of the effects can be beneficial. For instance, the cylindrical ground plane results in creeping waves around the curved surface, but this could be considered an advantage is some cases, since at low frequencies antennas on the top surface radiate into the lower hemisphere, obviating the need for a lower fuselage antenna in some cases.

4.3 Combination of Two Waves

In Chapter 1 we saw how the resultant of the interaction between two waves can be obtained by looking at the waveforms in the time domain as well as by considering the waves as phasors drawn to scale or by using simple trigonometry. We saw that it is difficult to predict the resultant of two waves, unless we know their amplitude as well as their relative phases.

We must remember that the radiation pattern shows the power, so that when we combine waves in the time domain we have to square their amplitudes to get the power. In this section we investigate the effect of varying the phases between two waves of the same frequency and equal amplitude and showing how the resultant power varies with the phase between the two waves in the time domain. We will also see, using trigonometry, how the resultant varies with the differential phase when the amplitude of one of the waves is one-half, one-quarter and one-eighth of the other.

A combination can give us a peak at one frequency but a null at another frequency. This could be due to the relative phases as well as the relative amplitudes of the two waves.

Table 4.1 Differences between a standard ground plane and the ground plane on the fuselage of an aircraft.

	Shape	Dimensions	Forward to rear dimension	Material of ground plane	Conductivity
Standard ground planes	Planar and usually circularly symmetrical	Diameters of 10 m for low-frequency antennas and 3 m for high-frequency antennas	Same	Same throughout	Good
Ground plane on the fuselage of an aircraft	Not planar (usually curved with dents, etc.) and uneven (doublers/triplers, etc.) and does not have circular symmetry	Much smaller than standard ground planes	Forward of the antenna is not the same as aft	A mixture of materials such as Al and composite materials	Does not necessarily have good conductivity

Table 4.2 Comparison between the radiation pattern of a monopole antenna on a standard ground plane and on an aircraft fuselage.

Monopole antennas on	Symmetry of radiation pattern in azimuth plane	Shape of radiation pattern in the horizontal plane	Position of peak of radiation pattern	Smoothness of radiation pattern
Standard ground planes	Symmetrical	Generally omni-directional for narrowband blades	Near the horizon	Fairly smooth
Aircraft fuselage	Asymmetrical	Not omnidirec-tional	Peak tilted usually away from the horizontal	Interference pattern (sometimes called beating) is obtained.

4.3.1 Combination in the Time Domain

This method can be used even when the two waves have different frequencies and we get the actual shape of the resultant shape in the time domain, that is, the variation of the electric field and/or power with time.

If we plot two waves of the same frequency and amplitude that are in phase we would get a resultant of twice the amplitude, as shown in Figure 4.2.

If we now look at the combination in terms of power we can see that the resultant (in Figure 4.3) is four times the power of either of the individual waves, because the power is proportional to the square of the electric field. In terms of dB this is 6 dB when the linear power is 4 times.

If we consider two waves that have a 45° phase difference, as shown in Figure 4.4, we can see that the amplitude of the resultant is about 1.8 times that of the individual waves and the power is about 3.4 times that of the individual waves. The phase difference of the resultant is 22.5° compared with that of the first wave.

As we increase the phase difference between the two waves to 65° (Figure 4.5) the amplitude of the resultant wave is about 1.7 and the power is about 2.8 times that of the individual waves, and its phase is 32.5° compared with that of the first wave.

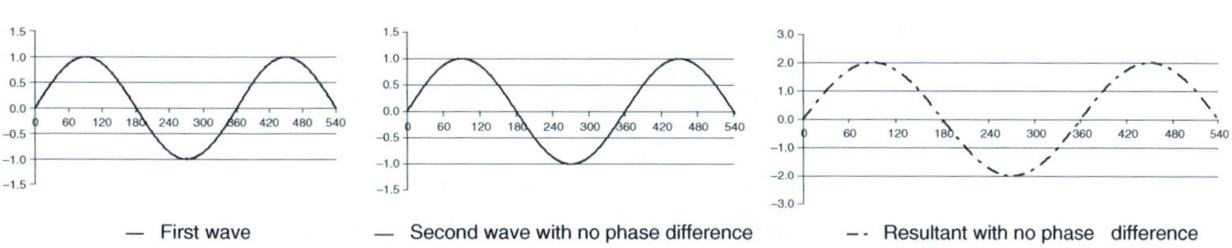

Figure 4.2 Combination of two waves that are in phase (i.e. no phase difference) in the time domain.

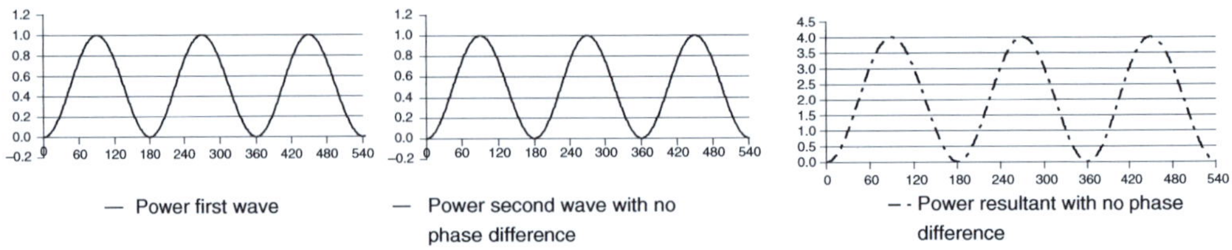

Figure 4.3 Power combination of two waves in phase (i.e. no phase difference).

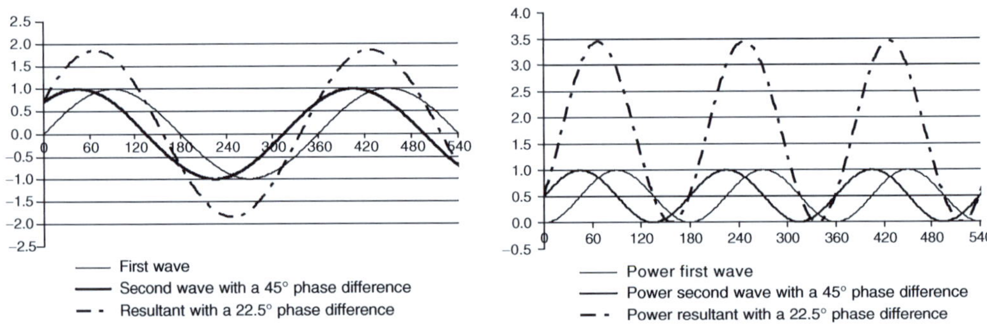

Figure 4.4 Combination of two waves with a relative phase of 45°.

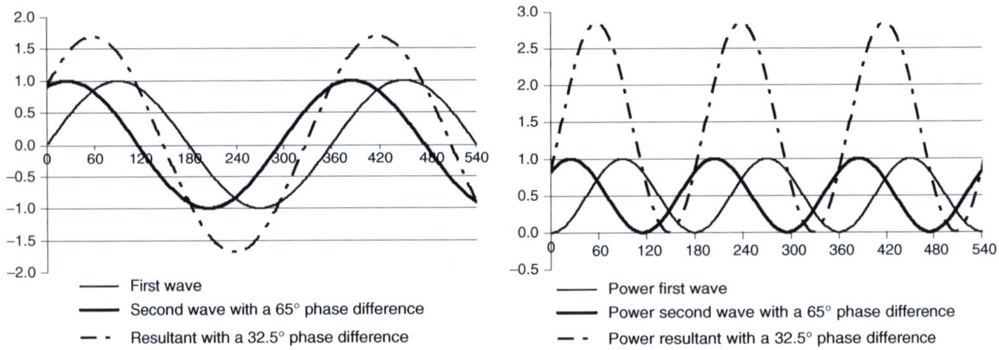

Figure 4.5 Combination of two waves with a relative phase of 65°.

As the phase between the two waves increases to 90° the amplitude of the resultant is about 1.4 and the power is about twice that of the individual waves (Figure 4.6) and its phase is 45° compared with that of the first wave.

As the phase between the two waves increases to 110° the amplitude of the resultant is about 1.15 and the power is about 1.3 times that of the individual waves (Figure 4.7) and its phase is 55° compared with that of the first wave.

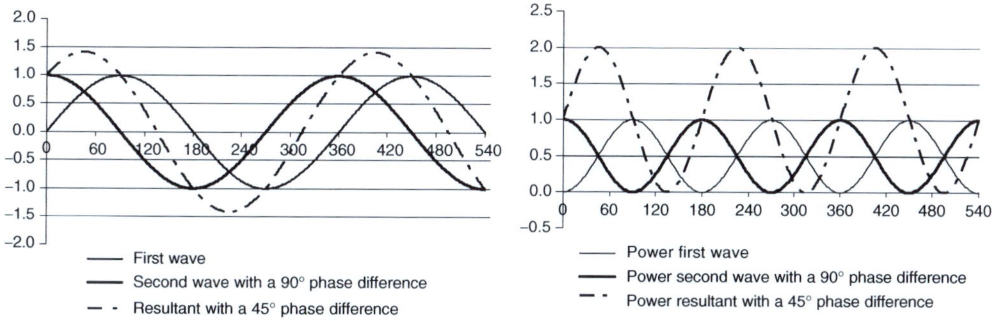

Figure 4.6 Combination of two waves with a relative phase of 90°.

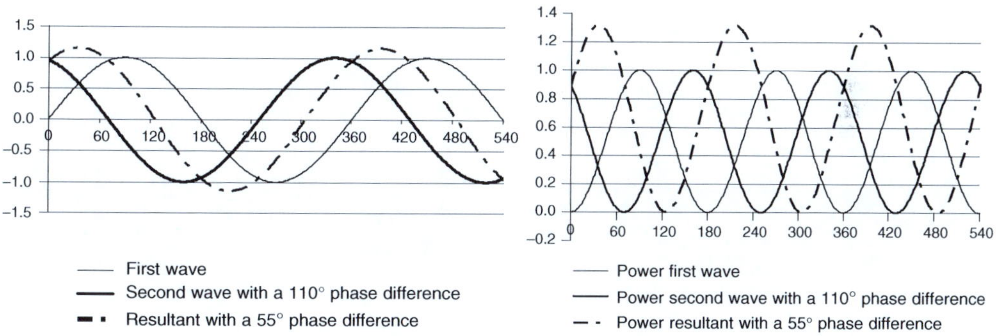

Figure 4.7 Combination of two waves with a relative phase of 110°.

At a phase difference of 120° (Figure 4.8), the resultant wave has the same amplitude as that of the individual waves, the power is equal to that of the individual waves, and its phase is 60° compared with that of the first wave.

As we increase the phase between the two waves of equal amplitude, the amplitude and power of the resultant obtained are higher than those of either of the individual waves until the phase difference is 120°, and then as the phase difference is increased to values greater than 120° the resultant wave (Figure 4.9) has an amplitude less than that of the individual waves (0.84 for a 130° phase difference) and the power is much less than that of the individual waves (0.7 for a 130° phase difference). We expect the resultant power to be lower than the resultant amplitude, since the square of a number less than 1 is smaller than the number, whereas in the case of numbers greater than 1 the square is greater than the number. This trend of a decreasing resultant (see Figures 4.9–4.12) is followed until we have a phase difference up to 180°, at which point the resultant is zero.

As the phase difference between the two waves increases to 135° (Figure 4.10) and 150° (Figure 4.11), the amplitude (and power) of the resultant decreases further until at 160° the amplitude of the resultant is about 0.35 and the power is about 0.12 times that of the individual waves (Figure 4.12).

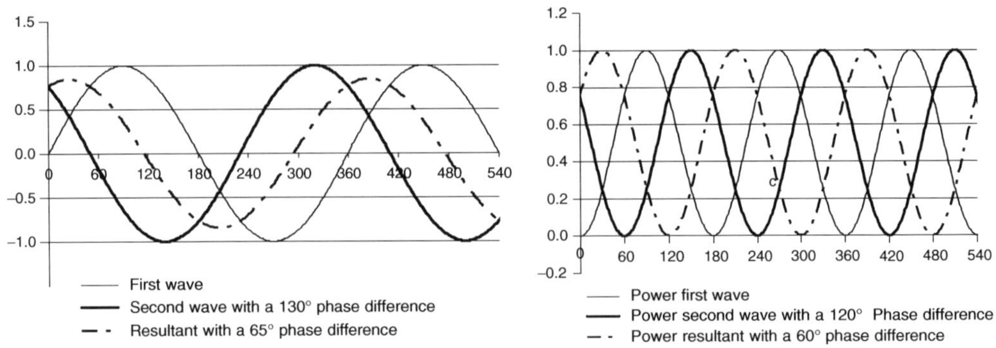

Figure 4.8 Combination of two waves with a relative phase of 120°.

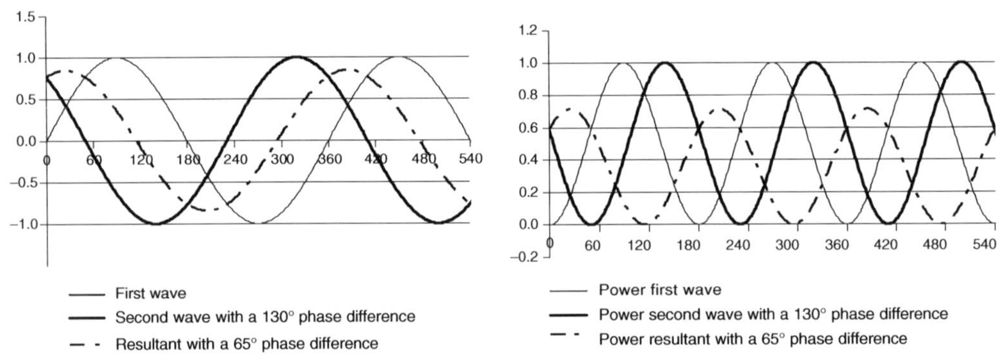

Figure 4.9 Combination of two waves with a relative phase of 130°.

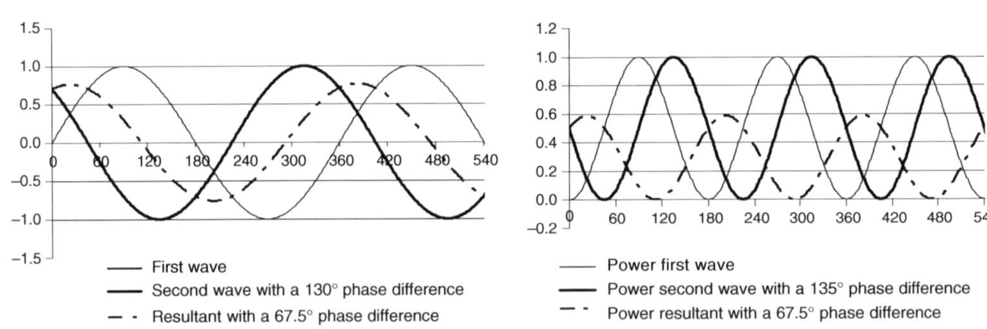

Figure 4.10 Combination of two waves with a relative phase of 135°.

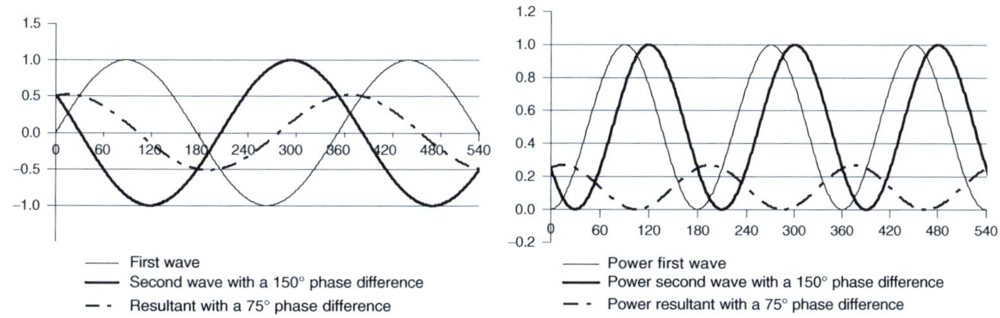

Figure 4.11 Combination of two waves with a relative phase of 150°.

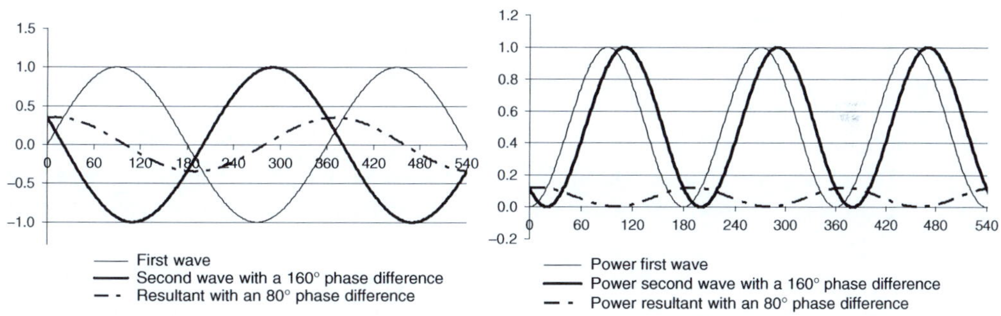

Figure 4.12 Combination of two waves with a relative phase of 160°.

A combination of two waves that are out of phase by 180° will give no resultant, that is, total destructive interference.

4.3.2 *Combination of Two Waves Using Trigonometry*

All the above resultant wave amplitudes can be calculated using the cosine formula, as described in Section 1.3.2.2. The relevant parameters are shown in Figure 4.13. However,

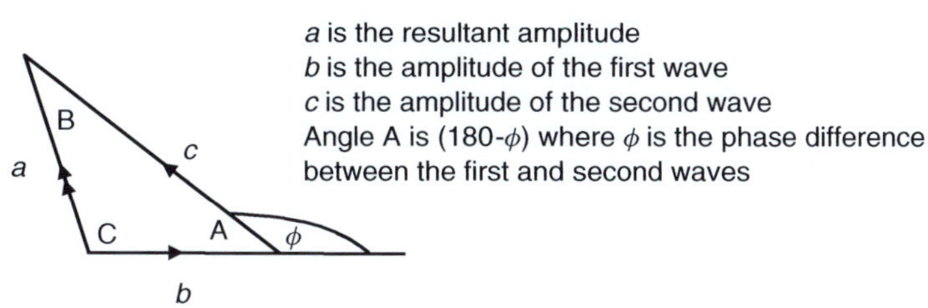

Figure 4.13 Calculation of the resultant amplitudes using the cosine formula.

as stated in Section 1.3, using phasors instead of the waveforms in the time domain, we only obtain the amplitude of the resultant wave and this does not give us the exact shape of the resultant wave. Phasors cannot be used if the individual waves are not the same frequency. The phase of the resultant wave can also be calculated in each case, but any ambiguity has to be checked as explained in Section 1.3.2.2.

All the resultants shown in Section 4.3.1 were obtained using unity (1) as the amplitude of the first as well as the second wave. In this section the resultants are calculated for waves of the same amplitude as well as for the cases where the amplitude of the second wave is a half, a quarter and an eighth that of the first wave.

As shown in Table 4.3, when the amplitudes are equal, the resultant power is zero ($-\infty$ in dB) when the phase difference is 180°. However, when the amplitude of one wave is half, a quarter and an eighth that of the other, the resultant powers are 0.25, 0.563 and 0.766 (−6.02, −2.5 and −1.16 in dB) when the phase difference is 180°.

If the resultant power is plotted against the phase difference for the four cases, we obtain the plots shown in Figure 4.14. There are six cases that would give us the same resultant power of 2 dB. This could occur with phase differences of 100° and 260° for waves of equal amplitude, for phase differences of 70° and 290° for waves where one has half the amplitude of the other, and for phase differences of 0 and 360° for waves where one has a quarter of the amplitude of the other. What can also be observed from Figure 4.14 is that, for the case where the second wave is half that of the first wave, the resultant is lower than that obtained for waves of equal amplitude, between 0 and 135°

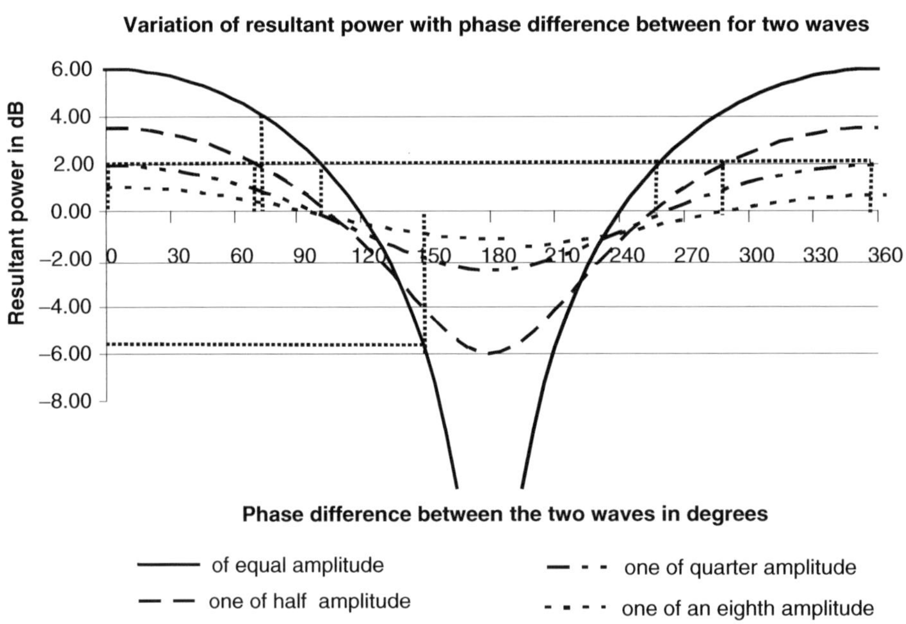

Figure 4.14 Variation of the resultant power with phase difference between two waves, for waves of equal amplitude as well as for the cases when one wave is half, a quarter and an eighth the amplitude of the other.

Table 4.3 Resultant of two waves of equal amplitudes as well as for the cases when one wave is a half, a quarter and an eighth of the amplitude of the other.

Amplitude of first wave	Amplitude of second wave	Phase difference ϕ between waves	Amplitude of resultant squared	Power of resultant	Amplitude Half that of first wave	Amplitude of resultant squared	Power of resultant	Amplitude 1/4 that of first wave	Amplitude of resultant squared	Power of resultant	Amplitude 1/8th that of first wave	Amplitude of resultant squared	Power of resultant
b	c	$180 - A$	a^2	dB	c	a^2	dB	c	a^2	dB	c	a^2	dB
1	1	0	4	6.02	0.5	2.25	3.52	0.25	1.563	1.94	0.125	1.266	1.02
1	1	30	3.732	5.72	0.5	2.116	3.26	0.25	1.496	1.75	0.125	1.232	0.91
1	1	60	3.000	4.77	0.5	1.750	2.43	0.25	1.313	1.18	0.125	1.141	0.57
1	1	90	2.000	3.01	0.5	1.250	0.97	0.25	1.063	0.26	0.125	1.016	0.07
1	1	120	1.000	0.00	0.5	0.750	−1.25	0.25	0.813	−0.90	0.125	0.891	−0.50
1	1	150	0.268	−5.72	0.5	0.384	−4.16	0.25	0.629	−2.01	0.125	0.799	−0.97
1	1	180	0.000	$-\infty$	0.5	0.250	−6.02	0.25	0.563	−2.50	0.125	0.766	−1.16
1	1	210	0.268	−5.72	0.5	0.384	−4.16	0.25	0.629	−2.01	0.125	0.799	−0.97
1	1	240	1.000	0.00	0.5	0.750	−1.25	0.25	0.813	−0.90	0.125	0.891	−0.50
1	1	270	2.000	3.01	0.5	1.250	0.97	0.25	1.063	0.26	0.125	1.016	0.07
1	1	300	3.000	4.77	0.5	1.750	2.43	0.25	1.313	1.18	0.125	1.141	0.57
1	1	330	3.732	5.72	0.5	2.116	3.26	0.25	1.496	1.75	0.125	1.232	0.91
1	1	360	4	6.02	0.5	2.25	3.52	0.25	1.563	1.94	0.125	1.266	1.02

(as well as between 225° and 360°). However, as the phase of the second wave increases from 135° to 180°, this trend is reversed (i.e. the resultants become greater than those for the cases of waves of equal amplitude). This is apparent for the other cases as well where the second wave has amplitudes of a quarter and an eighth of the second wave, although the crossover points occur at different phase differences. The crossover points are when the resultant power for waves of different amplitudes changes from being less than that for waves of equal amplitude to being greater than that for waves of equal amplitude. This demonstrates the difficulty of tracing the exact sources of the resultant radiation pattern of an antenna. The problem is magnified in the case of multiple combination of waves from different interactions.

When the frequency is doubled the wavelength is halved, so that for the same physical distance the phase difference will have doubled. Thus, referring to Figure 4.14, if the phase difference between two waves of equal amplitude is 75° at one frequency the resultant is about 4 dB, but if the frequency is doubled the phase difference is also doubled (to 150°) and the resultant is about −5.7 dB. Thus at one frequency we would get a peak, whereas at double that frequency we get a dip/null.

4.4 Measurements on Scaled Test Bodies

In order to analyse the radiation patterns on aircraft, we will examine the results of measurements on a 1/15th scaled model of a Fokker 100 airframe and a scaled cylinder with diameter that when scaled up 12 times is nearly the diameter of a single-aisle aircraft.

4.4.1 Fokker 100 Scaled Model Used for Measurements

Measurements were performed on a 1/15th subscale model of the Fokker 100 airframe, under the EU IPAS research project [2]; see Figure 4.15. The measurement frequency was scaled up 15 times; thus if the measurement frequency was 3.6 GHz the aircraft frequency would be 240 MHz, which is very near the distress frequency of 243 MHz.

Figure 4.15 Fokker 100 scaled model provided by NLR for measurements. Reproduced by kind permission of NLR.

The length of the 1/15th scaled model of the Fokker 100 fuselage was 220 cm and the fuselage diameter was 21.5 cm. The wing span was roughly 194 cm. Its weight was of the order of 15 kg. The origin of coordinates was taken as the front of the nose along the longitudinal axis of the circular fuselage.

The radiation patterns of antennas were measured at different positions on the scaled model, as shown in Figure 4.16. All antennas were installed on the spine of the fuselage (equal distances from the wings) along its vertical plane of symmetry as shown in Figure 4.16b. The first position was on the upper fuselage forward of the wings and 9.75 m from the front of the nose for the full-scale aircraft. The second position was on the upper fuselage above/between the wings and 16.8 m from the front of the nose. The third position was on the upper fuselage between the wings and the engines and 20.85 m from the front of the nose. The fourth position was on the lower fuselage behind the nose landing gear and 10.93 m from the front of the nose. Table 4.4 gives the positions of the antennas from the front of the nose and the positions of the wings and engines are given in order to explain the radiation patterns obtained.

4.4.2 *Cylinder Used for Radiation Pattern Measurements*

Under the IPAS EU project [2], a cylinder was constructed that corresponded to a scaled-down version of the fuselage of a single-aisle aircraft. The cylinder had a diameter of 0.406 m, and was scaled up 12× for the measurement frequencies. The radius of curvature was selected, so that the scaled-up value (4.872 m) was near that of the single-aisle aircraft which has a fuselage diameter of 5.64 m. However, it was not possible to have a length

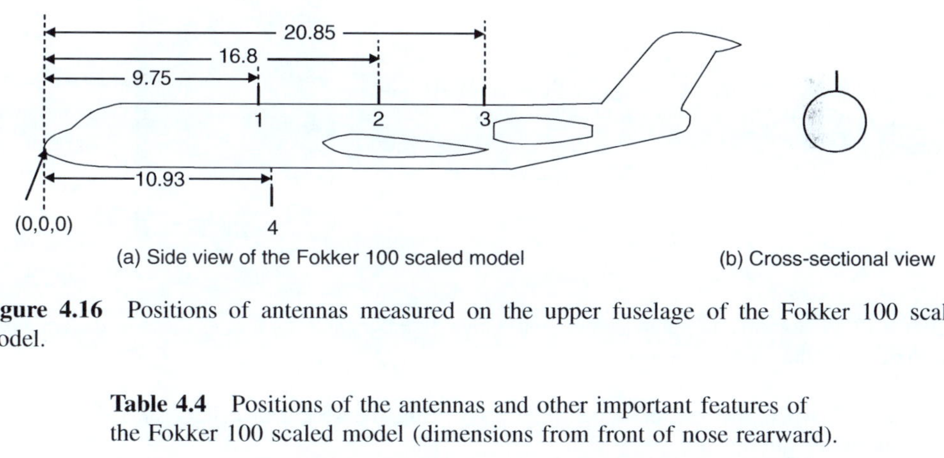

(a) Side view of the Fokker 100 scaled model (b) Cross-sectional view

Figure 4.16 Positions of antennas measured on the upper fuselage of the Fokker 100 scaled model.

Table 4.4 Positions of the antennas and other important features of the Fokker 100 scaled model (dimensions from front of nose rearward).

	Scale model in mm	Full scaled aircraft in m
Position 1	650	9.75
Start of wing	948	14.22
Position 2	1120	16.80
Position 3	1390	20.85
Position 4	728.4	10.93
Start of engines	1488	22.32

of 3.625 m corresponding to the single-aisle aircraft fuselage (which had a length of 43.5 m) because of the restriction on the length imposed by the measurement facility and the unmanageable weight that would have resulted if a 3.625 m long cylinder had been used. Thus the aspect ratio of the cylinder was 6.16 instead of the 7.71 of the single-aisle aircraft. The ripples obtained in the roll plane depend on the circumference of the fuselage, and thus the roll plane plots obtained on the cylinder are similar to those that would be obtained on the fuselage of an aircraft of these dimensions as long as the antennas are not installed above/between the wings. The azimuth plots are similar to those that would be obtained for antennas near the middle of the lower fuselage (on the vertical plane of symmetry) of an aircraft of these dimensions or if antennas were installed on the upper fuselage, electrically far from the tail fin or nose. The pitch plane plots, however near azimuth, would not be similar to those obtainable on an aircraft, since the cylinder was shortened and had radar absorbing material (RAM) covering one end and a sharp end face at the other end, which would not be the case for an airframe.

Six antennas were installed symmetrically on diametrically opposite sides of the cylinder as shown in Figure 4.17a. One end of the cylinder, near positions 1 and 4, was capped with a flat metal disc and the other end was covered in 'egg-box' RAM. This cylinder was also used for coupling measurements described in Chapter 6. Figure 4.17b shows the cylinder mounted vertically for roll plane radiation pattern measurements in the BAE Systems cylindrical near-field test facility in Great Baddow (UK). The dimensions of the cylinder are shown in Table 4.5, which also shows the dimensions of the single-aisle aircraft fuselage for comparison.

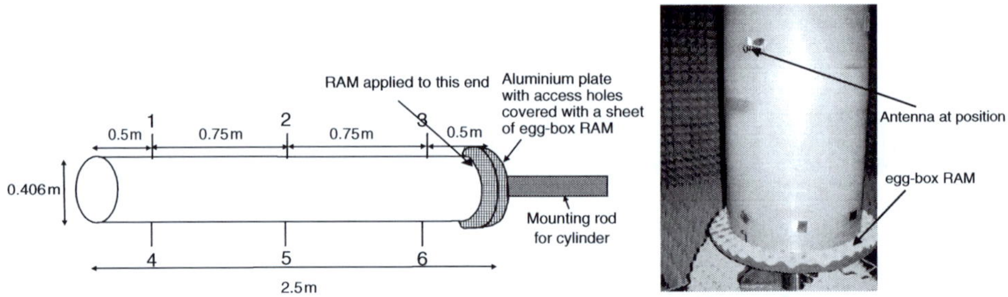

(a) Schematic of the cylinder (b) Photograph of the scaled cylinder

Figure 4.17 Diagram and photograph of the scaled cylinder showing the positions of the antennas. See Plate 6 for the colour figure.

Table 4.5 Comparison between dimensions of the measurement cylinder and single-aisle aircraft fuselage.

	Actual cylinder	Scaled-up cylinder 12×	Single-aisle aircraft fuselage	Ratio of scaled-up cylinder compared with single-aisle aircraft
Diameter in m	0.406	4.87	5.64	0.864
Length in m	2.5	30	43.5	0.690
Aspect ratio	6.16	6.16	7.71	–

4.5 Effect of Frequency on the Radiation Pattern

At high frequencies, if the antenna is installed electrically far from obstacles, the radiation pattern of the antenna is relatively unaffected by the obstacles. This is especially the case for directive antennas that do not require a ground plane, for example an array of antenna elements, reflector system, and so on. This occurs because the radiation at electrically far distances can be considered to be rectilinear propagation (like rays of light) and the second order effects such as diffraction and multiple reflection, have a much smaller influence on the radiation pattern. The line of sight (LOS) obscuration and first-order specular reflection can easily be investigated using simple trigonometry.

At low frequencies, the obstacles are electrically nearer and effects such as diffraction and multiple reflection have a much greater influence on the radiation pattern.

When we refer to the effect of frequency on the radiation pattern, we must consider the distance in terms of wavelength, that is, the electrical distance for the radiated wave as well as for the creeping wave. Since most antennas are installed on the fuselage and these have circular or elliptical cross-section, the creeping wave plays a crucial role in the radiation pattern.

4.6 Effect of Distance from Obstacles

When we are considering the effect that obstacles have on the radiation pattern, we have to consider the physical as well as the electrical distance. The physical distance gives us a first-order effect, since if we think of the radiation as rectilinear propagation (like rays of light) we know that if there is an obstacle in LOS of the antenna, this will act as a blockage, and specular as well as diffuse reflection will occur at flat and curved surfaces. As we move the antenna nearer to the obstacle the angular blockage will increase. This does not of course account for the wave attributes of the radiation which takes into account diffraction, and creeping waves, as well as higher-order interactions.

In the case of the effect due to the electrical distance, the best way to observe this is to keep the physical distance constant and vary the frequency.

In order to see the effect the distance from an obstacle has on the radiation pattern we can look at the radiation patterns of an antenna placed at three different positions on the upper fuselage of the Fokker 100 scaled model, described in Section 4.4.1.

4.6.1 Effect of the Physical Distance from Obstacles

As a first-pass activity, 3D obscuration plots can be obtained using a CAD package such as CATIA. A 3D external profile of the airframe is required and the obscuration plots are obtained by assuming that the antenna is a point source.

The radiation patterns do not show us the effect of the physical distance in isolation, since the wave effects will also be incorporated. However, if we ignore the ripples in the radiation pattern, and visualize the smooth pattern obtained without second-order effects, we can observe the effect that an obstacle has on the radiation pattern by moving the antenna to different distances from it. As the antenna on the Fokker 100 scaled model is moved further aft, the antenna's azimuth radiation pattern is more affected by the tailplane. This is noticeable in the patterns of Figure 4.18 for an antenna at the aircraft

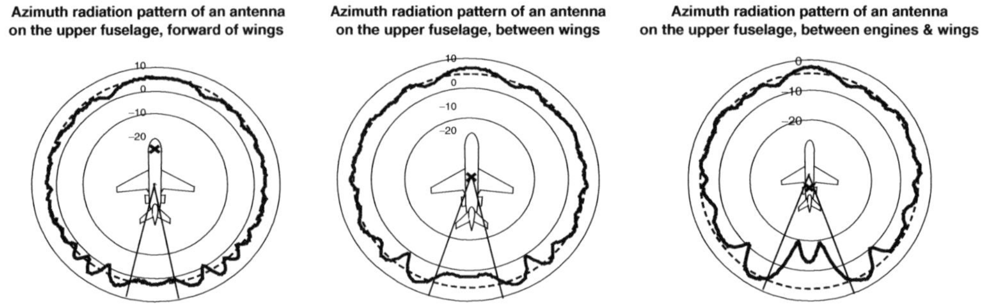

(a) Forward of wings (b) Above/between wings (c) Between wings and engines

Figure 4.18 Azimuth radiation patterns for an antenna at three different positions on the upper fuselage at 120 MHz showing the increasing angular blockage as the antenna is moved nearer the tail fin.

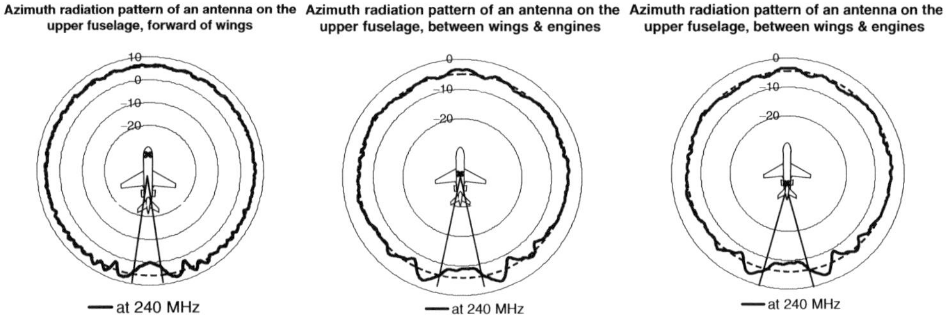

(a) Forward of wings (b) Above/between wings (c) Between engines and wings

Figure 4.19 Azimuth radiation patterns for an antenna at three different positions on the upper fuselage at the 240 MHz showing the increasing angular blockage as the antenna is moved nearer the tail fin.

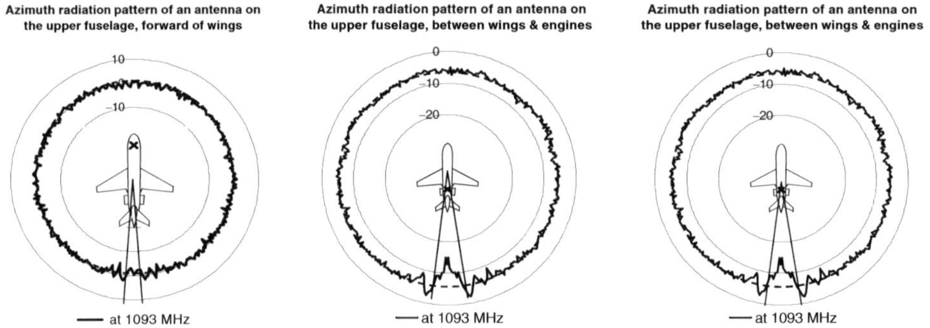

(a) Forward of wings (b) Above/between wings (c) Between wings and engines

Figure 4.20 Azimuth radiation patterns for an antenna at three different positions on the upper fuselage at the 1093 MHz showing the increasing angular blockage as the antenna is moved nearer the tail fin.

frequency of 120 MHz. The dotted circle is the approximate average relative gain pattern we would expect to get for an idealized antenna on an infinite ground plane. It can be assumed that if the tail fin was not blocking the LOS of the antenna, then the level would have been the approximate average that it is over the rest of the angular sector. Note that the angular blockage (denoted by the solid lines) is increased as the antenna is moved nearer the tail fin.

We would expect the blockage to be the first-order effect that we would get if we think of the EM radiation in terms of rays, where we are only considering the direct paths and first-order reflection. The peak at the position of the tail fin (in the third plot of Figure 4.18) is due to the wave properties of the radiation. We have seen from Figure 4.14 how if the phase difference is changed the resultant could be a positive or a negative value in decibels. The path length has been changed and hence the phase difference has also changed as the antenna is moved nearer the obstacle.

The other smaller ripples in the radiation pattern are due to the wave properties of the radiation, such as diffraction and creeping waves.

Note also that in the case of position 3, the forward gain is enhanced by the larger ground plane in front of the antenna.

We can see that as the frequency is increased the same trend of increasing angular blockage (as the antenna is moved nearer the tail fin) occurs for the higher frequencies as shown in Figures 4.19 and 4.20.

4.6.2 *Effect of the Electrical Distance from Obstacles*

The effect is demonstrated for different electrical distances from an obstacle, if we keep the physical position of the antenna fixed and just increase the frequency. Doubling the frequency doubles the electrical distance, since there are now twice as many wavelengths for the same distance.

The electrical distance from an obstacle has a marked effect on the far-field radiation pattern if the obstacle acts as a parasitic radiator and re-radiates, or if the specular radiation from it combines with the direct ray and causes constructive or destructive interference. There are of course third- and higher-order effects caused by interactions between waves that are diffracted or reflected (at one or more surfaces) with each other or with the direct wave.

We have seen in Section 4.3 how the resultant of two waves depends on the phase as well as the relative amplitudes of the waves. Figure 4.14 and Table.4.3 show that when the waves have the same amplitude the power of the resultant wave is four times (6 dB) greater than the power of each individual wave. However, when the amplitude of the second wave is one-eighth that of the first wave, the power of the resultant wave is only 1.27 times (1.02 dB) greater than the power of the first wave.

4.6.2.1 Effect of Tail Fin

If we examine the radiation patterns of Figure 4.21 at three different frequencies, for an antenna mounted at position 1, on the Fokker 100 1/15th scaled model on the upper

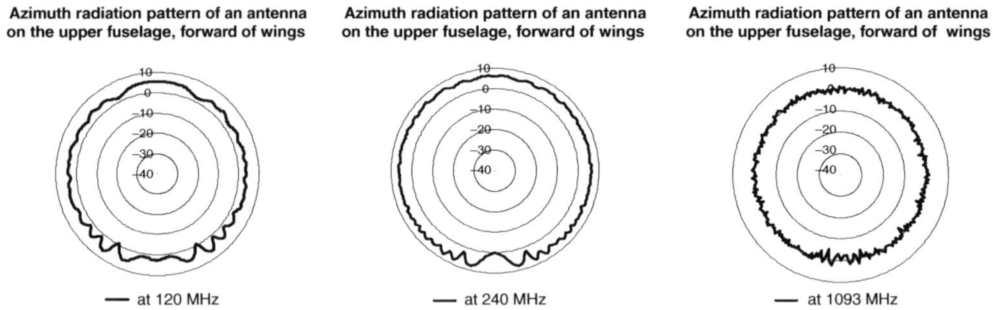

(a) 120 MHz (b) 240 MHz (c) 1093 MHz

Figure 4.21 Measured azimuth radiation patterns at three different frequencies for an antenna installed on the upper fuselage (of a scaled model) at position 1, forward of the wings and 9.75 m from the nose.

fuselage and forward of the wings, we can see the more marked effect of the lower frequencies. At 120 MHz the antenna is about 7.5 wavelengths from the tail fin, whereas at 240 MHz the antenna is about 15 wavelengths from the tail fin. At 1093 MHz the antenna is even further at about 67.5 wavelengths from the tail fin.

It can be seen from the azimuth radiation patterns that the effect of the tail fin is more pronounced (resulting is more blockage as well as deeper ripples to the rear) at the lower frequencies since the electrical distance from the fin is less at these frequencies. At the highest frequency of 1093 MHz the effect of the tail fin is hardly noticeable. We can also see the slight peak in the forward (nose) direction for the 120 MHz plot which is absent at the two higher frequencies. This is because the reflected/diffracted wave is a similar level to the direct wave and possibly almost in phase with it. At the higher frequencies the reflected/diffracted wave may only be half the amplitude of the direct wave or less. Referring to Figure 4.14 and Table 4.3, we can see that if the phase difference is 50° for waves of equal amplitude the resultant would be 5.7 dB, but at double the frequency when the phase difference is doubled to 100° the resultant is 0.3 dB. Even if we assume that the second wave has half the amplitude for both frequencies, the resultant would still be 2.77 dB at 120 MHz and 0.3 dB at 240 MHz.

The fact that we are dealing with waves that have magnitude as well as phase is also manifested in a slight peak rearward at 180°, as can be seen in Figure 4.21a. This is where the tail fin would be expected to cause a blockage and hence a reduction, resulting in a small null. However, we can see a slight peak appears at 120 MHz. In the case of the 240 MHz radiation pattern, a dip occurs in this region.

As we move the antenna further back to position 2 which is 16.8 m from the nose, we can see the greater effect in the spatial domain (see Figure 4.22), manifested by the wider null rearward due to the tail fin. Note the null aft for position 2 in the case of the 1093 MHz plot of Figure 4.22c, which was absent when the antenna was at position 1 (see Figure 4.21c).

The effect of the wings is not noticeable in the azimuth radiation patterns, since we can see that the patterns around 90° and 270° are very similar to those obtained at position

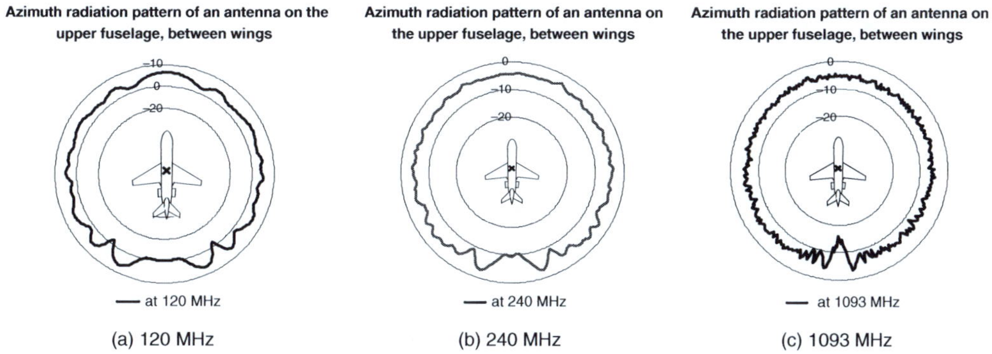

Figure 4.22 Measured azimuth radiation patterns at three different frequencies for an antenna installed on the upper fuselage (of a scaled model) at position 2, between the wings and 16.8 m from the nose.

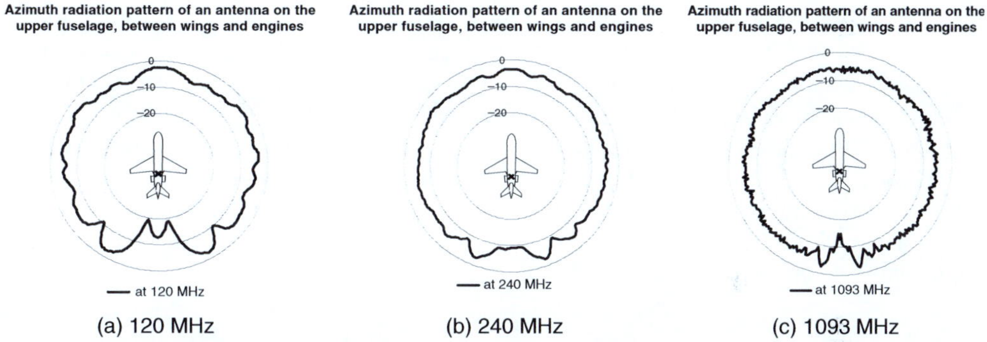

Figure 4.23 Measured azimuth radiation patterns at three different frequencies for an antenna installed on the upper fuselage (of a scaled model) at position 3, between the wings and engines and 20.85 m from the nose.

1. This is because any specular reflections off the wings or waves diffracted at the wings do not appear in the azimuth plane.

As the antenna is moved to the third position 20.85 m from the nose, the most marked difference is observed at the lowest frequency, where there is a peak aft, with nulls on either side of the peak (see Figure 4.23a). At the higher frequencies the patterns are very similar to those obtained with the antenna at position 2.

4.6.2.2 Effect of Wing Roots

The effect of the obstacles at different frequencies can also be seen in the roll plane patterns of Figure 4.24. The deep nulls at 110° and 250°, are caused by the wing roots. We can see that at the lowest frequency of 120 MHz there are very deep nulls possibly primarily due to LOS blockage and specular reflection of the EM waves where they are

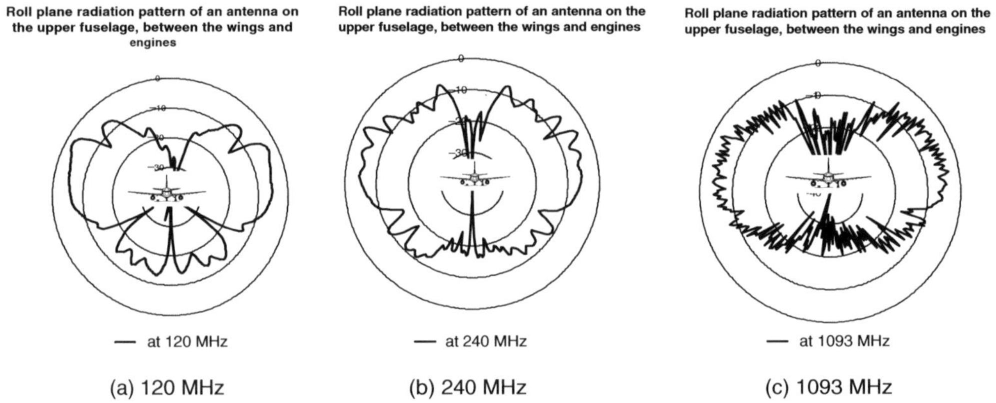

Figure 4.24 Roll plane radiation patterns for an antenna at three different frequencies on the upper fuselage at the position 3.

reflected off the wing roots, as well as multiple reflections, scattering and diffraction, at that particular angle. The antenna could not have been moved further aft, to avoid the wing roots, because of the presence of the engines. The wing roots are about 3 m away from the antenna at position 3, and at the lowest frequency of 120 MHz this distance corresponds to about 1.2 wavelengths away, whereas at 240 MHz it corresponds to about 2.4 wavelengths. As the frequency is increased further to 1093 MHz, the wing roots are about 11 wavelengths away and their effect is scarcely discernible, as can be seen from Figure 4.24c. Note that the number of ripples and nulls is greatly increased. The multiple ripples are sometimes referred to as beating.

4.7 Effect of Wings on the Radiation Pattern

From the patterns of Figure 4.25, at the frequency of 120 MHz the following observations can be made:

1. In the first position the antenna pattern resembles a monopole with a very small ground plane since the full-scale fuselage diameter of approximately 3.15 m is of the order of one wavelength. The wavelength at 120 MHz is 2.5 m.
2. In position 2, above the wings, the radiation in the lower hemisphere is severely compromised.
3. In position 3, between the wings and the engines, there is some coverage in the lower hemisphere, but the presence of the deep nulls (around 120° and 240°) indicates some multipath destructive interference because the antenna is above the wing roots, as discussed in Section 4.6.2.2. The other shallower nulls are most probably due to interactions after multiple reflections.

As the frequency is increased similar coverage results, but the aircraft no longer behaves like a monopole with a very small ground plane since the fuselage diameter is much greater than a wavelength. The number of ripples increases with frequency as expected.

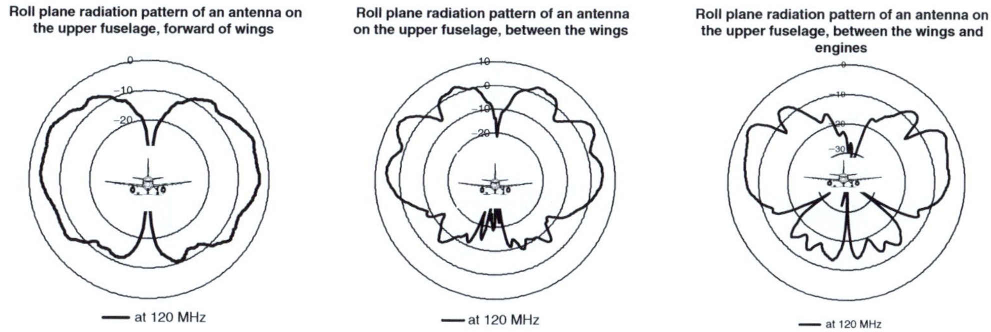

(a) Forward of wings (b) Above/between wings (c) Between wings and engines

Figure 4.25 Roll plane radiation patterns for an antenna at three different positions on the upper fuselage at the same frequency of 120 MHz.

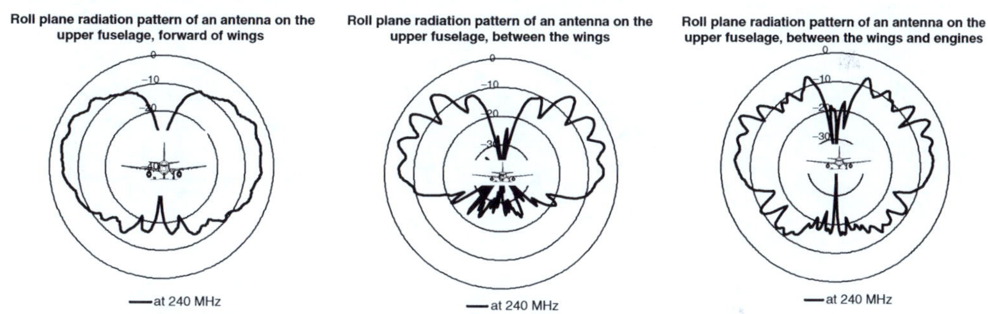

(a) Forward of wings (b) Above/between wings (c) Between wings and engines

Figure 4.26 Roll plane radiation patterns for an antenna at three different positions on the upper fuselage at higher frequency corresponding to aircraft frequency of 240 MHz.

The effect of the wing roots is not so pronounced at 240 MHz (Figure 4.26) because the wing roots are about 2.5 wavelengths away at this frequency. In position 2 between the wings we can see that the radiation in the lower hemisphere is severely restricted. In position 3 between the wings and the engines although the pattern in the lower hemisphere has many ripples, there is nevertheless radiation into the lower hemisphere.

At the highest frequency the radiation into the lower hemisphere is restricted for all three positions, with position 2 showing the smallest amount of radiation into the lower hemisphere (Figure 4.27).

Figure 4.28 shows the juxtaposed radiation patterns at the three frequencies for an antenna installed in position 2 between the wings. This clearly shows that although the radiation in the lower hemisphere is restricted, there is nevertheless some coverage at the lowest frequency of 120 MHz.

It should be stated that the above cases are for monopoles that are omnidirectional antennas. In the case of directive antennas (apart from radar antennas) that do not have obstacles in their main beam, the effects of the wings and other obstacles on the installed

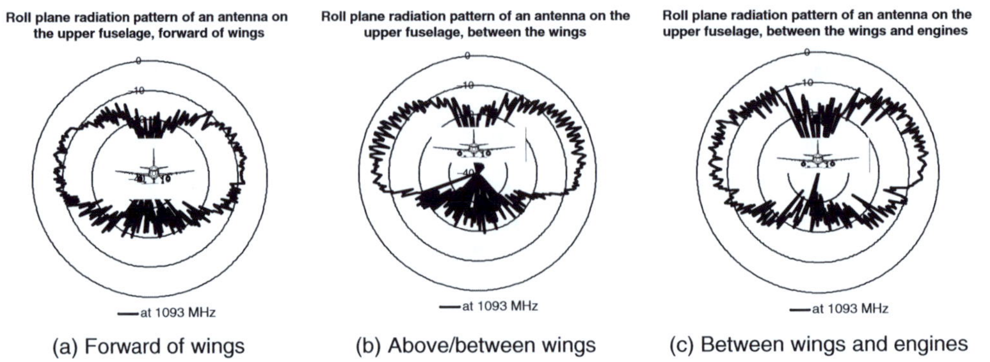

Figure 4.27 Roll plane radiation patterns for an antenna at three different positions on the upper fuselage at highest frequency corresponding to aircraft frequency of 1093 MHz.

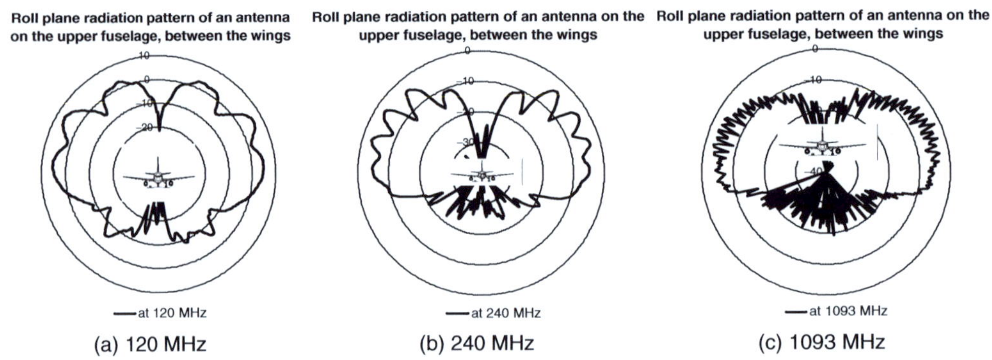

Figure 4.28 Measured roll plane radiation patterns at three different frequencies for an antenna installed on the upper fuselage (of a scaled model) between the wings.

performance would be negligible. However, in the case of radar antennas, the reflection from on-board obstacles, even if they are not in LOS of the main beam, can have a significant adverse effect. This is because false returns can cause confusion with regard to the positions and range of targets.

4.8 Effect of the Curved Ground Plane and the Electrical Dimensions of the Fuselage

Most fuselages of commercial aircraft are of circular cross-section. We can see in the azimuth radiation patterns of Figures 4.21–4.23 that as the frequency is increased the number of ripples also increases. In general, it can be said that for installed monopoles the higher the frequency the greater the number of ripples.

As the frequency is increased, the wavelength is reduced and around the same structure we would expect more wavelengths and hence more interactions between direct and reflected/diffracted/creeping waves.

4.8.1 *Effect of the Curved Fuselage on the Roll Plane Patterns*

In the case of the roll plane patterns, as the frequency increases, the number of ripples and nulls also increases, indicating more cases of near-destructive interference between the direct and first-order reflected waves (as well as creeping/diffracted waves), travelling in opposite directions around the fuselage in this case, due to the shorter wavelength. The presence of deep nulls would be due to waves of near-equal amplitude and in antiphase (i.e. 180° out of phase). If the nulls are not very deep it could be either the interaction between two waves that are of different amplitudes that are around 180° out of phase, or near-equal amplitudes that are more than 120° out of phase.

To investigate the effect that the fuselage diameter has on the radiation pattern in isolation (without effect of the wings), we must look at the radiation pattern on a cylinder. Measurements were undertaken on a scaled cylinder. Thus if the dimensions are scaled down by 12, the frequencies have to be scaled up by 12. The details of the scaled cylinder are given in Section 4.4.2.

If we scale down the size, we must scale up the frequency (or scale down the wavelength) by the same ratio. Since a 1/12th scaled cylinder is used the measurement frequencies of 1.458 and 2.916 GHz correspond to aircraft frequencies of 121.5 and 243 MHz.

The measurement frequency is 1.458 GHz so the wavelength is 0.2058 m. Thus the cylinder diameter of 0.406 m corresponds to ~2 wavelengths. The number of nulls and peaks obtained in the roll plane radiation patterns depends on the electrical circumference of the cylinder, that is, the circumference in terms of wavelength. In this case the roll plane patterns are like the ones we would expect to obtain on a single-aisle aircraft (with a fuselage diameter of ~5 m) at a frequency of 121.5 MHz, for an antenna placed forward of the wings. As the frequency is increased to 2.916 GHz (the wavelength is 0.1028 m) the diameter corresponds to ~4 wavelengths and the number of nulls increases. We would not get the same number of ripples on a much smaller or a much larger fuselage. On a fuselage of a much smaller diameter, such as a small corporate aircraft, we would get a much smoother pattern and it would appear more like a monopole with a very small ground plane.

We can see, for instance, that as we double the frequency from 121.5 to 243 MHz the ripples on the opposite side of the antenna are also doubled. However, we can only investigate the effect on a cylinder or scaled model if the scaled-up dimensions represent the correct dimensions of the real aircraft. If we look at the roll plane radiation pattern (Figure 4.29) at the low frequencies, we can see that the fuselage behaves like a small rod ground plane and the radiation creeps around it and in the plane transverse to the fuselage; the radiation pattern is similar to a monopole with a very small ground plane. Thus if an antenna is on top of the fuselage, a substantial amount still appears in the lower hemisphere, assuming that there are no wings.

The patterns have been inverted in Figure 4.30 and superposed on top of the original patterns in order compare the extent of the radiation in the lower hemisphere with that in the upper hemisphere. The heavier lines show the roll plane patterns and the lighter lines show the inverted patterns. It can be seen that, in the case of the lowest frequency of 121.5 MHz, a large part of the radiation appears on the side of the cylinder opposite the antenna, whereas at the higher frequencies, the amount is much less.

If we compare the roll plane patterns on the cylinder with those obtained on the forward fuselage of a Fokker 100, we can see that the patterns are similar but the number of nulls

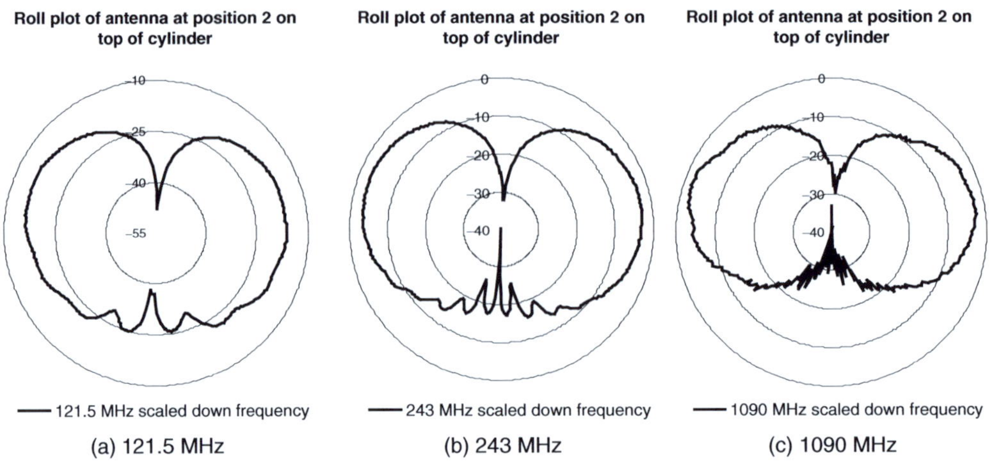

Figure 4.29 The measured roll plane radiation patterns at three different frequencies for an antenna installed on the scaled cylinder.

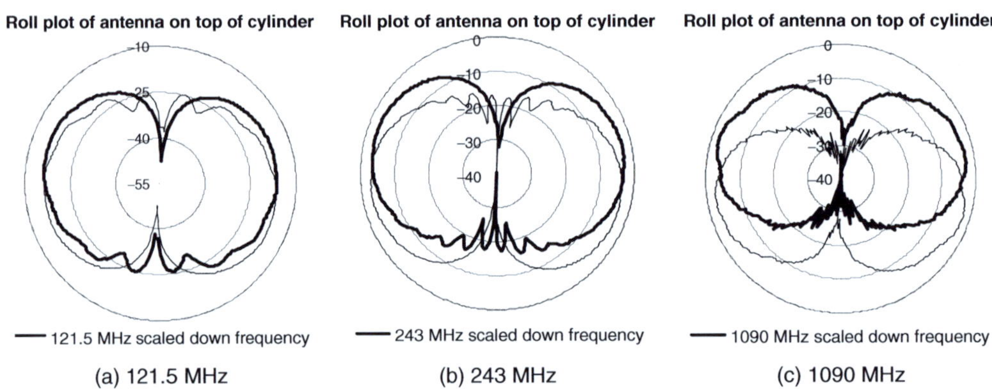

Figure 4.30 The measured roll plane radiation patterns at three different frequencies for an antenna installed on the scaled cylinder, showing the extent of the radiation in the lower hemisphere. The lighter plots are the inverted patterns.

differs (Figure 4.31). At the lowest frequencies the cylinder radiation pattern shows more ripples because the scaled-up diameter of the cylinder is 4.87 m whereas that of the Fokker 100 is 3.224 m. Thus the physical circumferences of the cylinder and Fokker 100 are 15.30 and 10.13 m respectively, and the electrical circumferences are shown in Table 4.6. It can be seen that at the lowest frequencies the patterns are similar apart from the extra ripples in the lower hemisphere due to the larger circumference of the cylinder. However, the presence of interactions with the parts of the airframe is manifest in the undulating pattern obtained over the rest of the angular sector in the case of the Fokker 100 compared with the smooth pattern obtained in the case of the cylinder. In the case of the 240 and 243 MHz plots the effects are magnified and we can see that there are roughly

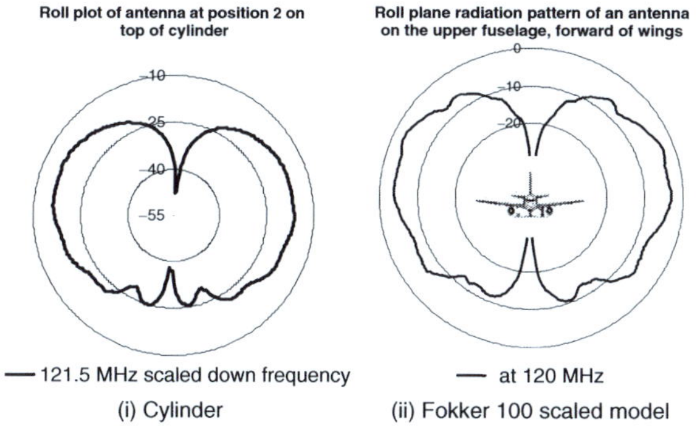

(a) Comparison between the roll plane radiation patterns on the cylinder at 121.5 MHz and the Fokker 100 at 120 MHz

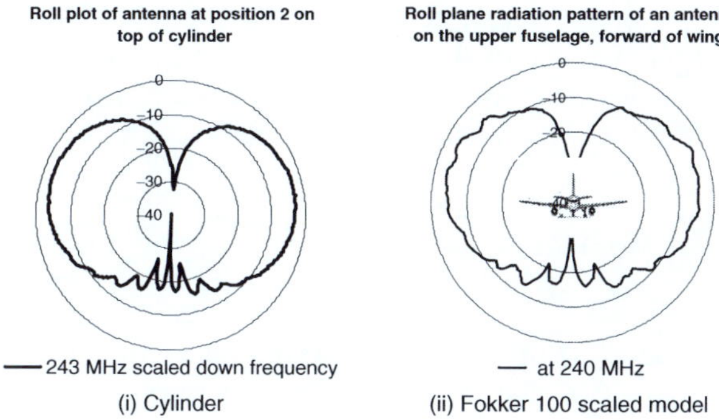

(b) Comparison between the roll plane radiation patterns on the cylinder at 243 MHz and the Fokker 100 at 240 MHz

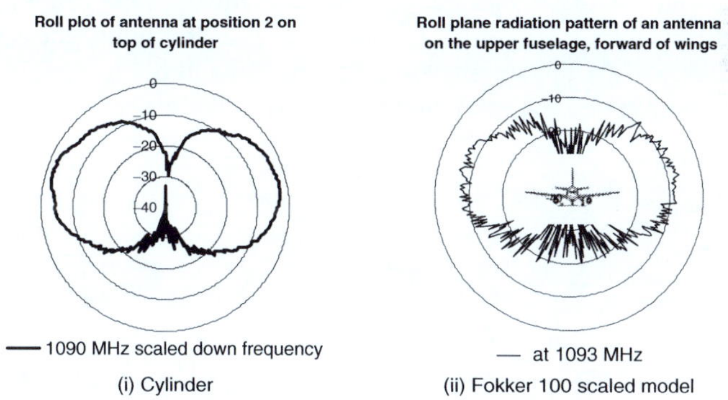

(c) Comparison between the roll plane radiation patterns on the cylinder at 1090 MHz and the Fokker 100 at 1093 MHz

Figure 4.31 The measured roll plane radiation patterns of an antenna at the centre of a cylinder of diameter 0.406 m and on the Fokker 100 forward of the wings.

Table 4.6 Comparison between the scaled-up electrical circumferences of the cylinder and Fokker 100 scaled model.

Frequency in MHz	λ in m	Elect circumference	
		Cylinder	Fokker 100
120	2.50		4.05
121.5	2.47	6.20	
240	1.25		8.10
243	1.235	12.39	
1090	0.28	55.59	
1093	0.27		36.90

one and a half times as many ripples in the lower hemisphere which is what would be expected since the circumference of the cylinder is ≃1.5 times that of the Fokker 100.

At the highest frequencies the Fokker 100 pattern shows the multiple interactions with the airframe as opposed to the relatively smooth pattern of the cylinder.

4.8.2 Effect of the Curved Nose Cone and Tail on the Pitch Plane Patterns

The nose cone has a more complex curved surface than the fuselage, with curvature in more than one plane. Because of this surface the waves could travel along more than one path and then interact with each other before radiation into the far field. Measurements were undertaken on the Fokker 100 scaled model on the upper fuselage at positions 1 (forward of the wings), 2 (above the wings) and 3 (between the wings and the engines) as indicated in Figure 4.16. If we look at the pitch plane radiation patterns in Figures 4.32–4.34, we can see the smoother patterns at the lower aircraft frequency of 120 MHz (Figure 4.32) and the more spiky patterns at the highest frequency of 1093 MHz

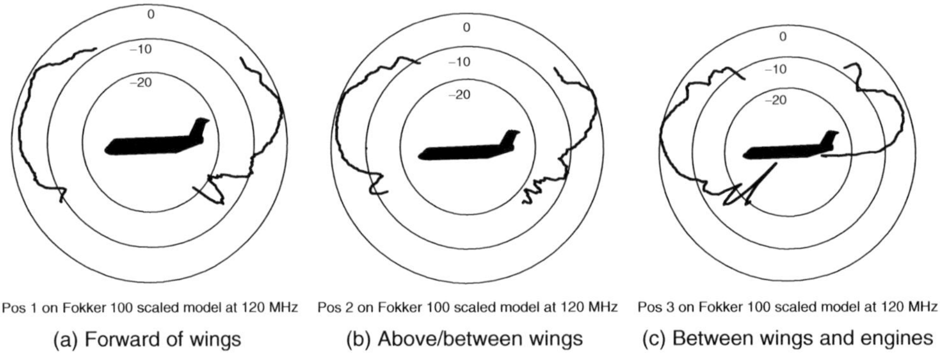

Figure 4.32 Measured pitch plane radiation patterns of an antenna at three different positions on the Fokker 100 at an aircraft frequency of 120 MHz.

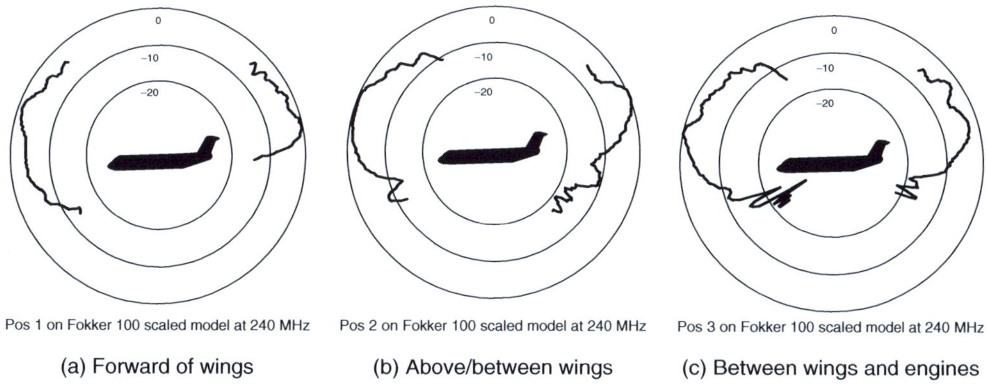

Figure 4.33 Measured pitch plane radiation patterns of an antenna at three different positions on the Fokker 100 at an aircraft frequency of 240 MHz.

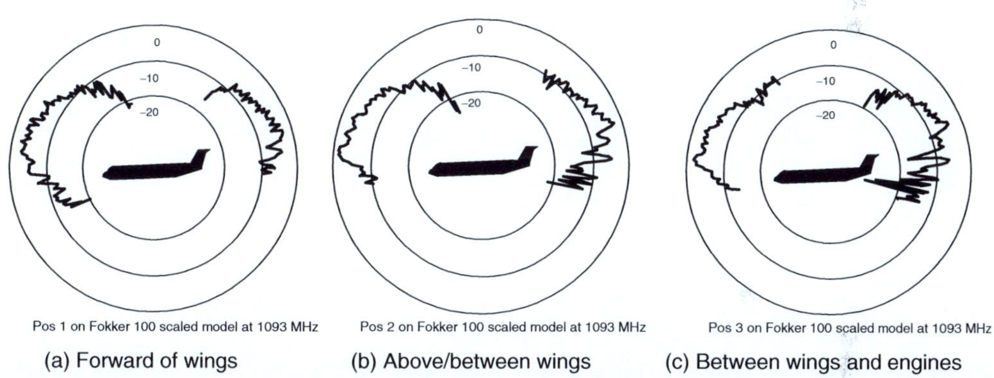

Figure 4.34 Measured pitch plane radiation patterns of an antenna at three different positions on the Fokker 100 at an aircraft frequency of 1093 MHz.

(Figure 4.34). The pitch plane patterns are incomplete because the measurements were undertaken in a cylindrical near-field facility that was not high enough in the vertical direction, to approximate a spherical facility.

Rather surprisingly, as the antenna is moved further back from position 1 to position 3, the coverage around the tail to the lower hemisphere is reduced. This could be due to the measurement system since the full spherical coverage could not be obtained in the cylindrical near-field facility used. There is most probably a deep null, and coverage is obtained at angles below the horizon at larger angles of depression. Very similar coverage is obtained in positions 1 and 2.

At the higher frequency of 240 MHz (Figure 4.33) the patterns show the expected trend of greater coverage in the forward lower hemisphere with the antenna at the front of the fuselage and greater coverage in the lower rear hemisphere as the antenna is moved further aft. The reason for better coverage at the higher frequencies in the cylindrical near-field facility is that the chamber is electrically larger (in the vertical plane) at the higher frequencies – see Section 8.11.2 for details of the near-field facility.

At the highest frequency of 1093 MHz (Figure 4.34) there is limited coverage in the lower hemisphere when the antenna is on the front fuselage but barely any coverage as the antenna is moved further back. The increased beating in the pattern at this frequency is as expected.

4.9 Radiation Patterns on Cylinders in the Absence of Obstacles

If there were no obstacles on the airframe the azimuth radiation patterns would be like those shown in Figure 4.35 for an antenna placed on the curved surface of the cylinder and equidistant from the flat ends, in position 2, as shown in Figure 4.17. The patterns are quite smooth and symmetrical with the number of ripples increasing at the higher

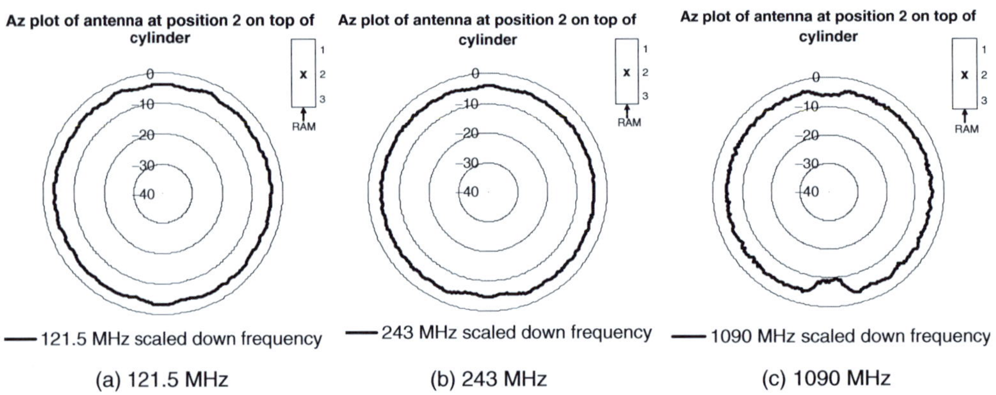

Figure 4.35 The measured azimuth radiations patterns for an antenna at position 2 on a scaled cylinder.

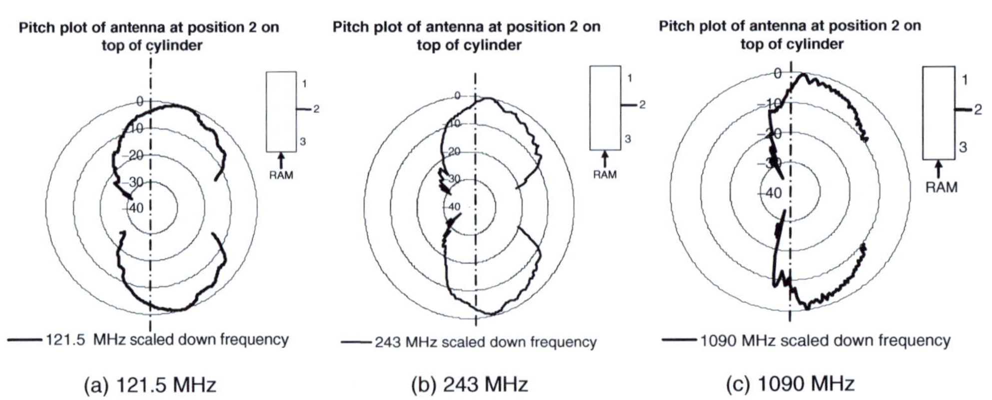

Figure 4.36 The measured pitch plane radiations patterns for an antenna at position 2 on the scaled cylinder.

frequencies as expected. At the highest frequency of 1093 MHz, the ground plane is very large in terms of wavelengths and thus we would expect there to be enhancement, manifested in higher gain forward and aft. However, in this case there are dips forward and aft and they are of different shapes and angular extent. This is most probably due to the diffraction at the sharp edge at the forward end (0°) of the cylinder, which increases at these higher frequencies. The bigger sector blockage at the RAM end in the rear direction (around 180°) is most probably due to absorption (and hence LOS blockage) caused by the presence of the RAM at this end of the cylinder. If we compare these patterns to the azimuth plots obtained on the Fokker 100 shown in Figure 4.22, we can see they are very similar apart from the effect of the tail in the rear and the additional ripples due to the other obstacles on the scaled model.

The pitch plane plots of Figure 4.36 show how the radiation 'wraps' around the end faces of the cylinder at the lower frequencies of 121.5 and 243 MHz, although this occurs to a lower extent compared to the roll plane patterns of Figure 4.30. Figure 4.36 is incomplete because the measurements were undertaken in a cylindrical near-field facility that was not high enough in the vertical direction, to approximate a spherical facility.

References

[1] Macnamara, T.M. (1995) *Handbook of Antennas for EMC*, Artech House, Boston, ISBN 0-89006-549-7.

[2] Verpoorte, J. (2006) Verification of modelling tools by measurements and computation on scaled aircraft, Deliverable D16 Version 3, IPAS-TR_3/NLR/JV 061020-3, 20 October. Installed Performance of Antennas on AeroStructures (IPAS), an EU Specific Targeted Research Project, November 2003 to January 2007, Contract No AST3-CT-2003-503611.

5

Antennas Used on Aircraft

5.1 Introduction

This chapter covers the most common types of antennas used on aircraft. It describes the near and far fields of an aperture antenna and the derivation of the far-field radiation patterns from the Fourier transform of the aperture illumination. It also covers the most common types of aperture illumination and shows the relative benefits of the different illuminations by tabulating the characteristics of these illuminations. This will enable the reader to deduce the characteristics of an aperture antenna from the radiation pattern, since suppliers are often reluctant to divulge details of their antennas.

The most common antennas used for the different systems are dipoles, helixes, horns, loops, monopoles, notches, patches, spirals and reflectors. The types of antenna radiating elements are listed in Table 5.1. Because aircraft antennas have to be aerodynamically suitable, it is not always easy to recognize the type of radiating element inside the aerodynamic enclosure. For instance, the 'blade' antenna may contain a wire or complex tracks on a printed circuit board.

5.2 Near and Far Fields of an Antenna

The plane wave only exists in the far field of the antenna. In the transition region between the antenna and this far-field region, there exists a near field which is very complex in nature and where constructive and destructive interference effects cause deep nulls and peaks in the electric and magnetic fields. The near-field region extends to a distance of $\lambda/2\pi$ for wire antennas and to $2D^2/\lambda$ for aperture antennas with a maximum aperture dimension of D. In practice, because an antenna on a structure would illuminate parts of the structure, the dimension D should be taken as the diameter of the smallest sphere that encloses all appreciable secondary radiating parts.

5.2.1 Far Field for Wire Antennas

The far-field distance for wire antennas is obtained by considering the ratio of the induction field to the radiation field. It can be shown that this ratio is given by

$$\frac{\text{Induction field}}{\text{Radiation field}} = \frac{\lambda}{2\pi R}. \tag{5.1}$$

Table 5.1 Types of antennas used for aircraft systems.

System	Frequency range in MHz		Polarization	Type of antenna
	Lowest	Highest		
ADF	0.19	1.8	Vert	Loop
Comms – Military	30	88	Vert	Monopole
DF	30	407	Vert	Loop
Distress maritime	2.182	–	Vert	Monopole
Distress – Civil	121.5	–	Vert	Monopole
Distress – Civil	243	–	Vert	Monopole
Distress – Satellite	406.028	–	Vert	Monopole
DME	960	1215	Vert	Monopole
EL	121	243	Vert	Monopole
ESM	500	40 000	All	Monopoles, spirals
GPS L1	1565	1586	RHCP	Patches and patch arrays
GPS L2	1217	1238	RHCP	Patches and patch arrays
GLONASS (Russian GPS)	1602	1615.5	RHCP	Patches and patch arrays
GLONASS (Russian GPS)	1240	1256	RHCP	Patches and patch arrays
HF	2	30	Horiz & vert	Monopoles, loops
ILS – Glideslope	329	335	Horiz	Folded dipoles/loops
ILS – Localizer	108	112	Horiz	Monopoles, dipoles
ILS – Marker	74.75	75.25	Horiz	Loops
MLS	5031	5091	Vert	Monopole and Yagi
RadAlt	4200	4400	Horiz	Rect. Waveguide horns, patches
Radar S-band	2000	4000	Vert	Slotted waveguides
Radar X-band	8200	12 000	All	Reflectors and slotted waveguides
Radio Broadcast FM	88	108	Vert	Monopole
SatCom Military UHF	240	310	Vert and circular	Monopole and crossed dipoles
SatCom Military SHF	7200	8400	Circular	Reflector, helical arrays
SSR Tx 1030 Rx 1090	1030	1090	Vert	Monopole
Telemetry	1500	6400	Vert	Monopoles, dipoles, reflectors
TACAN	960	1215	Vert	Monopole
TCAS Tx 1030 Rx 1090	1030	1090	Vert	Monopole and monopole arrays
UHF	225	400	Vert	Monopole
UMTS	1800	1900	Vert	Monopole
VHF Civil	108	137	Vert	Monopole
VHF Maritime	150	174	Vert	Monopole
VHF Military	108	152	Vert	Monopole
VOR en-route navigation	112	118	Horiz	Monopoles, dipoles
VOR terminal	108	112	Horiz	Monopoles, dipoles
Weather Radar C	5500		All	Reflectors and slotted waveguides
Weather Radar X	9330	9354	All	Reflectors and slotted waveguides

The distance of the far field is taken as the position when the induction field is equal to the radiation field. Thus using Equation 5.1 and equating it to 1, we can see that this gives $R = \lambda/2\pi$. However, to be on the safe side, this distance is usually doubled, giving the far-field distance as λ/π.

5.2.2 *Far Field for Aperture Antennas*

In the case of aperture antennas we obtain the far-field distance by considering the power density. In the region where interference effects predominate, the power density (the power per unit area) does not decrease monotonically (i.e. reduce consistently) as the distance from the antenna increases, but follows an oscillatory pattern. This is because the beam alternately divides and re-forms in an oscillatory manner. This beam broadening also results in a reduction in the gain in the near field, instead of the gain being inversely proportional to the distance. This fact is used in determining whether an antenna is in the far field. If the receive antenna is in the near field of a transmitting antenna the power it receives (from the direct wave only) will increase as it is moved away, instead of decreasing. The power density and gain variation with distance depends on the illumination across the aperture of the antenna. For aperture antennas the region of the near field which extends to $D^2/4\lambda$ is sometimes called the Rayleigh region, after the eighteenth-century physicist Lord Rayleigh, although it is more commonly known as the Fresnel region. At this distance P (shown in Figure 5.1) the phase of the waves from A and B at the edges of the aperture are 180° out of phase compared with the wave from the centre of the aperture. This means that the path differences to the Fresnel distance, between the edges of the aperture and the centre of the aperture, are $\lambda/2$ since 180° corresponds to $\lambda/2$.

However, a plane wave cannot be assumed to be present in the region directly beyond this distance, and the Fresnel region in practice is the portion of the near field that extends up to $2D^2/\lambda$.

A plane wave can be assumed to be propagated beyond the Fresnel distance (R_f) of $2D^2/\lambda$. The Fraunhofer region exists at distances greater than $2D^2/\lambda$ and in this region the power density decreases monotonically and follows the inverse square law. The electric field is inversely proportional to the distance from the antenna in this region. The Fraunhofer or far field is considered to be the point at which the path differences between points on the aperture are negligible.

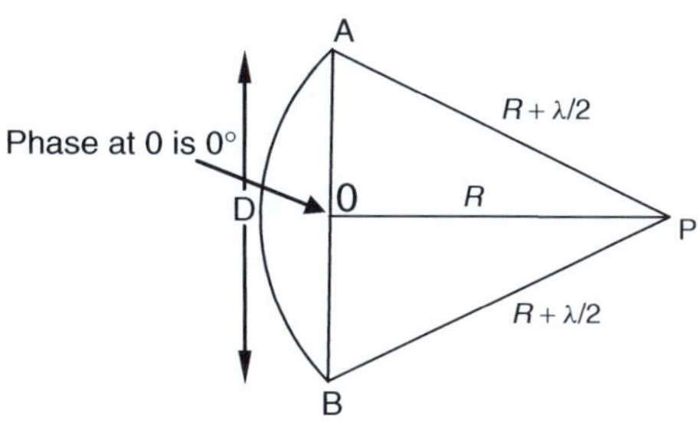

Figure 5.1 The Fresnel distance for a reflector antenna.

In order to qualitatively understand the oscillatory nature of the power density in the near field, it is useful to think of the aperture in terms of the wave-like nature of EM radiation as postulated by Huygens, for physical optics. The aperture can be considered to be constituted of a number of elemental areas. Each elemental area can be regarded as a secondary source which gives rise to a spherical wavelet. The wave amplitude at any point from the aperture is obtained by considering the superposition of these elemental wavelets taking into account their relative phase differences when they arrive at the point in question. Constructive and destructive interference may occur as in the case of light waves. In the case of a linear aperture of length 5λ that is uniformly illuminated as shown in Figure 5.2a, the variation of the field pattern as the distance from the antenna increases is shown in

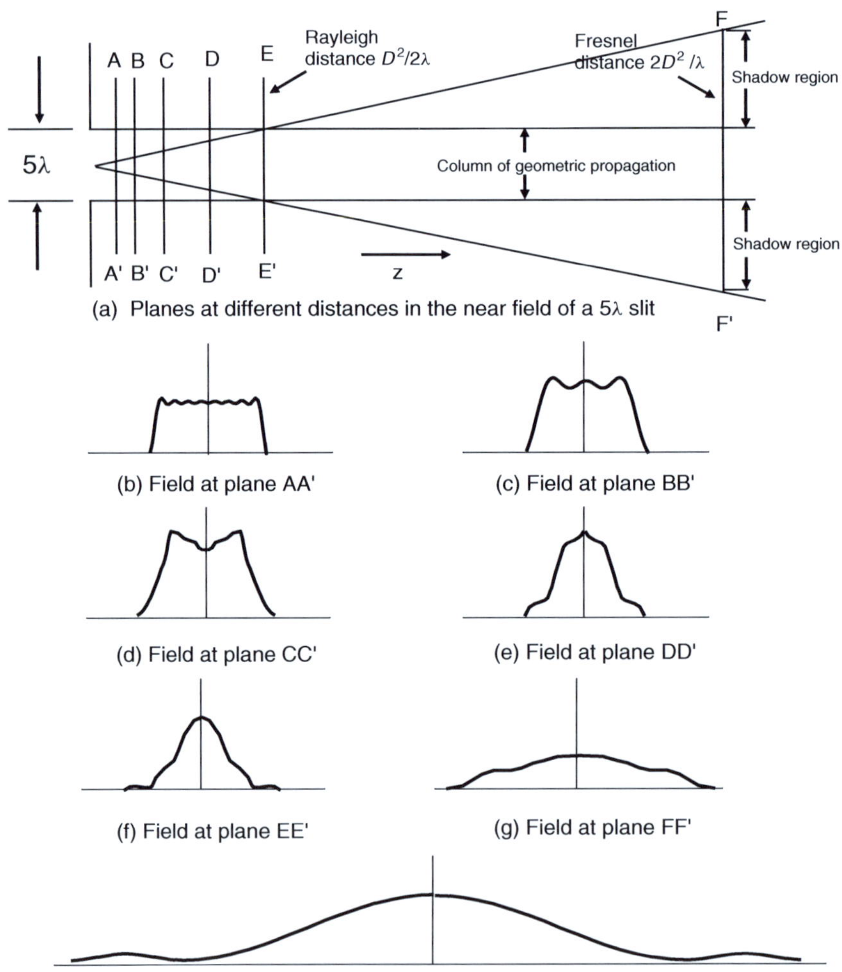

(a) Planes at different distances in the near field of a 5λ slit

(b) Field at plane AA'

(c) Field at plane BB'

(d) Field at plane CC'

(e) Field at plane DD'

(f) Field at plane EE'

(g) Field at plane FF'

(h) The far field pattern

Figure 5.2 The transition from the near-field to the far-field pattern for a 5λ aperture antenna.

Figure 5.2b–g. It can be seen that as the distance increases, the field diverges from the column of radiation propagated geometrically from the aperture and parallel to the z axis. The field progressively diffuses into the shadow region. The Fresnel region is characterized by the onset of this diffusion outside the boundaries of the column of radiation propagated geometrically from the aperture beyond the plane EE′ ([1], p. 173). The radiation pattern is shown in Figure 5.2h. However, the boundaries between the three regions cannot be sharply defined and the transition from the Fresnel to the Fraunhofer approximations is a gradual one which is determined by the acceptable errors in the approximations that are made.

For circular apertures D is the diameter, but in general, reflector antennas used for surveillance and weather radar have elliptical apertures with major and minor axis a and b, respectively. This is because a narrow beam is required in the horizontal plane to provide greater resolution as the radar scans in the azimuth plane. In this case the Fresnel region for one plane is different from that in the other plane. In the plane containing the major axis a the Fresnel field region extends up to a distance of $2a^2/\lambda$, whereas in the plane containing the minor axis b the Fresnel field region extends up to a distance of $2b^2/\lambda$. The larger Fresnel distance of $2a^2/\lambda$ should be taken as the correct value.

In many cases the Fresnel distance is taken as $4D^2/\lambda$. In terms of the gains for antennas, it is useful to compare the gains obtained at $2D^2/\lambda$ and at $4D^2/\lambda$ with that obtained at infinity. If the gain at infinity is G_0, the gain at $2D^2/\lambda$ is $0.94G_0$, whereas at $4D^2/\lambda$ the gain is $0.99G_0$ ([1], p. 199). Therefore the far-field distance of $4D^2/\lambda$ is a more suitable distance for antenna gain measurements, that is, for radiation patterns.

5.3 Antennas on Aerostructures

The radiation patterns in manufacturers' data sheets are those obtained for antennas on standard ground planes or in free space. Real antennas are greatly affected by the structure on which they are installed, and the patterns obtained are unique to the structure on which they are installed. Thus the radiation patterns obtained on one airframe cannot be assumed to the same on a different airframe.

A quarter-wave monopole (that has been 'tuned') tends to radiate in the shape of a doughnut, so that the majority of the power is radiated in the horizontal plane. However, this only applies when it is on an ideal ground plane of reasonable size. This doughnut shape of radiation is greatly distorted when it is installed on an aircraft. Most communication, guidance/landing and navigational antennas are monopoles or variations of monopoles. The radiation patterns obtained for antennas at different positions on the fuselage of an aircraft can be seen in Chapter 4.

For satellite communications an antenna is used that would radiate upwards when placed on the aircraft. The radiation pattern is less affected by the structure in these cases, since the main beam is directed away from the aircraft and other obstacles and the higher frequencies used in these cases mean that the antenna is electrically further away from obstacles and hence the effect of the latter is reduced. For antennas placed on the upper fuselage flat plate, arrays of 'patches' and in some cases helixes and crossed dipoles antennas are used.

For weather and search radar, 'aperture' antennas are used. Optical system aperture antennas radiate through their 'mouths' like the reflectors of torches. They are usually mounted in the nose of the aircraft. Flat plate slotted waveguide antenna arrays are now becoming more commonly used instead of reflector antennas. These arrays enable phasing to be applied so that the beams can be scanned without physically moving the antenna, although to obtain large angles of scan the arrays are usually physically moved as well. In some cases physical scanning is used in the azimuth plane and electronic scanning in the elevation plane since the angular coverage required in the vertical plane is usually much less than that required in the azimuth plane.

The antenna of each system is required to provide the optimum spatial coverage at the correct frequency, as well as provide adequate gain to attain the correct range for its system.

The spatial separation between antennas is calculated to provide the optimal isolation between systems in the spatial domain. Where the RF interoperability cannot be achieved by spatial separation, separation in the time and frequency domains is used as well as power reduction of transmitters and/or sensitivity reduction of receivers. More detail on these measures is given in Chapter 6.

5.4 Polar Radiation Patterns

Polar radiation patterns usually only show the relative gain at each angle and not the absolute gain. As mentioned in Chapter 1, quite often the radiation pattern is shown without any scales on the plot and the scales on these patterns are usually shown without a sign, but because the values nearer the centre are greater than those further away, we can deduce that the negative signs have been omitted.

The whole shape of a polar plot can look quite different if the minimum levels are varied. Because the radiation pattern of a monopole is dependent on the size and shape of the ground plane the radiation patterns of dipoles are shown below instead of monopoles, since dipoles do not have ground planes.

In the case of a polar plot especially, like that shown in Figure 5.3a, the scale is very important in defining the shape. For instance, the pattern can look completely different if plotted with different scales, (from 0 to $-100\,\text{dB}$) as shown in Figure 5.3b.

5.5 Dipoles

Dipoles are usually fed by a coaxial cable, although twin wires are sometimes used at very low frequencies. The inner conductor is connected to one arm of the dipole and

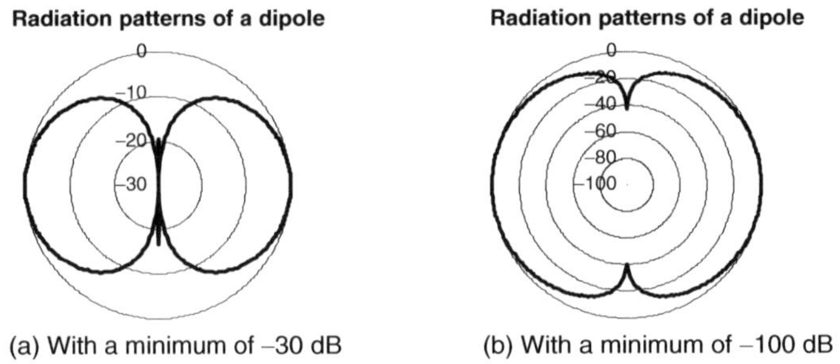

Figure 5.3 The radiation pattern of a tuned half-wave dipole plotted using different scales.

Table 5.2 Characteristics of different size dipoles.

	Small/Hertzian dipole $\lambda/8$	Half-wave dipole	One wavelength dipole
Resistance (Ω)	3.125	73	$\sim$185
HPBW ($^\circ$)	90	78	47
Peak directivity (dBi)	1.96	2.16	2.42

the outer conductor is connected to other arm of the dipole. They have to be mounted away from metallic areas and therefore have to protrude from the sides of an aircraft. These are, therefore, not suitable for mounting on aircraft since they would disturb the air flow. However, some adaptations of dipoles are used on aircraft especially if they can be mounted horizontally as in the case of the ILS localizer antennas. These are sometimes installed on the nose of the aircraft or on either side of the tail fin. They are often bent dipoles and are called ram's horns because of their appearance. In some cases the VOR Localizer antenna is combined with the VHF antenna and is mounted above the VHF blade. It looks like a horizontal boomerang on top of the blade. Crossed dipoles are used in SatCom antennas for communication with high-angle satellites. They are circularly polarized, whereas the blades on which they are mounted are vertically polarized for communication with low-angle satellites. The characteristics of different size dipoles are shown in Table 5.2.

5.5.1 *Small Dipoles*

When we refer to 'small dipoles' we mean *electrically* small dipoles. Small dipole antennas are sometimes defined as those that are shorter than $\lambda/8$ in length. These small antennas have much lower gains than tuned ones and, more importantly, their spatial radiation patterns are different. Thus we cannot assume that a small antenna placed at the position of a half-wavelength tuned antenna will just give us reduced gain at all angles. At some angles the gain will greater and at some angles it will be less. Additionally, the positions of the nulls are at different angular positions.

Manufacturers would normally use capacitors and inductors for these broadband antennas to ensure that the antennas are not electrically longer than $\lambda/2$ at the higher frequencies.

An isotropic antenna is effectively a point source and, as its name suggests, radiates isotropically, that is, its radiation pattern is a sphere with the same level of radiation at all angles.

A theoretical very small dipole is called a Hertzian dipole and has an HPBW of 90° and a maximum gain of 1.96 dBi (linear gain of 1.5).

5.5.2 *Resonant Dipoles*

The smallest length of tuned or resonant dipole is approximately half a wavelength long and is called a half-wave dipole. It has a doughnut-shaped radiation pattern with the peak occurring at the horizon for a vertical dipole. The HPBW of a tuned half-wave dipole is 78° and the maximum gain is 2.16 dBi (linear gain of 1.64).

The impedance of a dipole normally has a reactive as well as a resistive component. At the first resonant point of 0.5λ the reactive component is zero and thus the impedance of the antenna is purely resistive, and its resistance is 73 Ω.

The total impedance X of a dipole is given by

$$X = 30\left\{2S(kL) + \cos(kL)[2S(kL) - (2kL)] - \sin(kL)[2C(kL) - C(2ka^2/L)]\right\}, \tag{5.2}$$

where

S and C are the sine and cosine integrals
a is the radius of the wire
L is the length of the dipole
and $k = 2\pi/\lambda$.

The reactance is dependent on the length as well as the thickness of the wire from which the dipole is made. The variation of reactance with the electrical length of a thin dipole with a radius-to-length ratio (a/L) of 0.001, is shown in Figure 5.4a. It can be seen that the reactance of the half-wave dipole is zero when the length is just under 0.5λ – that is, when we have a tuned dipole. The variation of resistance with length is shown in Figure 5.4b (see [2], p. 103).

For the ideal tuned half-wave dipole in free space, the typical doughnut shape is obtained, as shown in Figure 5.5a, with the roll and pitch planes being the same and the azimuth plot of Figure 5.5b being a circle, that is, the antenna is said to be omnidirectional in the azimuth plane. Note that these show relative and not absolute gain levels. The highest or peak gain for the particular cut, is usually adjusted to the 0 dB value, and this is known as normalizing to the peak gain. However, in the case of Figure 5.5b this is shown at a lower level for clarity, so that the pattern does not coincide with the 0 dB circle.

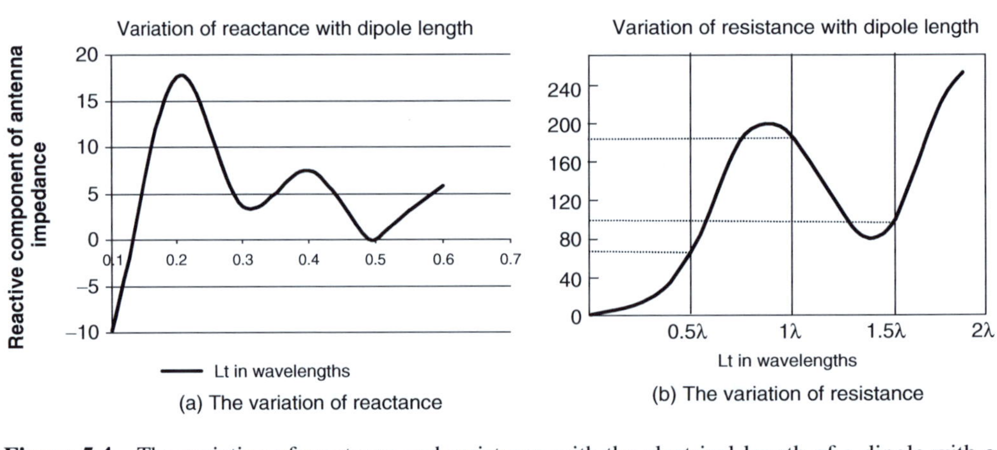

Figure 5.4 The variation of reactance and resistance with the electrical length of a dipole with a radius-to-length ratio (a/L) of 0.001.

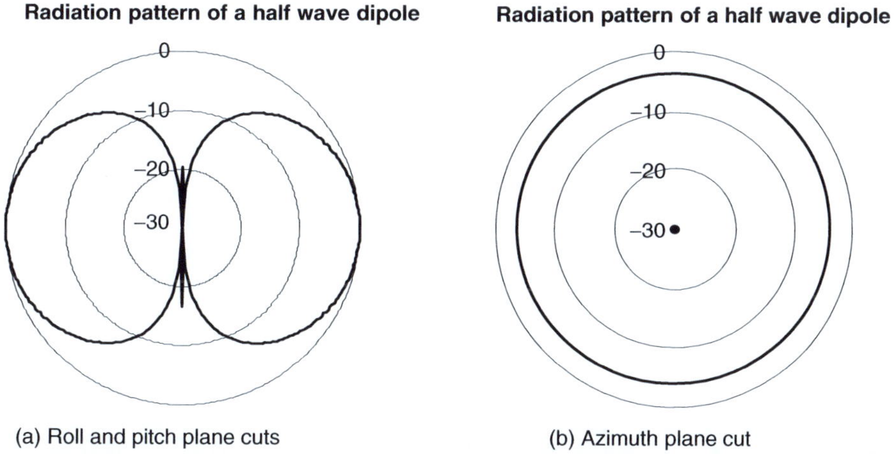

Figure 5.5 The principal plane cuts of the radiation pattern of a half-wave dipole.

5.5.2.1 Higher-Order Resonant Dipoles

We tend to talk about dipoles, for instance, without mentioning their electrical lengths. We assume that the dipole is half a wavelength ($\lambda/2$) long. Dipoles could be very small (much smaller than $\lambda/2$), or longer than $\lambda/2$. For an antenna of fixed length, as the frequency is increased the physical length of the antenna becomes equal to a value greater than half a wavelength and then becomes one, one and a half wavelengths, two wavelengths, and so on. The radiation patterns for one-wavelength and one-and-a-half-wavelength dipoles are shown in Figure 5.6. The HPBW decreases to 47° for a full-wave dipole (length λ). Note how the familiar doughnut shape of the half-wave and full-wave dipole splits into two slanted doughnuts at one and a half wavelengths. This of course could cause severe problems for the aircraft systems connected to these antennas, since the antennas are no longer omnidirectional in the azimuth plane, but instead have significant nulls in this plane.

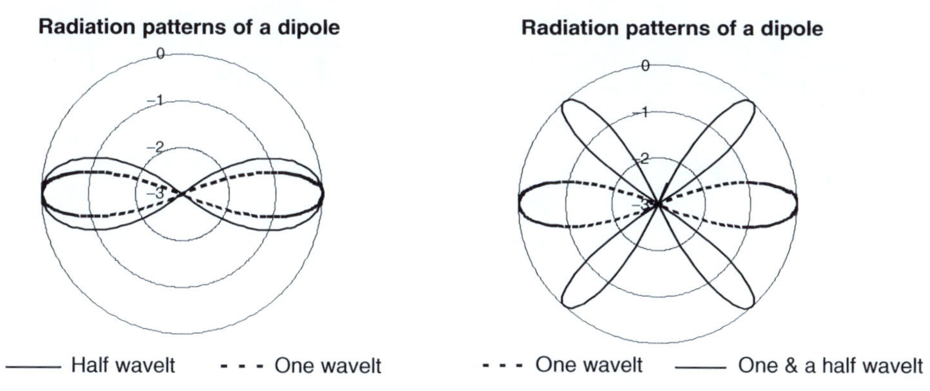

Figure 5.6 One-wavelength and one-and-a-half-wavelength dipoles and their radiation patterns.

5.5.3 *Broadband Dipoles*

Broadband passive antennas that cover several octaves will behave like electrically small antennas at the low frequencies, a tuned half-wave dipole at the midband frequencies, and like a one-and-a-half-wave or larger dipole at the highest frequencies. A dipole is made more broadband by increasing its thickness. A linear dipole with an *l/d* ratio of 5000, and a bandwidth of 3%, can have its bandwidth extended to 30% by keeping its length constant, but having its *l/d* ratio decreased to 260 ([3], p. 333). The physical length of a thin dipole is usually less than the electrical length and varies between 91% and 98% of the physical length.

5.5.4 *Crossed Dipoles*

Crossed dipoles are used instead of flat spiral antennas to transmit and receive circular polarization since they can handle higher powers than flat spirals, which tend to be implemented on printed circuit boards. Circular polarization can be generated by two waves of equal magnitude at right angles to each other, if they are 90° out of phase with each other, that is, they are in phase quadrature. Thus, if we have two dipoles at right angles to each other and one of them is fed by a 90° phase shifter we would get circular polarization.

Consider the horizontal and vertical waves shown in Figure 5.7a which are 90° out of phase, At point A the vertical wave is zero so that the resultant of the two waves is the horizontal vector of one unit, as depicted by the double-arrowed vector shown in Figure 5.7b. At point B, one-eighth of a period later, the horizontal and vertical vectors are 0.707 units so that their resultant is one unit (using Pythagoras) at 45° to the horizontal. We can see from Figure 5.7b that the resultants retain a value of 1 at all points and over one period of time produce a rotating vector that has uniform magnitude of one unit, that is, we have circular polarization.

On aircraft crossed dipoles are used for military UHF SatCom. One version of this type of antenna consists of a crossed dipole mounted horizontally above a blade. The

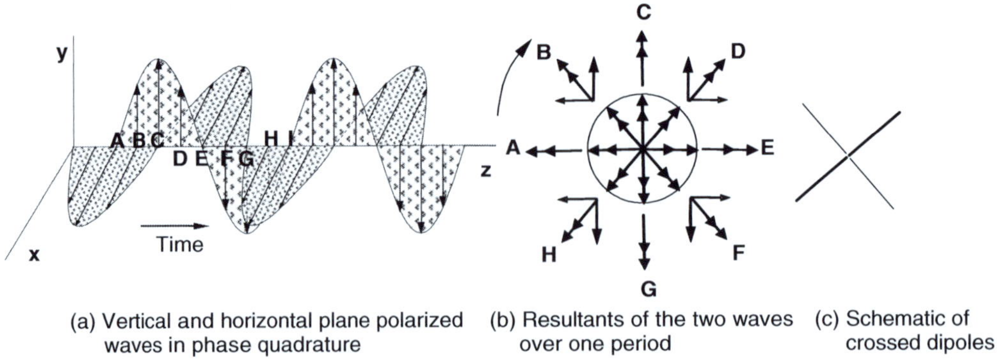

Figure 5.7 Circular polarization obtained by two plane waves in space and time quadrature.

blade/monopole is used for communication with low-angle satellites and the crossed dipole is used for high-angle satellites.

5.5.5 Dipole Arrays

If more than one radiating element is used to form the beam of the antenna then we have an array. Dipole arrays are used to increase the gain of antennas as well as to increase their frequency bandwidth. Yagis are used to increase the gain and bandwidth of a single dipole as well as to produce a directional antenna. Log periodic arrays are used primarily to increase the bandwidth, since they are frequency independent (FI) antennas. However, they also produce a directed beam. They are not normally used as airborne antennas, although they may be used for special applications.

5.5.5.1 Yagis

Although this antenna is commonly known as a Yagi, it was actually designed by Professor Uda of Japan, but the paper was first presented in English by his student Mr Yagi, who also performed most of the measurements. It should therefore strictly be called an Uda–Yagi antenna. In principle it consists of a single resonant/tuned fed dipole (also known as a driven dipole) with a number of slightly shorter parasitic dipole elements called directors in front of it, as shown in Figure 5.8a. Parasitic elements are ones that are not directly fed but have currents induced in them in the same way as currents from the primary of a transformer induce a current in the secondary. The parasitic elements re-radiate like the directly fed radiating dipole. A reflector dipole which is slightly longer than the driven element is placed behind the driven element. All elements are parallel to each other and in the same plane. The antenna radiates in the end fire direction, that is, from the reflector towards the shorter directors. The radiation pattern in the plane of the array, the xy plane, is shown in Figure 5.8b, and that in the yz plane is shown in Figure 5.8c. In the case of uniform Yagi arrays all the elements have the same circular cross-section,

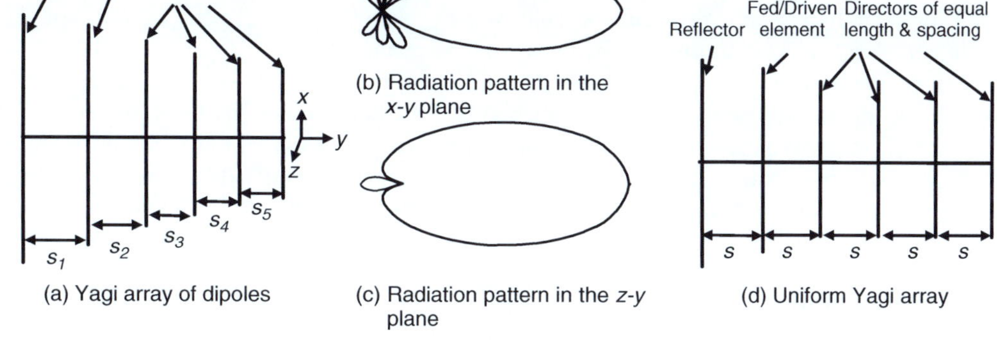

Figure 5.8 The Yagi–Uda dipole array.

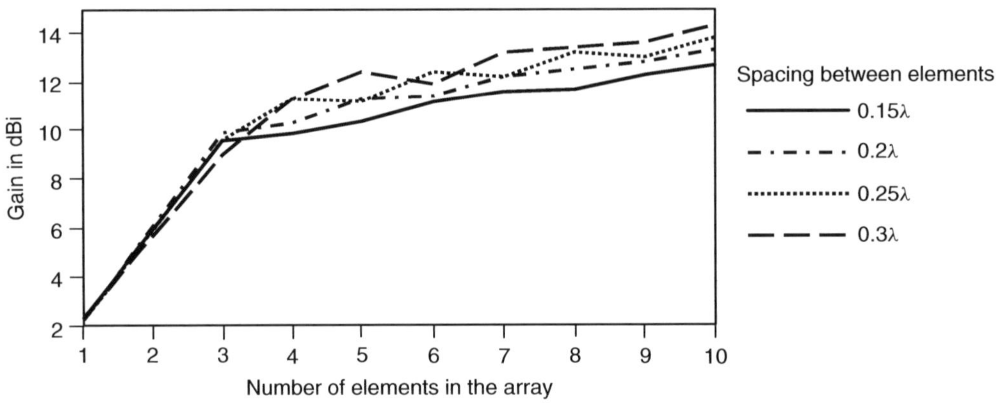

Figure 5.9 Variation of gain with number of elements for uniform Yagis.

the spacings between the elements are the same and all directors have the same physical length, as shown in Figure 5.8d. Although the directors are made the same length for ease of construction, they do not carry equal currents.

The gain of the Yagi can be increased by using more directors, but once the number of parasitic elements is above about 12 the increase in gain is limited ([4], Vol. 2, p. 765). Computations have been undertaken for the gains obtainable for uniform Yagis with differing numbers of elements with different wire diameters and spacings [5]. These are plotted in Figure 5.9 for a wire diameter of 0.0025λ using the spacing as a parameter. It can be seen from the graph that there is a linear increase in the gain for all the spacings up to three elements, but above this number the gains start to level out with the largest spacing showing the best increase in gain.

Comparison of measured results and computed values shows that, as in the case of single dipoles, the elements behave as though they are slightly longer than their physical lengths, due to an 'end effect'. Thus, in order to obtain optimum performance, the elements are slightly shortened at the design frequency. The input resistance of the antenna varies with the number of elements used, but a four-element array with a spacing of 0.2λ has an input resistance of about 45 Ω. Increasing the number of elements reduces the radiation resistance, which could result in a mismatch to the transmission line feeding the antenna.

A broader bandwidth Yagi can be achieved by using shorter directors and a longer reflector element. The shorter elements would be resonant at the upper end of the frequency band and the longer reflector element is designed for the low end of the band. However, the broadband operation is obtained by sacrificing the gain. Thus, for instance, an antenna with five elements (one reflector and three directors) can have its bandwidth increased to 60%, but its gain would be reduced to around 6 dBi ([4], Vol. 2, p. 765). Increasing its gain to 10 dBi reduces its bandwidth to about 9%.

5.6 Monopoles

As its name suggests, a monopole only has one arm and requires a ground plane. The coaxial cable feeding the monopole has the inner conductor connected to the monopole and the outer conductor connected to the ground plane. These are the most common type of antennas used on aircraft. They are usually mounted inside aerodynamically shaped enclosures and are commonly referred to as blades. Blades are basically monopoles encapsulated in a covering to provide better aerodynamic properties.

The smallest resonant monopole is approximately a quarter of a wavelength long and radiates the maximum power. At resonance the impedance of a thin monopole is purely resistive and has a value of 36.8 Ω. It has zero reactive impedance, that is, it does not have an inductance or capacitance. The value of this resistance varies by a small amount depending on the gap between the inner and outer conductor of the coaxial line feeding the antenna. It has been shown experimentally that the actual length of a tuned quarter-wave monopole is 0.236λ when the ratio of the diameter to length (d/l) of the conductor is 0.00318. Thus at 100 MHz the length of the monopole would be 0.708 m and its diameter would be 9.5 mm. Since tuned monopoles only have a real impedance (the imaginary part being zero), they are easily matched to the characteristic impedance of the RF line (usually coaxial cable).

As in the case of dipoles, monopoles can be made more broadband by making the diameter of the conductor larger. Length-to-diameter ratios (l/d) of 10 to 10^4 (i.e. d/l ratios of 0.001 to 0.1) are common [6]. A thick monopole has a length greater than 0.25λ and its resistance is less than 37 Ω.

5.6.1 *Ground Plane Dependence*

The radiation pattern of the monopole depends on the ground plane on which it is installed. When the ground plane is electrically small (in terms of wavelengths) the outer edge of the ground plane diffracts the incident radiation in all directions. The currents on the top and bottom surfaces of the ground plane are equal in magnitude but must be opposite in direction since the net current at the edge is zero. This outer edge diffraction becomes more significant as the size of the ground plane is reduced since the edge is nearer the monopole where the currents are larger. Edge diffraction can alter the input impedance by more than 100% and the gain by more than 6 dB compared with a monopole on an infinite ground plane.

For monopoles with infinite and zero ground planes, the peak directivity occurs at the horizon, that is, at elevation angles (θ) of 0° (zenith angles of −90°). In general, however, it can be said that as the size of the infinite ground plane is decreased the peak directivity occurs at larger elevation angles (>0°), that is, the pattern tends to tilt upwards. However, this variation is not monotonic, that is, it does not increase consistently. The gain at the horizon is 1.76 dBi in the absence of a ground plane, with a large but finite ground plane it is −1.249 dBi, whereas it is 4.77 dBi in the case of an infinite ground plane [6]. The smoothed curve obtained for the variation of the peak directivity with the radius of the ground plane is shown in Figure 5.10.

5.6.2 *Top Loaded Monopoles*

On an aircraft it is aerodynamically preferable to have low-profile (i.e. short) antennas. Thus, many blades are ‘top loaded’. This top loading can be embodied by attaching a

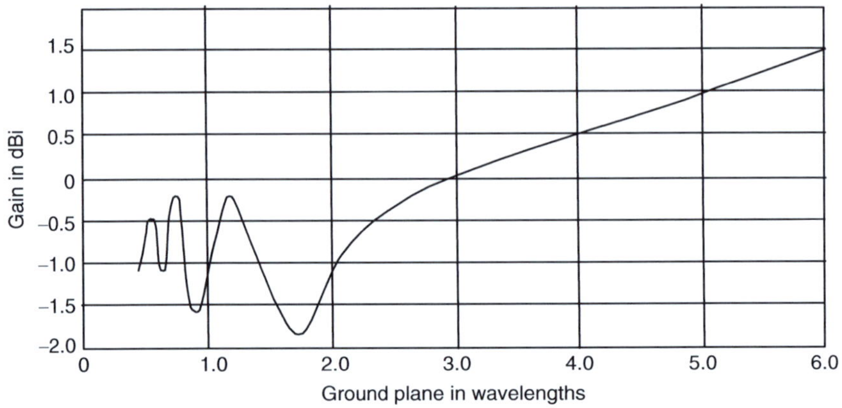

Figure 5.10 Variation of the peak directivity for a tuned monopole, with the electrical diameter of the ground plane.

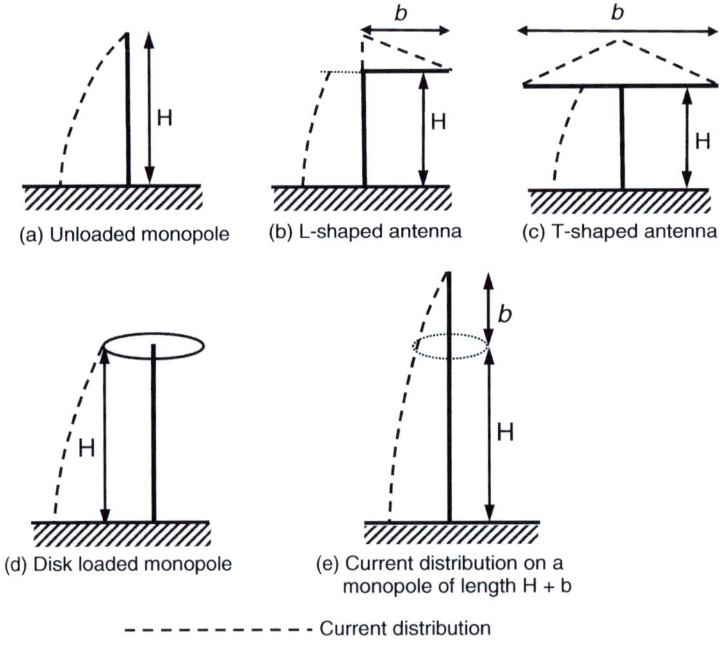

Figure 5.11 Top loaded monopoles.

disc at the free end, by bending the free end into the shape of an L or by using a T-shape as shown in Figure 5.11.

Top loading can increase the effective length by a factor of 2; this increases the gain of the antenna. Without any loading the current distribution of the unloaded monopole is as shown in Figure 5.11a. In the case of the L or T antennas, shown in Figure 5.11b and c, the bent sections draw the same amount of current that would be drawn by an additional

Figure 5.12 Chelton 16-21 VHF comms antenna 118–136 MHz and Chelton 19-85 VOR/GS 108–118, 331–333 MHz Ref IPAS [8].

length b connected to the antenna. This additional length b is approximately equal to the physical length of the horizontal portion of the L and T antennas. Top loading by use of a disc changes the current distribution that would be present if the monopole was extended by a length b equivalent to the effective increased length of the top loaded monopole (see [7], p. 512).

Although top loading gives an increased effective length and thus improves the total gain, these horizontal sections introduce a component of horizontally polarized radiation and usually result in a reduction in XPD. In the case of a vertical monopole, the XPD is the ability of the antenna to reject the horizontal or cross-polar radiation. Monopoles that are greater than one wavelength also suffer from the same break-up of lobes as dipoles.

Antenna manufacturers tend to use reactive circuits in radiating elements to prevent the electrical length of the antenna attaining these lengths.

Figure 5.12 shows two blades manufactured by Chelton. The first one is a top loaded blade used at VHF and the second one is a combined VOR and glideslope antenna.

5.6.3 Small Monopoles

As in the case of dipoles, when we refer to 'small' monopoles we mean electrically small monopoles. Thus at HF these small monopoles could be several metres long. The aircraft HF band spans 2–30 MHz, which is nearly four octaves. The wavelength varies between 150 and 10 m. A tuned (quarter-wave) monopole at the lowest frequency would have to be 37.5 m long and, since most commercial airliners are around 50 m long, it is impractical to install tuned monopoles on an aircraft at the low end of the band. Even at the high end of the band a tuned monopole would be 2.5 m long. Because of this large length small antennas are used at this frequency. Electrically small antennas have imaginary (reactive) parts of impedance and in the case of HF require tuning circuits in the form of separate units called HF tuners/couplers that have to be used in conjunction with the antenna. These tuners work on the principle of conjugate matching to the antenna. Thus if the antenna has an impedance of $Z = R + jX$, the tuner applies an impedance of $-jX$ to tune out the reactive part of the impedance. The reactive part of the impedance varies in both sign and magnitude over the frequency band. A negative reactance is capacitive and a positive reactance is inductive. Because of the low frequency and power requirements, the dimensions of the HF tuners are large, require cooling and have to be placed very near the antenna radiating element. In addition, the radiated emissions from the HF tuners could cause problems to the other electronic equipment on board.

At HF vertical polarization is mainly used for line-of-sight communications between aircraft or between an aircraft and the ground, whereas horizontal polarization is used for long-distance communications. In the latter case, the waves are reflected (using the principle of total internal reflection) by the ionosphere the troposphere (see Chapter 1 for further details). Because HF uses horizontal as well as vertical polarization, the antennas are usually installed at an angle to the vertical, preferably at 45° where this is achievable. An antenna inclined at 45° to the vertical would result in a 3 dB loss for both horizontally and vertically polarized waves. However, if the antenna were vertical the loss would be infinite (in theory) for horizontally polarized waves. In practice the loss would about 15 dB for real antennas depending on the XPD of the antenna.

One of the main advantages of using low frequencies is that the space loss or attenuation is much less than it would be at higher frequencies. For instance, the loss over a distance of 300 m is 28 dB at 2 MHz but 46 dB at 16 MHz, that is, 64 times more. This means that the power required would also need to be 64 times more at the higher frequency to transmit or receive a wave over the same distance.

5.6.3.1 Long-Wire Antennas

Long-wire antennas are 'small' monopoles that are only used for HF and are often attached between the fuselage and the tail fin or between the wings and the tail fin, and are therefore at angles approaching 45°. This makes them suitable for both horizontal and vertical polarization. Because of this configuration, they can be about 30 m long (depending on the aircraft size) and hence have higher gain than towel rails, loops or notches. The higher gain permits longer ranges (distances) for the system, since an increase of 6 dB would effectively double the range.

However, they are not aerodynamically suitable, since they cause drag and have to be tensioned and examined regularly for wear. They are easily installed as retrofits, but the installation of the HF tuner may not be achievable at the optimum position.

HF antennas have to transmit as well as receive signals and thus have to be capable of handling high power. The gain of the antenna is very important in determining the maximum range of the system.

The tuner is sometimes installed externally into the dorsal fin or on the upper fuselage. In the latter case, because the tuner is outside the airframe, the radiated emissions from the tuner cause fewer problems to the other electronic equipment inside the fuselage.

5.6.4 Broadband Monopoles

Broadband passive antennas that cover several octaves will behave like an electrically small antenna at the low frequencies, a tuned quarter-wave monopole at the midband frequencies, and like a half-wave or larger monopole at the highest frequencies.

The broadband monopoles such as V/UHF blades that span several octaves are often on printed circuit boards and are shaped to give broadband performance. In general broadband antennas will have reduced gains over the entire frequency band and the antenna will only act as a quarter-wave monopole over a very narrow band at the centre or high end of

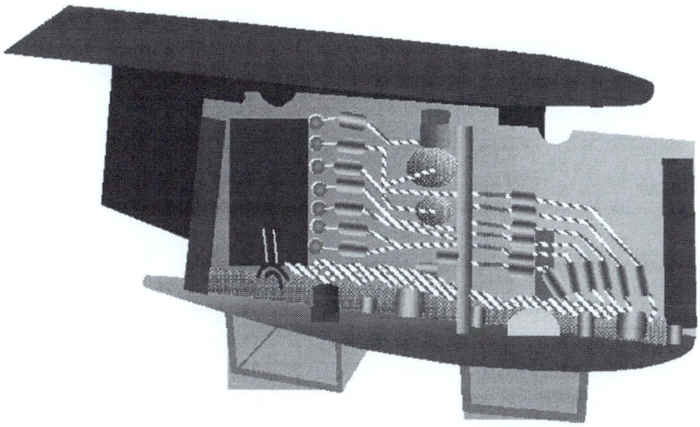

Figure 5.13 A UHF blade antenna showing the complex circuitry on the printed circuit element – manufactured by Chelton – IPAS [8]. See Plate 6 for the colour figure.

their frequency range. They may also have more than one element and thus have more than one RF connector. Thus for instance, a V/UHF antenna may also contain a D band element such as a DME antenna.

Blades are also available as 'active' antennas. These are passive blades used with tuning circuits to improve the performance over specific frequency channels. The various tuning components are switched (usually by positive-intrinsic-negative (pin) diodes) through logic units that are controlled by the radios. The logic units could be built into the same LRU as the antenna, or they could be implemented as separate LRUs and mounted within the fuselage. One typical active antenna is shown in Figure 5.13.

Active antennas are several orders of magnitude more expensive than their passive counterparts and of course have to be compatible with the radios used.

5.6.5 Tuned Monopoles

At the higher frequencies and narrow frequency bands the blades are tuned quarter-wave monopoles that could be simple rods/wires approximately a quarter of a wavelength long. These include systems such as DME, TCAS, IFF and MLS. Whip antennas are also used as dedicated antennas for the distress frequencies of 121.5 and 243 MHz.

5.6.5.1 Trailing-Wire Antenna

This is effectively a tuned antenna whose length is adjusted so that it is a quarter of a wavelength (resonant monopole) at the frequency it is receiving or transmitting. This antenna provides the highest gain of the aircraft antennas used at HF.

The wire are wound onto a drum inside the airframe, and the correct length (a quarter of a wavelength at the operating frequency) is unwound. However, because of the time taken to unwind the correct length this antenna cannot be used when the frequency is being changed continuously, as in the case of frequency hopping systems. The drum used for the wire is also rather bulky and heavy, and necessarily has to be positioned on the tail so that, when unwound, the wire does not get entangled with parts of the airframe or with any excrescence. These wires have also been known to be left trailing when the crew forget to wind them in. However, they are used on specialized aircraft such as the EC-135E version of the ARIA.

5.6.6 *Monopole Arrays*

Monopole arrays used on aircraft tend to be:

1. circular element arrays used for beam switching systems;
2. linear arrays used in combinations of two, in-phase comparison systems;
3. circular element arrays used for amplitude comparison systems;
4. Yagi arrays used to obtain a narrow beam as in the case of dipole Yagis; and
5. directional arrays, such as TCAS and ESM arrays.

5.7 Loops and Notches

A tuned loop has a circumference that is a multiple of a half or full wavelength. If the loop has a circumference of 1λ and is shaped into a *circular* loop, the radius would be $r = \lambda/2\pi$ and thus the area enclosed is πr^2 which is equal to $\lambda^2/4\pi$ or $\lambda^2/12.4 = 0.79\lambda^2$. If the loop of 1λ is shaped into a *square* loop, each side would be $\lambda/4$ and thus the area enclosed is $\lambda^2/16 = 0.625\lambda^2$. Thus it can be seen that the area enclosed by a circular loop is larger than that enclosed by a square loop of the same length. Since the gain of the loop is proportional to the area enclosed, the circular loop gives the highest gain.

The small horizontal loop behaves like a vertical dipole and has its strongest signal in the plane of the loop, and nulls in the axis perpendicular to the plane of the loop. The loop does not have to be circular or have an axis of symmetry and can be a printed circuit track.

In the case of an aircraft, the loop antennas used at HF are towel rails. Loops are also used for ADF, DF, VOR and ILS. HF towel rails are effectively loops that are installed externally on the rear or lower fuselage. They look like towel rails, hence the name. They are electrically quite small and have to be conjugatively matched over most of their operating frequency range. They usually have two or three masts, and in some cases the tuner may be installed inside one of the masts. In the latter case, because the tuner is outside the airframe, the radiated emissions from the tuner cause fewer problems to the other electronic equipment inside the fuselage. They also cause drag but are easier to maintain than the long-wire antennas. The whole airframe has currents induced on it and thus re-radiates when the HF is being transmitted. For this reason many other systems have to be switched off when the HF system is operating.

The frequency band of the ADF system is 100 kHz to 1.8 MHz, This corresponds to wavelengths from 3000 to 166.7 m, so that tuned quarter-wave monopoles cannot be used on even a very large aircraft.

The DF system works in the VHF and UHF frequency band. In the case of DF the antenna is a loop that is rotated physically or electronically to detect the direction of broadcast stations and NDBs. This antenna is usually contained in a cylinder of height around 10 inches and of circular cross-sectional diameter around 20 inches. It is often mounted on the lower rear fuselage.

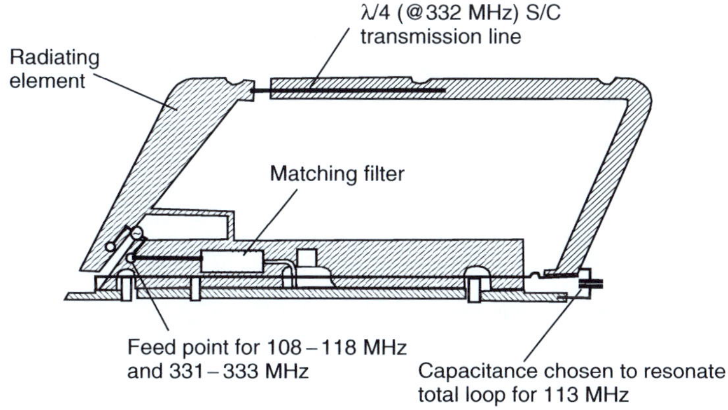

Figure 5.14 Chelton 19-85 VOR/glideslope schematic of radiating antenna elements – IPAS [8].

The ILS marker is available as a small loop, almost rectangular, and because the aircraft flies over the ground marker, the gain does not have to be very high. The signal from the ground station is horizontally polarized, thus the antenna is mounted on the lower fuselage with its longer side parallel to the ground and its plane perpendicular to the ground. The gain of the loop is of the order of −18 dBi. This antenna is mounted externally or could be a 'suppressed' antenna mounted within the envelope of the external airframe but recessed rather than conformal.

The VOR localiser loop is sometimes combined with the glideslope antenna as shown in Figure 5.14. The exterior profile is shown in Figure 5.12. At the lower VOR frequency band (108–118 MHz) the whole loop is effectively used, whereas at the higher glideslope frequency band (329–335 MHz) only part of the loop is used since the wavelength is much shorter.

Notches are the complements of loops in that the metal part of the loop is replaced by air or a material of low dielectric constant such as fibreglass. According to Babinet's

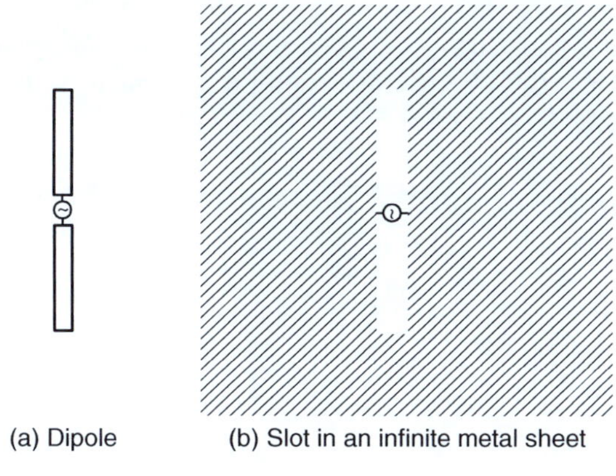

Figure 5.15 Babinet's principle of equivalence between dipoles and slots.

principle, a slot cut in an infinite metal sheet and excited at its centre radiates in the same way as the equivalent dipole (Figure 5.15). However, it should be noted that whereas the dipole is fed across its vertical arms, the slot is excited at rights angles, that is, it is fed horizontally. This causes the electric and magnetic fields to be interchanged ([1], p. 168).

The notches used as aircraft antennas are usually embedded in the dorsal fin, the leading edge of the tail fin, or other suitable appendages.

5.8 Helixes

Although the helical antenna can produce several different types of radiation depending on the electrical diameter, the most common mode of radiation of the helical antenna is the axial mode – see Figure 5.16.

The pitch angle ψ is related to the spacing and the diameter of the helix by the expression

$$\tan\psi = \frac{S}{\pi D},$$

where

ψ is the pitch angle in degrees
S is the spacing in m
D is the diameter of the helix in m.

This results in a radiation pattern similar to that of a square waveguide horn and gives a peak on boresight, and occurs when the circumference of one turn is of the order of one wavelength, that is, the diameter is λ/π or 0.32λ. The axial mode has greater impedance bandwidth and provides circular polarization. A left hand wound helix receives left hand circularly polarized waves – see [9], Chapter 13. By varying the diameter of the helix, the bandwidth can be increased to 1.7 times of that of a uniform diameter helix. The helix usually has a ground plane or is cavity-backed. Cavity-backed helixes are preferred, since they reduce back radiation and the forward gain is enhanced, resulting in an improved front-to-back ratio.

By adding two additional tapered turns at the free (outer diameter) end of the helix, the reflected currents can be suppressed, resulting in a much better impedance match at the antenna input terminals.

An empirical approximation for the HPBW derived from measured data for helixes with pitch angles of between 12° and 15°, circumferences C between 0.67λ and 1.3λ and

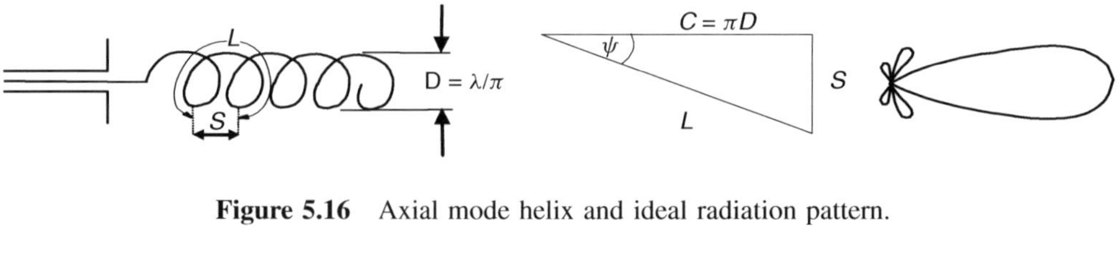

Figure 5.16 Axial mode helix and ideal radiation pattern.

of at least three turns is given by

$$\theta = \frac{52}{\frac{C}{\lambda}\sqrt{\frac{NS}{\lambda}}} \tag{5.3}$$

where

θ is the HPBW in degrees
C is the circumference of the helix in m
S is the spacing between adjacent turns in m
N is the number of turns.

When the circumference is between 0.67λ and 1.3λ the terminal impedance R_h is nearly resistive ([10], p. 7-6) and given (within 20%) by the empirical relation

$$R_h = \frac{140C}{\lambda}, \tag{5.4}$$

where C is the circumference of the helix. The variation of this terminal impedance with frequency is less pronounced for helixes with a larger number of turns. Increasing the number of turns results in a narrower beam since this is equivalent to increasing the number of elements in an array, or increasing the aperture of an antenna.

The gain is given by the approximate relation ([10], p. 7-5)

$$G = 15\left(\frac{C}{\lambda}\right)^2 \frac{NS}{\lambda},$$

where

C is the circumference of the helix in m
S is the spacing between adjacent turns in m
N is the number of turns.

For axial mode helixes, $C = \lambda$, the gain is given by

$$G = 15\frac{NS}{\lambda}.$$

5.9 Flat/Planar Spirals

Spiral antennas have radiation patterns similar to patch antennas, but they are broadband and can cover several octaves. They are usually used as passive receiving antennas for ESM.

The first practical bidirectional planar spiral was constructed in 1958 by John D. Dyson [11] at the University of Illinois ([12], p. 697). An ideal equiangular or log spiral is an FI (Frequency Independant) antenna – that is, its characteristics such as boresight, HPBW

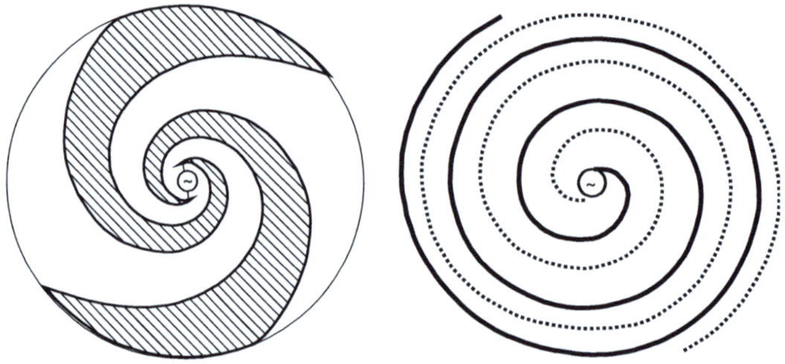

Figure 5.17 Schematics of equiangular and Archimedean planar spirals.

and gain are constant over the frequency band. In practice this cannot be attained and all these characteristics vary slightly with frequency. An equiangular spiral can be recognized by the fact that the width of the spiral, as well as the distance between the turns, increases as the radial distance from the centre increases, that is, the spiral appears to 'open out' (see Figure 5.17a).

Most planar spirals used on aircraft have two arms so that a rotation of 180° will superimpose the spiral onto itself ([9], p. 14-4). Spirals that have two arms wound in the same sense (i.e. both clockwise or anticlockwise) radiate in the first or fundamental mode (which is also known as the sum pattern mode) and have a single main lobe coincident with main (z) axis of the spiral, that is, perpendicular to the plane of the spiral. All the other modes obtained from spirals with four or more arms (wound in the same sense) have a null on the main axis and two lobes either side of the axis, with the angle of the lobe increasing in proportion to the mode number (see [13], Chapter 6). These higher-order modes are known as difference pattern modes. However, dual polarized antennas have one pair of spirals wound clockwise and one set wound anticlockwise. Sinuous spirals that are dually polarized only have two arms, but these spirals are wound in a special way.

For mode 1 propagation the active region is confined to a radius of $\lambda/2\pi$ or 0.16λ or a diameter d_1 of 0.32λ. The radiation pattern of the spiral antenna rotates as the frequency is changed. When the frequency is changed from a frequency of (say) f to f/k_3 (where k_3 is a constant) the pattern is rotated through an angle of $\log_e k_3/a$ (where a is the expansion coefficient). This rotation is not observable for tightly wound spirals since the radiation patterns have circular symmetry. Tightly wound spirals also tend to have smoother and more uniform radiation patterns and exhibit smaller variations in HPBWs with frequency. Most spiral antennas used for airborne applications tend to be a combination of equiangular and Archimedean spirals. The latter, sometimes called arithmetic spirals, are not true FI antennas and must have many turns and be tightly wound to operate over wide bandwidths, whereas equiangular spirals can be constructed with just one or two turns that are loosely wound. However, because equiangular spirals are loosely wound their overall diameters tend to be large. The width of the arms of Archimedean spirals does not vary

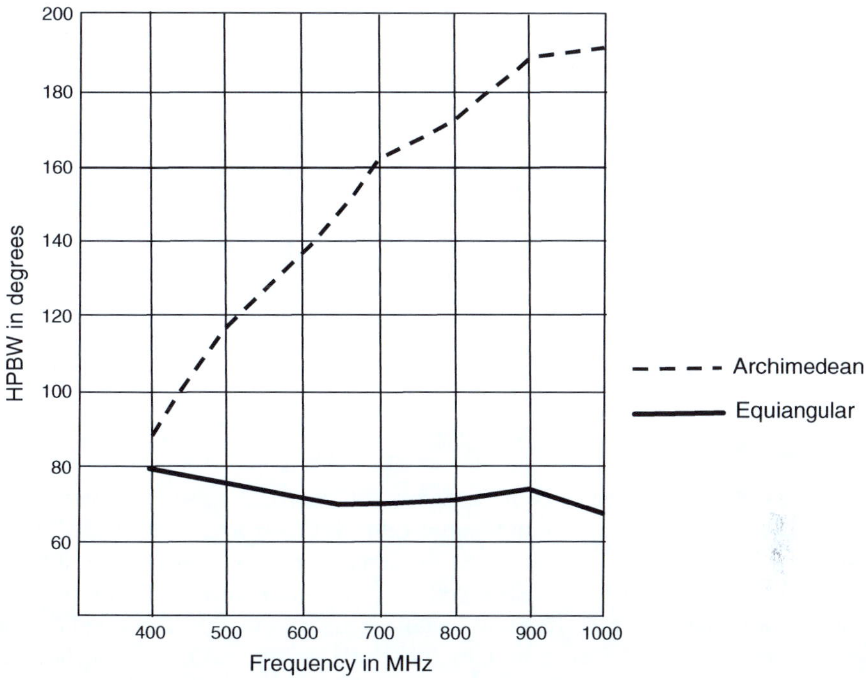

Figure 5.18 The variation of the HPBW with frequency for equiangular and Archimedean spirals [14].

with angle as is the case of equiangular spirals (see Figure 5.17b). One of the arms of the antenna is shown dotted for clarity. For airborne applications where space and mass are at a premium, Archimedean spirals provide better gain and wider bandwidth (although the variation of gain, HPBW, etc. with frequency is greater than their equiangular counterparts). The variation of the HPBW with frequency for equiangular and Archimedean spirals is shown in Figure 5.18. It can be seen that the HPBW of the Archimedean spiral increases with frequency, whereas that of the equiangular spiral is fairly constant over the whole frequency band of nearly 2.4 octaves.

A moderately wide bandwidth of between 5:1 and 10:1 can be obtained for bidirectional spirals ([9], p. 14-9). Airborne spiral antennas are usually unidirectional and implemented on printed circuit boards backed by a cavity with radiation absorbing material to absorb the back radiation. However, this absorption results in a reduction of the transmitted power that can be as high as 50%. In the case of receive antennas this dissipative power loss raises the antenna noise temperature, thus reducing the threshold of the receiving system.

Commercially available spirals that are a combination of Archimedean and equiangular spirals are optimized so that their gain and beamwidths are fairly constant over the frequency band, as shown in Figure 5.19.

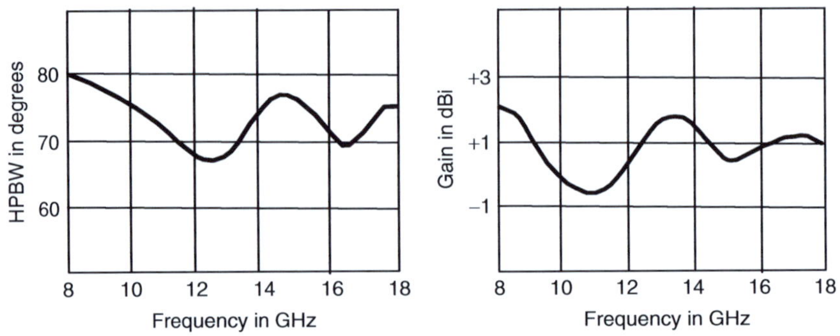

Figure 5.19 Typical variation of the HPBWs and gains of commercially available planar spirals.

5.10 Patches

Patch antennas usually consist of printed circuit boards and are used for systems such as GPS and RadAlt. They give a radiation pattern that varies from a pear-shaped to roughly hemispherical. Patch antennas are used on the upper and lower surfaces of airframes to radiate upwards and downwards, respectively. They can be shaped or conformed to the shape of the structure on which they are installed, and are therefore particularly suitable for use on aircraft. The GPS antenna currently used on aircraft consist of one or more circular patch antenna elements. Patch antennas can only handle about 5 W power and are more often used as receive antennas, although in the case of the RadAlt there is also a transmit antenna. Circular polarization is obtained from these patch antennas, by feeding them at a particular point.

5.10.1 Patch Arrays

The arrays of patch antennas are used on aircraft to form a narrower beam and also, in the case of military aircraft, to employ null steering. In these cases the beam pattern is formed so that there is a null in the direction of an unwanted interfering or jamming signal. These are known as controlled radiation pattern antennas (CRPA) and consist of seven element arrays consisting of a central patch surrounded by six elements arranged symmetrically around it as shown in Figure 5.20.

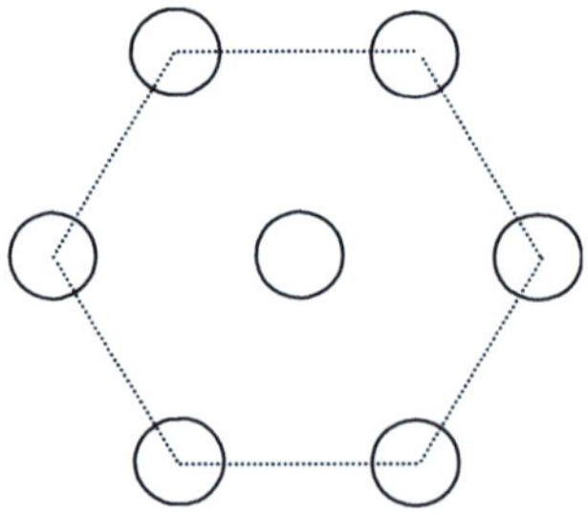

Figure 5.20 A GPS controlled radiation pattern antenna array.

5.11 Aperture Antennas

The most common aperture antennas used on an aircraft are reflectors, phased arrays and horns. These antennas are not affected to any significant extent by the ground plane, as long as the direct radiation from the antenna does not strike the ground plane. These antennas do not require a ground plane, as is the case for monopole antennas.

In the case of aperture antennas, the beamwidth is inversely proportional to the dimensions, so that the larger the antenna the smaller its beamwidth, assuming that the frequency is unchanged. The beamwidth depends on the electrical dimensions of the antenna, that is, its dimensions in terms of wavelength. Thus if the frequency of operation is doubled the wavelength is halved, and so an antenna of only half the physical dimensions would be required to obtain the same beamwidth.

The boresight gain, on the other hand, is proportional to the dimensions of the antenna, thus the larger the antenna the higher is its gain, assuming that its frequency is not changed. The gain depends on the electrical dimensions of the antenna, that is, its dimensions in terms of wavelength. Thus if the frequency of operation is doubled the wavelength is halved and so an antenna of only half the physical dimensions would be required to obtain the same gain.

5.12 Reflectors

Reflector antennas are based on optical systems which usually have one or two reflectors, fed by a horn, planar spiral, conical log spiral or a dipole. For broadband operation the feed could be a log-periodic array. A log-periodic array is an array of dipoles or monopoles that are used instead of single elements, to give broadband frequency operation.

In the case of the single reflector the feed is placed at the focus of the reflector which is usually parabolic as shown in Figure 5.21a. The feed, however, may be offset, so that it is not on the principal axis of the reflector as shown in Figure 5.21b.

In the Cassegrain system, two reflectors are used. The feed is placed at one focus of the hyperbolic sub-reflector, and the second focus of the sub-reflector is coincident with the focus of the main parabolic reflector, as shown in Figure 5.21c.

5.12.1 Antennas and the Fourier Transform

Fourier analysis was first suggested by J.B.J. Fourier (1768–1830). He showed that a square wave could be decomposed into a sum of sine waves. The Fourier transform is used to transform a function from one domain to another. It can be used to convert the waveform in the time domain (variation of the shape of the waveform) to the frequency domain, that is, the frequency components of the waveform.

The waveform in the time domain shows the amplitude variation with time and the frequency domain gives us the frequencies and their amplitude. An oscilloscope can give us a time domain output, by showing us the shape of the pulse, whereas a spectrum analyser shows us its characteristics in the frequency domain, by showing us its frequency content or spectrum, that is, the amplitude at each frequency of its spectrum.

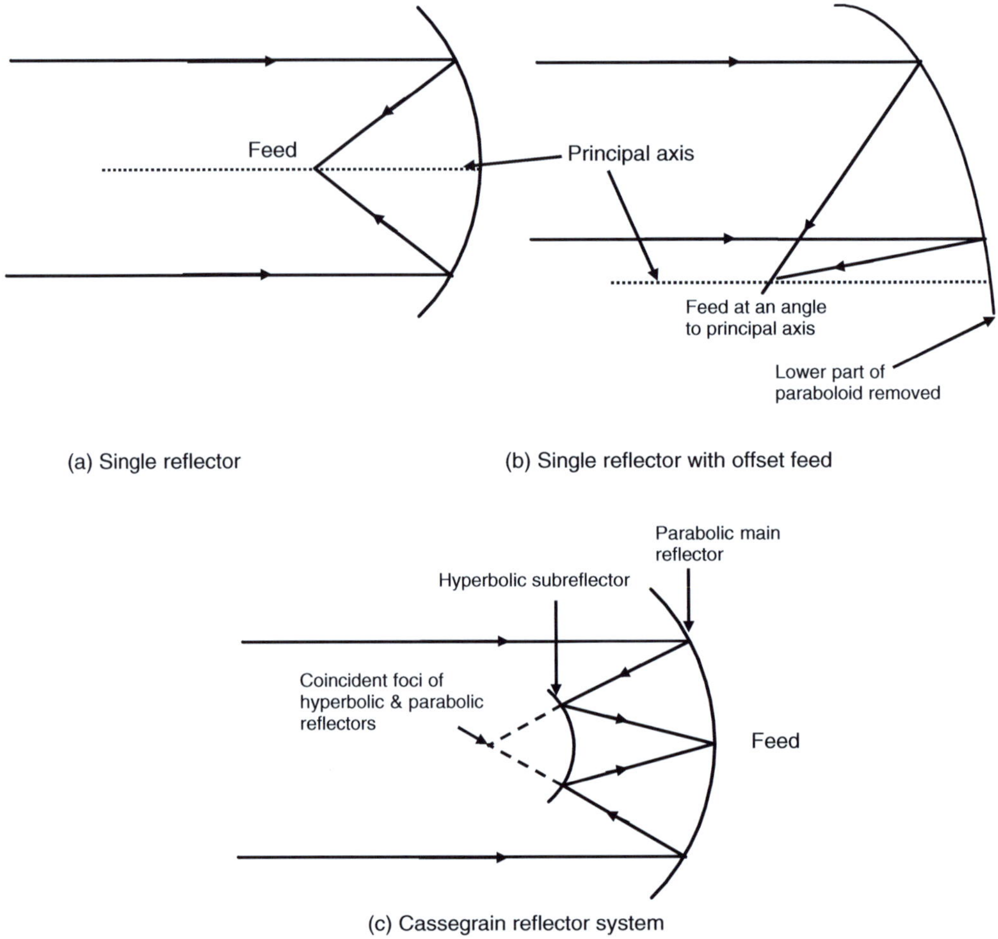

Figure 5.21 Reflector antenna system.

5.12.1.1 Fourier Transform of a Square Wave

Consider a square wave as shown in Figure 5.22a. This has a spectrum given by

$$F(x) = A\left(\sin \omega t + \frac{1}{3}\sin 3\omega t + \frac{1}{5}\sin 5\omega t + \frac{1}{7}\sin 7\omega t + \ldots + \frac{1}{(2n-1)}\sin(2n-1)\omega t\right), \qquad (5.5)$$

where

$\omega = 2\pi f$ is the angular frequency in radians per second (rad s^{-1})
A is a constant and
f is the frequency in Hz.

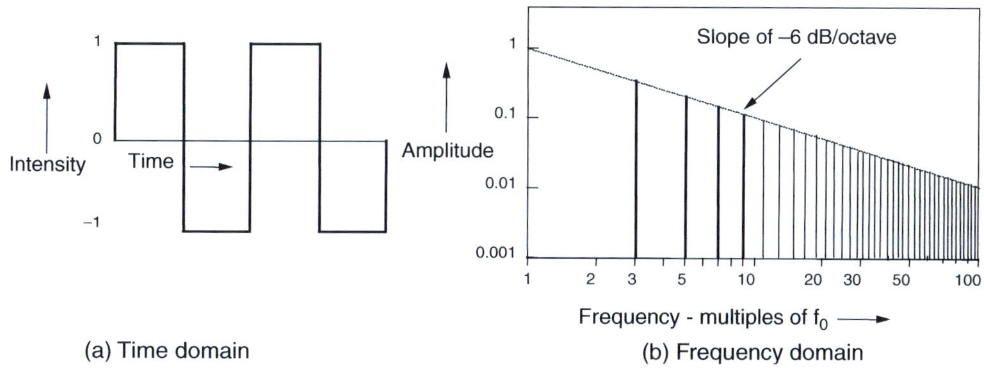

Figure 5.22 Time and frequency domains of a square wave.

The number of terms n is ideally infinite. If we input a pure square wave into a spectrum analyser we would see the series of lines shown in the right-hand plot of Figure 5.22b but extending to an infinitely large multiple of f.

The lowest frequency, f_0, is called the fundamental. The next frequency is called the first overtone. The first overtone is not necessarily the second harmonic ($2f_0$), it could be any multiple of f_0, and hence any harmonic depending on the shape of the initial wave (in the time domain). For instance, in the case of the square wave, the first overtone is the third harmonic $3f_0$ and not $2f_0$.

It is the harmonic content that gives a musical instrument as well as someone's voice its unique sound. A note on a violin sounds different from the same note on the piano because although they both have the same fundamental they have different overtones and/or the harmonics may have different amplitudes. Amplifiers that do not cover a wide enough frequency range cannot reproduce sound faithfully because they do not reproduce the higher overtones of a particular note.

The transformation from time to frequency domains can be undertaken mathematically by means of the Fourier transform.

A square wave with a ripple will have fewer terms than a true square wave with an infinite number of terms. If we have a square wave with a ripple as shown in the time domain of Figure 5.23 it will have a frequency spectrum with just the first three terms, as shown in the frequency domain of the figure.

As mentioned above the function of a square wave has an infinite number of terms and can be written as

$$f(\omega t) = \frac{4}{\pi}\left(\sin\omega t + \frac{1}{3}\sin 3\omega t + \frac{1}{5}\sin 5\omega t + \ldots + \frac{1}{\infty}\sin\infty\omega t\right) \tag{5.6}$$

where the constant A is now given the value of $4/\pi$. Thus a square wave contains many sine waves of different frequencies and amplitudes.

The Fourier transform of a square wave of frequency f can be written as a summation given by

$$f(\omega t) = \frac{4}{\pi}\sum_{n=1}^{\infty}\frac{\sin(2n-1)\omega t}{(2n-1)\omega t} \tag{5.7}$$

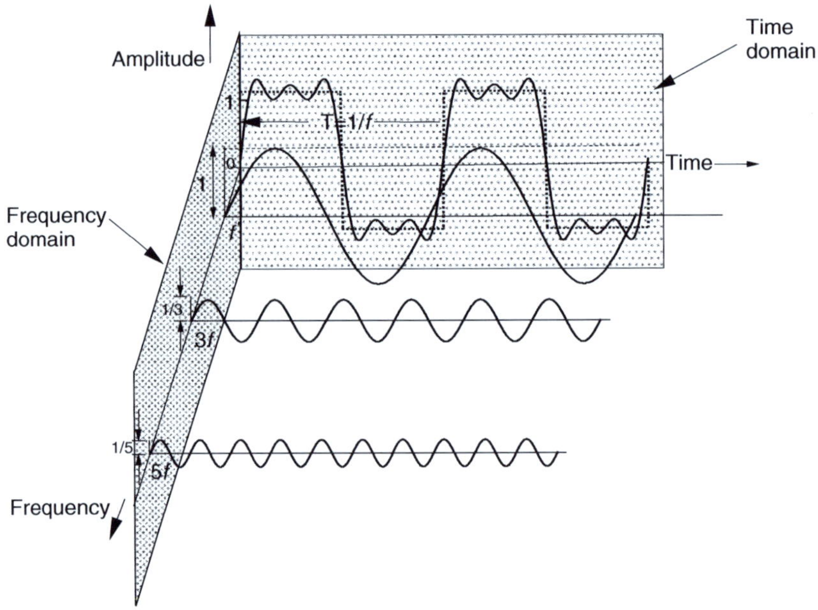

Figure 5.23 Time and frequency domains of a square wave with a ripple.

To get an exact square wave function $f(\omega t)$ we have to add an infinite number of terms of the series. We can see that if we just add the first three terms we would get the square wave with the a ripple as shown in Figure 5.23. This would have the equation

$$f(\omega t) = \frac{4}{\pi}\left(\sin \omega t + \frac{1}{3}\sin 3\omega t + \frac{1}{5}\sin 5\omega t\right). \tag{5.8}$$

If we add the first eight terms we would get the shape shown in Figure 5.24, which is starting to resemble the square wave more closely. This would have the equation

$$f(\omega t) = \frac{4}{\pi}\left(\sin \omega t + \frac{1}{3}\sin 3\omega t + \frac{1}{5}\sin 5\omega t + \frac{1}{7}\sin 7\omega t \right.$$
$$\left. + \frac{1}{9}\sin 9\omega t + \frac{1}{11}\sin 11\omega t + \frac{1}{13}\sin 13\omega t + \frac{1}{15}\sin 15\omega t\right). \tag{5.9}$$

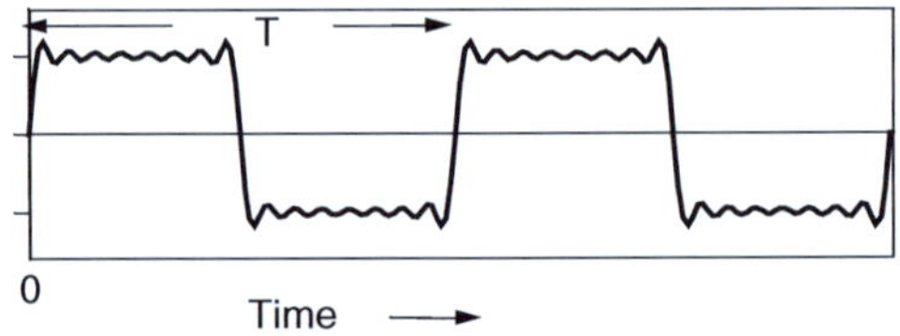

Figure 5.24 Summation of the first eight terms of a square wave with a ripple.

5.12.1.2 Fourier Transform of a Square Pulse to Obtain Radiation Patterns

In the case of antennas the Fourier transform is used to convert the electric field illumination across the aperture of an antenna into the electric field spatial pattern in the far field. The aperture illumination is the variation of the electric field over the aperture of an antenna, for example a reflector. The electric field can vary in both amplitude and phase. The radiation pattern is the spatial variation of power with the angle, and this is proportional to the square of this electric far-field pattern. Since the radiation pattern only gives the power relative to boresight, the constant of proportionality does not matter, and we can get the radiation pattern by squaring the Fourier transform of the aperture illumination and normalizing to the boresight gain. In the case of linear aperture distributions, the Fourier transform gives us the spatial electric field distribution. In the case of two-dimensional rectangular apertures, the Fourier transform can be applied across each of the two orthogonal axes, assuming that the field variations in these two axes are separable. The power on boresight will be the product of the square of these two transforms, or the sum if the powers are in dB. Linear interpolation between the two planes is often performed to give intermediate values, but quadratic interpolation produces more accurate results. The Fourier transforms may also be used for circular and elliptical apertures, to give approximate magnitudes of the radiation patterns.

Uniform illumination occurs when the electric field has the same constant value across the aperture, and the electric field also has the same phase. This would be like a square pulse as shown in Figure 5.25. Looking at a square pulse in the frequency domain, we see that it is made up of many frequencies that have a $\sin(x)/x$ spectrum.

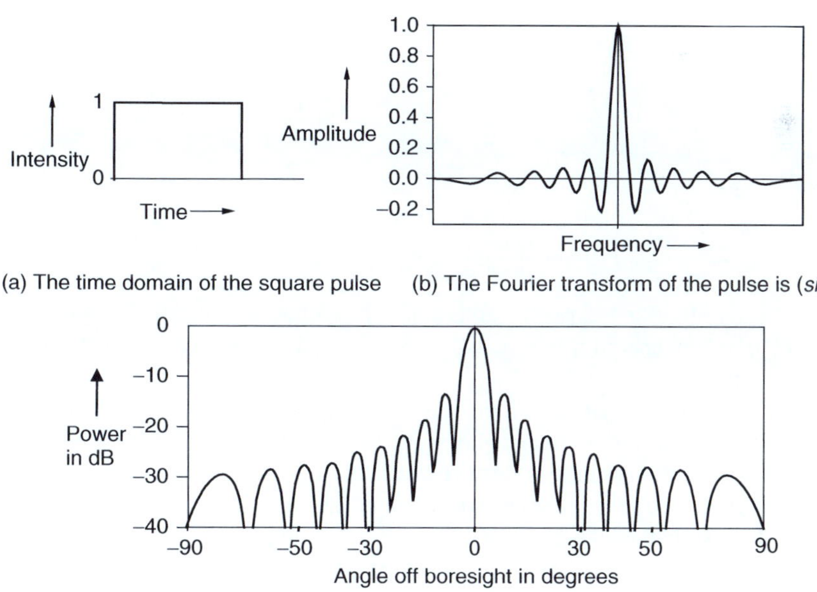

(a) The time domain of the square pulse (b) The Fourier transform of the pulse is (*sin u)/u*

(c) The square of the Fourier transform of the pulse

Figure 5.25 Time and frequency domains (Fourier transform) of a square pulse and the square of the Fourier transform that gives the radiation pattern.

When we see the shape of a pulse, we are looking at its characteristics in the time domain; that is, we see how its electric field intensity varies with time. This shape is the result of different frequencies each with a different amplitude which, added together, make up that unique shape. When we look at these different frequencies, we say that we are looking at looking at the frequency spectrum of the pulse. For instance, if the aperture illumination is uniform, the electric field has the same amplitude and phase. The electric field in the far field is as shown in Figure 5.25b. It can be seen that the Fourier transform gives positive as well as negative values. The radiation pattern is obtained by squaring the electric field to obtain a value proportional to the power. This also produces all positive values, as can be seen as in Figure 5.25c.

Aperture illuminations can be of many different types but uniform illumination aperture gives the highest gain on the peak of the radiation pattern and the narrowest beam. However, it also produces the highest level of sidelobes – just 13 dB before the main peak of the beam.

5.12.1.3 Uniform Illumination

If an antenna is uniformly illuminated the electric field across the aperture has the same magnitude and constant phase. The electric field can be normalized to the value at the centre of the aperture, and thus we can say that the electric field has a magnitude of unity across the whole aperture; that is, the electric field is given by $f(x) = 1$ for values of x from $-L/2$ to $+L/2$, and $x = 0$ at the centre of the aperture.

The Fourier transform F_t for uniform illuminated linear apertures is proportional to the spatial electric field variation in the far field and is given by

$$E = \frac{\sin u}{u} \tag{5.10}$$

where

$u = L\pi \sin\theta/\lambda$

L/λ is the length in wavelengths of the aperture

θ is the angle off boresight in degrees.

Squaring the electric field gives power and hence the radiation pattern. Radiation patterns give us the relative power and this is usually normalized to the peak power. That is, the peak of the beam is plotted as 0 dB and all other values are negative dB. The peak is normally at the electrical boresight of the uninstalled antenna, but this is not always the case. In the case of installed antenna patterns there may be a shift of the position of the peak compared to the uninstalled pattern, especially if the antenna is installed inside a radome, for example for radar antennas installed in the nose cone of the aircraft.

If we want to calculate the actual power (or power density) available at a particular point, we have to know the gain of the transmit antenna at boresight as well as the angle between the boresight and line joining the transmit antenna to the point. The electrical length of the antenna, L/λ, affects the gain at boresight as well as the number and position of the sidelobes, but the level of the first (near-in) sidelobe below peak is always -13 dB regardless of the size of the aperture. The width of the main lobe or its HPBW is inversely

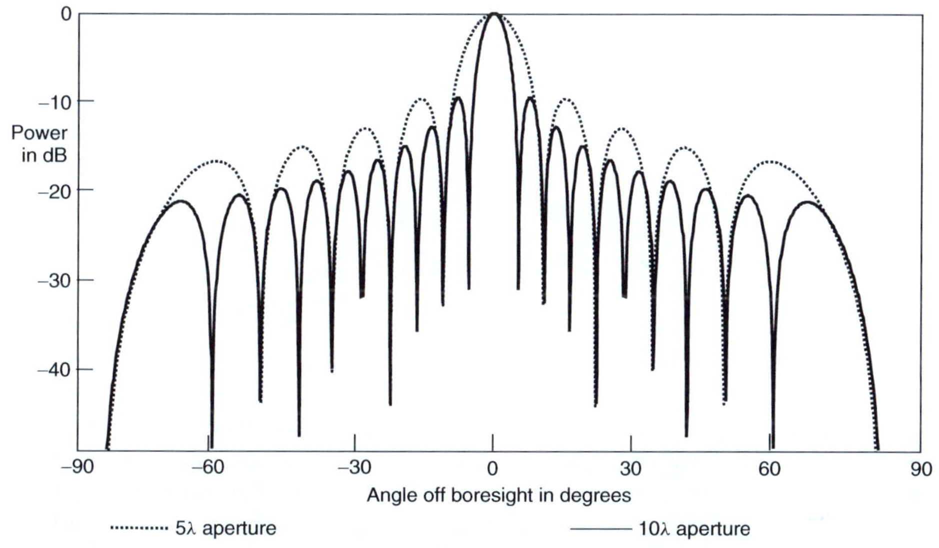

Figure 5.26 Radiation patterns for 5λ and 10λ apertures with uniform illumination.

proportional to the gain and the electrical length of the antenna, since the value of u in Equation 5.10 is larger. The HPBW will be narrower and the gain larger for longer electrical lengths. Thus it useful to specify the HPBWs and the position of the nulls normalized to (divided by) the electrical length of the aperture. The HPBW for uniform illumination is $50.8\lambda/L$, thus for a 10λ antenna the HPBW is 5.08°, whereas for a 5λ antenna the HPBW is 10.2°. The position of the first nulls on each side of the boresight is $57.3\lambda/L$, which corresponds to 5.7° for the 10λ antenna and to 11.4° for the 5λ antenna. The radiation pattern of two antennas of lengths 10λ and 5λ both with uniform illumination are compared in Figure 5.26. It can be seen that the number of sidelobes for the 10λ antenna is also almost double the number for the 5λ antenna. For uniform illumination the total number of sidelobes is $2(L/\lambda - 1)$. Thus a 5λ aperture has 8 sidelobes whereas the 10λ aperture has 18 sidelobes, and a 20λ aperture has 38 sidelobes. The boresight gain is shown as the same value in both cases since the gains are normalized to the respective boresight for each pattern. The absolute gain at boresight for the longer aperture will be larger than that for the shorter length. The rate at which the level of the far-out sidelobes decreases is inversely proportional to u (i.e. proportional to $1/u$).

An aperture antenna with uniform illumination has the maximum directivity, the narrowest beam but the highest sidelobe level (SLL). Tapering of the aperture illumination reduces the SLL at the expense of lower boresight gain and the broadening of the HPBW.

5.12.1.4 Cosine Illumination

In the case of cosine illumination where the aperture extends from $x = -L/2$ to $+L/2$, and $x = 0$ at the centre of the aperture, the electric field is given by the function

$$f(x) = \cos(\pi x/L)$$

for $-L/2 < x < +L/2$. The Fourier transform for cosine illumination is given by

$$E = \frac{L\pi \cos(u)}{2\lambda\left(\frac{\pi^2}{4} - u^2\right)} \tag{5.11}$$

where

$u = L\pi \sin\theta/\lambda$
L/λ is the length in wavelengths of the aperture
θ is the angle off boresight in degrees.

This type of illumination would appear across the aperture of a rectangular horn. Although the phase across the aperture of the horn would not usually be constant, this can be taken to be a fair approximation. The radiation pattern obtained by plotting the square of the Fourier transform is shown in Figure 5.27 for a 5λ antenna compared with the pattern for uniform illumination. It can be seen that the HPBW is 13.8° and the first nulls occur at 17° off boresight, whereas in the case of uniform illumination the HPBW is 10.2° and the first nulls occur at 11.4°, The boresight gain is 0.81 of that obtainable with uniform illumination and the HPBW is increased, but the SLL (of the first sidelobe) is reduced to -23 dB. The rate at which the level of the far-out sidelobes decreases is inversely proportional to the square of u (proportional to $1/u^2$). For cosine illumination the total number of sidelobes is $2(L/\lambda - 1) - 1$. Thus a 5λ long antenna has 7 sidelobes whereas a 10λ long antenna has 17 sidelobes. The odd number of sidelobes is manifest by the presence of half a lobe around +90° and −90° off boresight.

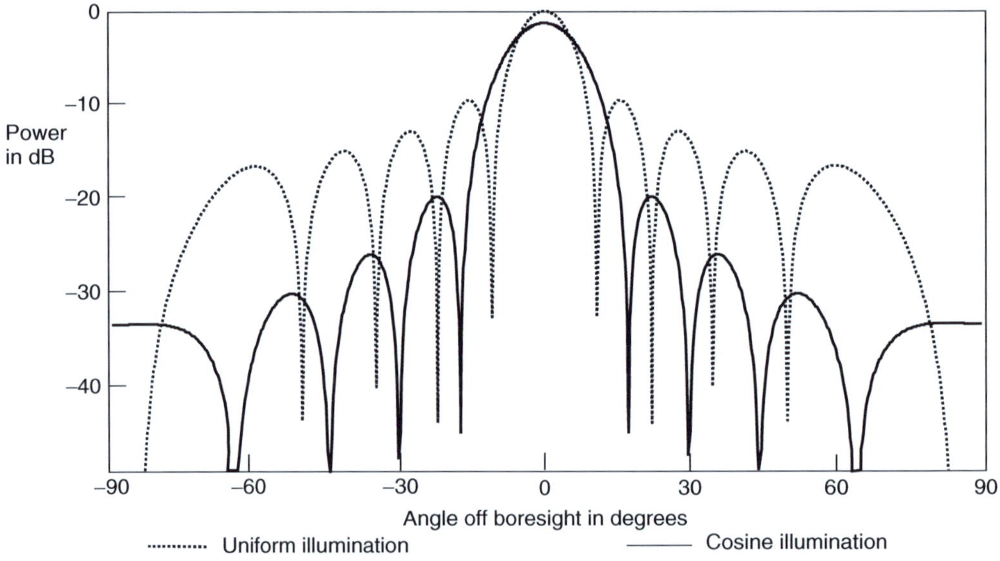

Figure 5.27 Comparison between the radiation patterns for uniform and cosine illumination for a 5λ aperture.

5.12.1.5 Cosine Squared Illumination

In the case of cosine squared illumination where the aperture extends from $x = L/2$ to $+ L/2$, and $x = 0$ at the centre of the aperture, the electric field is given by the function

$$f(x) = \cos^2(\pi x/L)$$

for $-L/2 < x < +L/2$. The Fourier transform for cosine squared illumination is given by

$$E = \frac{L\pi^2}{2\lambda(\pi^2 - u^2)} \frac{\sin(u)}{u} \tag{5.12}$$

where

$u = L\pi \sin\theta/\lambda$
L/λ is the length in wavelengths of the aperture
θ is the angle off boresight in degrees.

The radiation pattern for cosine squared illumination is obtained by plotting the square of the Fourier transform of Equation 5.12, as shown in Figure 5.28 for a 5λ antenna compared with the pattern for uniform illumination. It can be seen that the HPBW is 16.6° and the first nulls occur at 23° off boresight, whereas in the case of uniform illumination the HPBW is 10.2° and the first nulls occur at 11.4°. The boresight gain is 0.667 of that

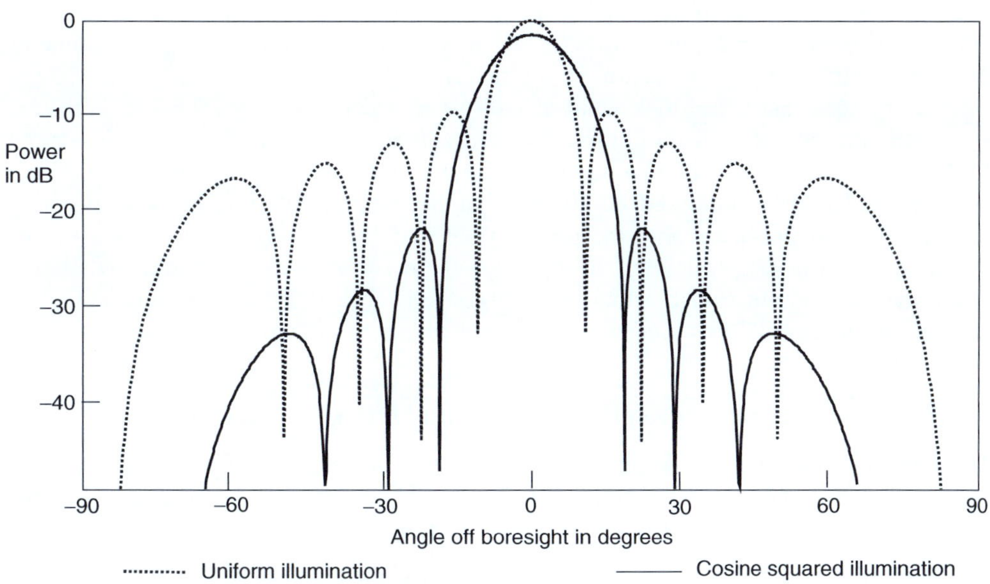

Figure 5.28 Comparison between the radiation patterns for uniform and cosine squared illumination for a 5λ aperture.

obtainable with uniform illumination and the HPBW is increased, but the SLL (of the first sidelobe) is reduced to $-32\,\text{dB}$.

The rate at which the level of the far-out sidelobes decreases is inversely proportional to the cube of u (proportional to $1/u^3$). For cosine squared illumination the total number of sidelobes is $2(L/\lambda - 1) - 2$. Thus a 5λ long antenna has 6 sidelobes whereas a 10λ long antenna has 16 sidelobes.

5.12.1.6 Triangular Illumination

In the case of triangular illumination where the aperture extends from $x = -L/2$ to $+L/2$, and $x = 0$ at the centre of the aperture, the electric field is given by the function

$$f(x) = 1 - |2x/L|$$

for $-L/2 < x < +L/2$. The Fourier transform for triangular illumination is given by

$$E = \frac{L}{2\lambda}\frac{\sin(u/2)}{(u/2)^2} \tag{5.13}$$

where

$u = L\pi \sin\theta/\lambda$
L/λ is the length in wavelengths of the aperture
θ is the angle off boresight in degrees.

The radiation pattern for triangular illumination is obtained by plotting the square of the Fourier transform of the Equation 5.13 and is shown in Figure 5.29 for a 5λ antenna compared with the pattern for uniform illumination.

It can be seen that the HPBW is 14.6° and the first nulls occur at 22.8° off boresight, whereas in the case of uniform illumination the HPBW is 10.2° and the first nulls occur at 11.4°, The boresight gain is 0.75 of that obtainable with uniform illumination (1.249 dB less) and the HPBW is increased, but the SLL (of the first sidelobe) is reduced to $-26.4\,\text{dB}$.

The rate at which the level of the far-out sidelobes decreases is inversely proportional to the square of u (proportional to $1/u^2$), as is the case for cosine aperture illumination. For triangular illumination the total number of sidelobes is $L/\lambda - 2$. Thus a 5λ long antenna has three sidelobes whereas a 10λ long antenna has eight sidelobes.

5.12.1.7 Comparison between Parameters with Different Illuminations

The directivities for all other illuminations can be calculated as a fraction of the directivity of an uniformly illuminated aperture, that is, the directivities can be normalized to that for uniform illumination. This is known as the gain factor. Similarly, the beamwidths can be normalized to that obtainable for uniform illumination.

The shape of the beam required depends on the application. For instance, for search radar it is best to have low sidelobes to avoid ambiguity of the direction and range of targets.

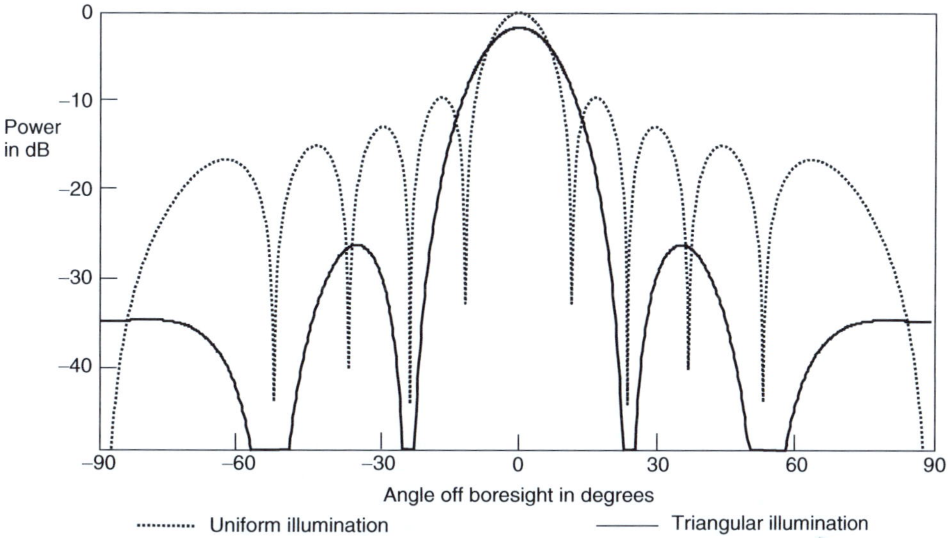

Figure 5.29 Comparison between the radiation patterns for uniform and triangular illumination for a 5λ aperture.

When assessing the power density available at a particular location, the details of the illumination of an antenna are not always available. However, if the HPBW and antenna dimensions are known it is possible to deduce the type of illumination across the antenna aperture and thus obtain the radiation pattern as well as the absolute directivity. The gain factors, HPBWs and SLLs (below boresight) for different aperture illuminations are shown in Figure 5.30, and in Table 5.3 the exact values and formulas are tabulated. The values of radiation patterns for different illuminations for a 20λ linear aperture are shown in Table 5.4.

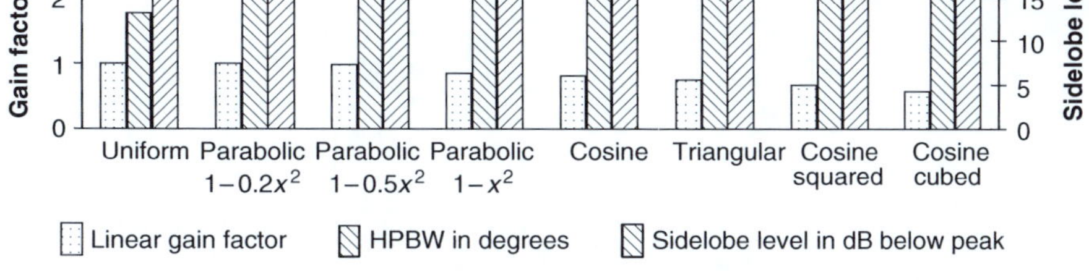

Figure 5.30 Comparison between the parameters for different aperture illuminations.

Table 5.3 Formulas for the radiation pattern parameters in the case of a 20λ linear aperture with different illuminations after Silver [1].

Type of illumination	Formula for HPBW	Gain factor	SLL below boresight in dB	Formula for position of first null	Position of first null in degrees	Formula for number of sidelobes to ±90°	Number of sidelobes to ±90°	Approx gain in dBi	Fourier transform
Uniform	$50.8\lambda/L$	1	13.2	$57.3\lambda/L$	2.87	$2(L/\lambda - 1)$	38	37.78	$(\sin u/u)$
Cosine	$68.8\lambda/L$	0.81	23.9	$85.9\tilde{\lambda}L$	4.30	$2(L/\lambda - 1) - 1$	37	36.87	$(L\pi/2\lambda)^*\cos u/((\pi/2)^2 - u^2)$
Cosine squared	$83.2\lambda/L$	0.67	33	$114.6\lambda/L$	5.73	$2(L/\lambda - 1) - 2$	36	36.02	$(L/2\lambda)^*(\sin u/u)^*(\pi^2/(\pi^2 - u^2)$
Triangular	$73.4\lambda/L$	0.75	27.7	$114.6\lambda/L$	5.73	$L/\lambda - 2$	18	36.53	$(L/2\lambda^*(\sin u/2)/(u/2)^2$

Table 5.4 Values of radiation patterns for different illuminations for a 20λ linear aperture after Silver [1].

Type of illumination	Aperture illumination	Relative gain in dB	Linear gain factor	HPBW in degrees	HPBW relative to uniform illumination	Position of first null in radians	Position of first null in degrees	Near-in SLL below peak in dB
Uniform	Uniform	65.91	1	2.521	1	0.050	2.865	13.2
Parabolic 1	$1 - 0.2x^2$	65.88	0.994	2.636	1.045	0.053	3.037	15.8
Parabolic 2	$1 - 0.5x^2$	65.77	0.97	2.779	1.102	0.057	3.266	17.1
Parabolic 3	$1 - x^3$	65.11	0.833	3.295	1.307	0.072	4.097	20.6
Cosine	$\cos(\pi x/2)$	64.99	0.81	3.438	1.364	0.075	4.297	23
Triangular	$1 - \lvert x \rvert$	64.66	0.75	3.667	1.455	0.100	5.730	26.4
Cosine squared	$\cos^2(\pi x/2)$	64.15	0.667	4.154	1.648	0.100	5.730	32
Cosine cubed	$\cos^3(\pi x/2)$	63.50	0.575	4.784	1.898	0.125	7.162	40

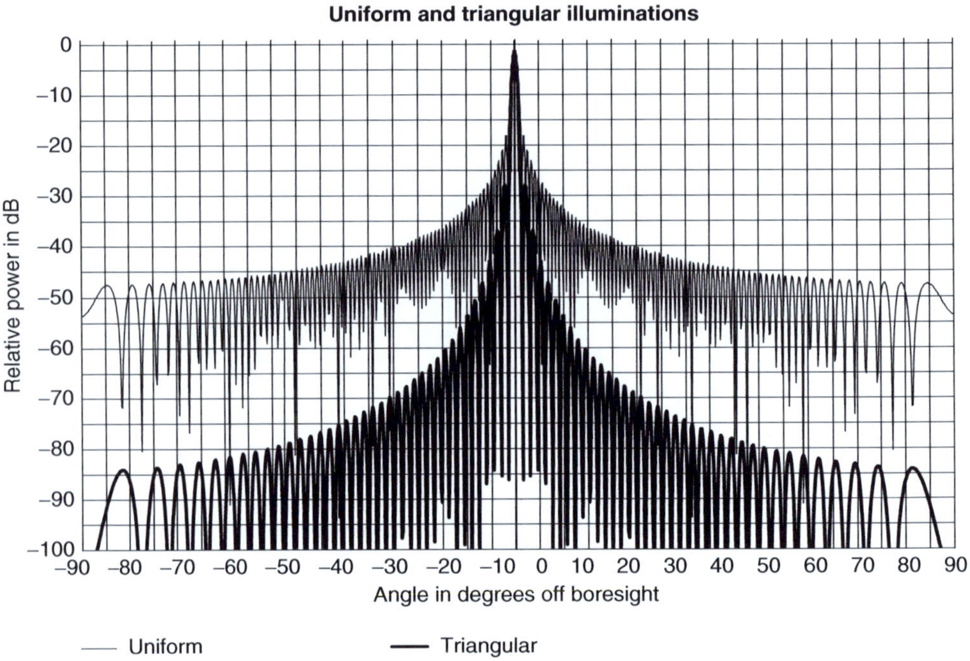

Figure 5.31 Comparison between the radiation patterns obtained for uniform and triangular illumination for a 75.8λ aperture.

For an array, the SLL can be reduced (by sacrificing some of the boresight gain. For instance, by applying a triangular amplitude taper to an array of length 9 m at 2.5 GHz, we can see that the SLL is greatly reduced. An aperture antenna of physical length 9 m would have an electrical length of 75.8 at a frequency of 2.5 GHz, as shown in Figure 5.31. This reduction in SLL is particularly important for radar systems, since sidelobes can give false radar returns. The ambiguity arises from the fact that it is difficult to distinguish the return from a distant object in the mainbeam/boresight, from a lower level return from a less distant object in a sidelobe. The SLL can be reduced further to −40 dB if cosine cubed illumination is used.

5.13 Waveguide Fed Antennas

The aircraft antennas that are fed by waveguides are pyramid horns and slotted waveguide arrays. A brief qualitative explanation of waveguide theory is given below, in order to understand rectangular waveguide fed antennas. We are accustomed in everyday life to think of EM waves being transmitted through free space or by a pair of wires. The latter is what is required to transmit the TEM waves that propagate in twin wires, twisted pairs and coaxial cables. However, the waves can also be transmitted using pipes of circular/elliptical and rectangular/square guides, known as waveguides.

5.13.1 Waveguide Theory

The EM wave propagates in the waveguide in a similar manner to the way light waves travel in an optic fibre. In order to understand the propagation of EM waves in a rectangular waveguide we have think of it as two waves excited at a point between the two side walls and reflected off the metal walls as they travel down the guide. Consider a rectangular waveguide of cross-section dimensions a and b, as shown in Figure 5.32a. In Figure 5.32b the rays are shown reflecting off the walls, whereas in Figure 5.32c the wavefronts that are perpendicular to the rays are shown. Dotted lines are used for the troughs, and the crests are denoted by full lines. The waves reflected off the side walls cross and where two troughs cross at point Q, for instance, the resultant is a trough of twice the magnitude of the trough of each individual wave, that is, a deep trough. Similarly, where crests cross (at P, for instance) the resultant crest is twice the magnitude of the individual crests. Thus the resultant wave travelling down the guide is of twice the amplitude of the individual waves. This is the guided wave, and the distance between two adjacent crests or between two adjacent troughs gives us the guide wavelength λ_g. Thus, the distance between P and Q, which is the distance between a trough and the crest, gives us half of the guide wavelength and PS is a quarter of this value. We can find the relationship between the guide wavelength and the wavelength of the individual incident waves by using simple trigonometry. We know that the wavelength of the individual

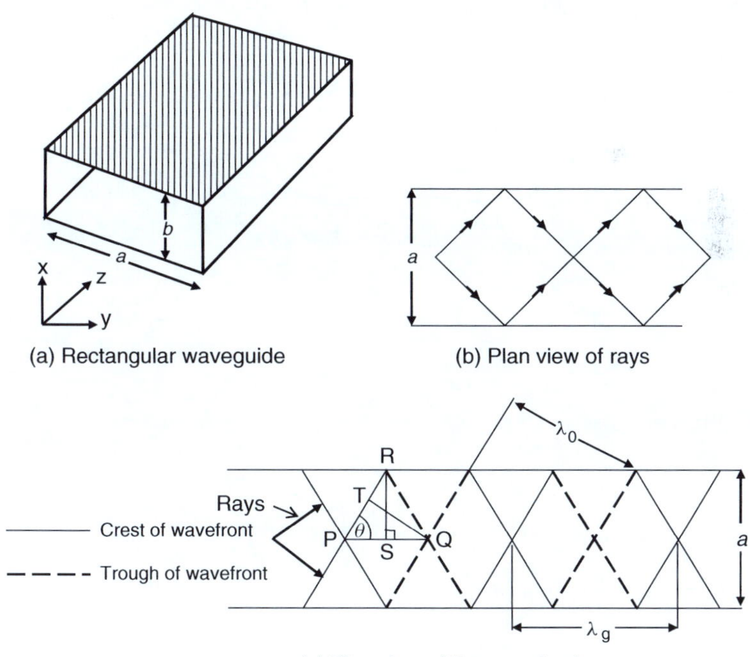

Figure 5.32 Propagation of waves in a rectangular waveguide.

waves is the same as the free space wavelength λ_0, and thus the distance denoted by TQ, which is the distance between a crest and a trough, is equal to $\lambda_0/2$ (half the free space wavelength). Since the width of the guide is a, and the distance RS is $a/2$.

In triangle PTQ we can see that the sine of angle TPQ is given by

$$\sin\theta = \frac{TQ}{PQ} = \frac{\lambda_0/2}{\lambda_g/2} = \frac{\lambda_0}{\lambda_g}. \tag{5.14}$$

Using Pythagoras' theorem, for triangle PRS, we get

$$\mathrm{PR}^2 = \mathrm{RS}^2 + \mathrm{PS}^2$$

$$\mathrm{PR}^2 = \left(\frac{a}{2}\right)^2 + \left(\frac{\lambda_g}{4}\right)^2.$$

Thus

$$\mathrm{PR} = \sqrt{\left(\frac{a}{2}\right)^2 + \left(\frac{\lambda_g}{4}\right)^2} \tag{5.15}$$

and

$$\sin\theta = \frac{\mathrm{RS}}{\mathrm{PR}} = \frac{a/2}{\sqrt{(a/2)^2 + \left(\lambda_g/4\right)^2}}. \tag{5.16}$$

But sin θ is also equal to λ_0/λ_g (see Equation 5.14). Equating Equations 5.14 and 5.16, we get

$$\frac{\lambda_0}{\lambda_g} = \frac{a/2}{\sqrt{(a/2)^2 + \left(\lambda_g/4\right)^2}} \tag{5.17}$$

Squaring gives

$$\frac{\lambda_0^2}{\lambda_g^2} = \frac{a^2/4}{(a/2)^2 + \left(\lambda_g/4\right)^2}$$

and cross-multiplying gives

$$\lambda_0^2\left[\left(\frac{a}{2}\right)^2 + \left(\frac{\lambda_g}{4}\right)^2\right] = \frac{\lambda_g^2 a^2}{4}. \tag{5.18}$$

Multiplying throughout by $4/\lambda_0^2\lambda_g^2 a^2$, we have

$$\frac{4}{\lambda_g a^2}\left(\frac{a^2}{4} + \frac{\lambda_g^2}{16}\right) = \frac{1}{\lambda_0^2}$$

$$\frac{1}{\lambda_g^2} + \frac{1}{4a^2} = \frac{1}{\lambda_0^2},$$

which gives

$$\left(\frac{1}{\lambda_g}\right)^2 = \left(\frac{1}{\lambda_0}\right)^2 - \left(\frac{1}{2a}\right)^2. \tag{5.19}$$

If $\lambda_0 = 2a$, then the right-hand side of Equation 5.19 becomes zero, and thus $1/\lambda_g = 0$, and thus the guide wavelength tends to infinity. Since the frequency is inversely proportional to the wavelength, the frequency tends to zero, that is, the wave does not propagate. The free space wavelength λ_0 at which this occurs is known as the cut-off wavelength $\lambda_c(=2a)$, and the corresponding frequency is known as the cut-off frequency f_c. Propagation can only occur at frequencies above this cut-off frequency, and thus the waveguide acts as a high-pass frequency filter.

In this type of wave the electric and magnetic fields are not perpendicular to each other and mutually perpendicular to the direction of propagation, as in the case of a TEM wave. This type of wave is what is known as a transverse electric field wave, and since it is the lowest-order mode of TE waves, it is designated as the TE_{10} mode. The subscript 1 refers to one half cycle variation of the electric field in the y direction and the subscript 0 refers to no variation of the electric field in the x direction. For a qualitative explanation of waveguide modes see Chapter 6 of [13]. The electric and magnetic fields in and along the sides of the waveguide are shown in Figure 5.33.

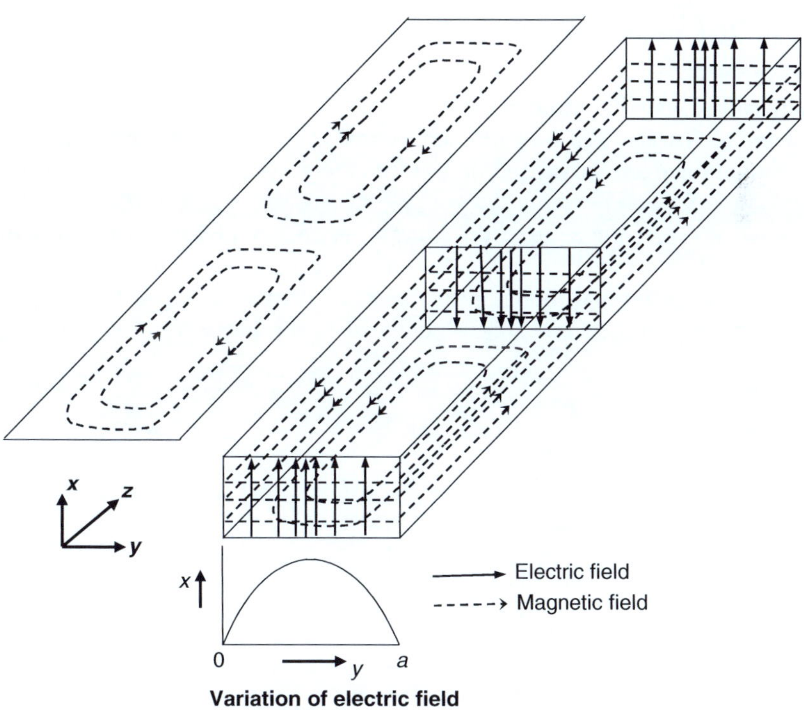

Figure 5.33 The electric and magnetic fields of a TE_{10} wave in a rectangular waveguide.

5.13.2 Horns

The horn could have a square, rectangular or circular open mouths. For airborne applications they are usually used as a feed for a reflector system or as 'suppressed' antennas since they are not suitable aerodynamically unless they are enclosed in a radome. Suppressed antennas are not like conformal antennas that are shaped to the airframe skin. These antennas do not protrude from the airframe because a hole is cut in the airframe and the antenna is installed inside the profile of the airframe, and a dielectric/fibreglass covering/radome is installed (over the antenna) flush with the aircraft's outer skin. Horns are usually fed by waveguides which are in turn are fed by coaxial cables using a waveguide-to-coaxial adapter. The two most common forms of airborne waveguide-to-coaxial adapter have a supported centre conductor since these can withstand the vibration encountered on aircraft as shown in Figure 5.34a,b. The doorknob adaptor can withstand higher power levels than the crossbar version.

The feed horns used in weather radar reflectors are usually circular waveguides since circular polarization is used in heavy rain/sea clutter and horizontal polarization is used in clear weather. Search radar usually employs linear polarization, which may be vertical or horizontal and vertical.

Older RadAlt systems have a pair of suppressed rectangular pyramidal horns, but in newer installations patch antennas are used. In these pyramidal horns the waveguide is flared in both planes as shown in Figure 5.34c.

5.13.3 Slotted Waveguide Array Antenna

Slotted waveguide arrays are sometimes (erroneously) called 'flat plate antennas'. Flat plate antennas are usually printed circuit arrays of antenna elements. However, radar antennas have to handle high powers that preclude the use of printed circuit arrays and therefore the flat plate antennas used in airborne radar systems are usually slotted waveguide arrays.

Most slotted array antennas used on aircraft are rectangular waveguides with slots in the smaller wall (height) of the waveguide. These are spaced at half-wavelength intervals. Bent probes are inserted and the angle of the hook varied to adjust the coupling of the electric field, by turning the screw head attached to the probe as shown in Figure 5.35.

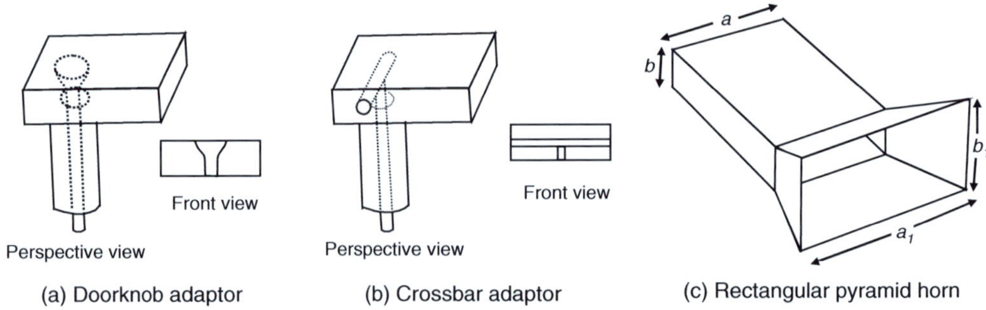

Figure 5.34 Rectangular waveguide-to-coaxial adaptors and pyramid horns.

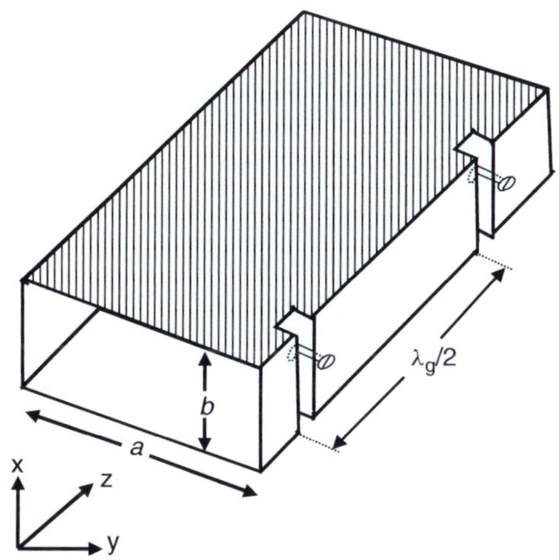

Figure 5.35 Slotted rectangular waveguide radiators.

The screw heads can then be sheared off so prevent any undesirable impedance effects ([1], p. 301). The phasing of the slots can be adjusted to give the desired illumination. Beam scanning can be attained by applying a phase gradient.

5.13.4 Phased Array Antennas

In the case of phased arrays the beam is formed by the combination of more than one element, but in addition the beam could be steered without physically moving the array. If we have a linear array this is equivalent to increasing the aperture of the antenna, so that we get a narrow beam.

Phased arrays are becoming increasingly commonplace with the emergence of new technologies and computer control. In this case the antenna may consist of one or two dimensional arrays of antenna elements, the amplitudes and/or phases of the individual elements are selected to produce the desired beam shape, and the beam is then steered by altering the phase gradient across the whole array of elements. Thus in order to scan the beam, a beam forming as well as a beam steering network is required.

The beam forming network could be a array of formed by having fixed phase shifters in the feed network. However, in the case of the beam steering network the phase is altered in equal increments across the elements so that a uniform phase gradient is applied across all the elements. Each phase gradient applied gives a different angle through which the beam is tilted.

If we have two elements, A and B, which are radiating in phase, then in the direction indicated by the arrows in Figure 5.36a, perpendicular to the line joining the elements, the wavefronts will be in phase and we have a plane wave. Now consider the case where the phase of element A is given a phase so that it lags behind that of B by ϕ degrees. The wave from A will have to travel a distance equal to a phase of ϕ degrees before it is in phase with the wave from B. It travels the distance AP (as shown in Figure 5.36b) which corresponds to the phase of ϕ degrees and if the distance between the two elements is

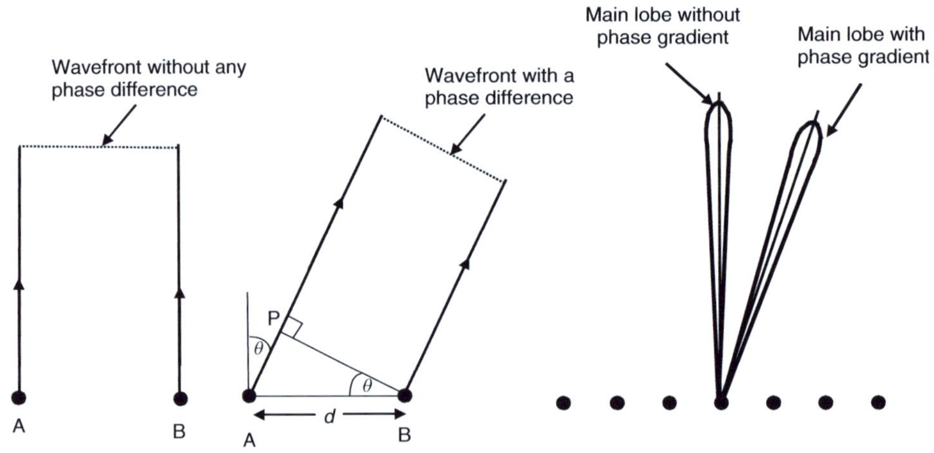

Figure 5.36 Phasing between two radiating elements and phased arrays.

d, then AP $= d \sin\theta$. The angle of the radiation is now at an angle of θ compared to the case then when the two elements were in phase.

Thus by applying a phase difference between the two elements the direction of the beam has been altered. By applying a phase difference across the array the beam can be steered in a direction depending on the phase applied. This is the principle of beam steering in phased arrays, as shown in Figure 5.36c. Thus phasing of the antenna elements can be used to shape the beam, and application of a phase gradient can be used to steer the beam, resulting in a scanning phased array.

Thus by altering the phase gradient in any or all directions across a two-dimensional array, we can have a scanning antenna, without the need for mechanically moving the array. This form of electronic scanning is limited to relatively small angles for planar arrays, and in general to get large sector coverage, phased arrays may also be mechanical rotated, with phasing used to produce the beam shape alone. However, large scan angles result in an increase of the sidelobes, and it is common to apply small angles of scan in the vertical direction (pitch) by the application of phase gradients, but use mechanical rotation of the array in the horizontal direction, as in the case of reflector antennas.

Electronic scanning has been used with circular arrays to give 360° coverage, as in the case of the TACAN ground station.

5.14 Model Numbers Used by Different Manufacturers

Some antennas, especially on military aircraft, have the manufacturer's name overpainted to match the airframe body. In addition, some aircraft publications and maintenance manuals give the model numbers of the antennas but do not state the manufacturer's name. It is useful in some cases, such as when an antenna has to be replaced urgently, to identify the manufacturer. It is also useful to find the equivalent antenna supplied by other manufacturers, although the physical form and fit may not permit replacement in some cases. Table 5.5 shows the model numbers of equivalent antennas for ATC/DME/TCAS, glideslope, ILS marker and VHF supplied by different manufacturers.

Table 5.5 The equivalent antennas for ATC/DME/TCAS, Glideslope, ILS Marker and VHF supplied by different manufacturers.

Manufacturer	Chelton	Collins	Commant	Dayton Granger	Dorne and Margolin	HR Smith	Sensor systems
ATC/DME/TCAS	10A9		C100-5	L10-16	DMN 50-6	10-203-IP	S65-5366-10L
			C100-2			10-203-2P	
			C100-3		DMN 50-3	10-203-3P	565-5366-10LC
	10A1	2377-1	C100-4	750147	DMN 50-4	10-203-4P	S65-5366-2L
Glideslope	17-21	37-P5		720036	DMN 25-2	10-204-1P	S41422-2
			720036-1	DMN25-1		10-205-4P	
	17-20N	37-P4				10-205-1P	S41422-6
	17-20					10-205-3P	S41422-5
Marker	37X-2	C1 118-1				10-208-2, 10-208-2FP, 10-208-7FP, 10-208-8FP, 10-208-9FP	
VHF	12-1	37-R2	C-108	72004 VF10-22 VC10-126	DMC 50-2, DMC 60-1	10-50-112,	S65-8280 S65-8282
					DMC 50-1	10-105-24	S65-8282-2
					DMC 50-17	10-105-20	
	16-21			VF-10-347		10-118-20	

References

[1] Silver, S. (1949) *Microwave Antenna Theory and Design*, MIT Radiation Laboratory Series, 1st edn, McGraw-Hill, New York.

[2] Wait, J.R. (1986) *Introduction to Antennas and Propagation*, Peter Peregrinus, London, ISBN 0-86341-054-5.

[3] Balanis, C.A. (1982) *Antenna Theory – Analysis and Design*, Harper & Row, New York.

[4] Rudge, A.W., Milne, K., Olver, A.D. and Knight, P. (1982) *The Handbook of Antenna Design*, vols. 1 & 2, Peter Peregrinus, Stevenage.

[5] Green, H.E. (1985) Antenna designer's notebook. *IEEE Antennas and Propagation Society Newsletter*, 11–13.

[6] Weiner, M.M. (1987) Monopole element at the centre of a circular ground plane whose radius is small or comparable to a wavelength. *IEEE Transactions on Antennas and Propagation*, **35** (5), 488–495.

[7] Jordan, E.C. (1953) *Electromagnetic Waves and Radiating Systems*, Constable and Company Ltd.

[8] Installed Performance of Antennas on AeroStructures (IPAS), an EU Specific Targeted Research Project, November 2003 to January 2007, Contract No AST3-CT-2003-503611.

[9] Johnson, R.C. and Jasik, H. (1984) *Antenna Engineering Handbook*, McGraw-Hill, New York.

[10] Jasik, H. (1961) *Antenna Engineering Handbook*, McGraw-Hill, New York.

[11] Dyson, J.D. (1959) The equiangular spiral antenna. *IRE Transactions on Antennas and Propagation*, **7** (2), 181–187.

[12] Kraus, J.D. (1988) *Antennas*, international edition, McGraw-Hill, New York ISBN 0-07-100482-3.

[13] Macnamara, T.M. (1995) *Handbook of Antennas for EMC*, Artech House, Boston, ISBN 0-89006-549-7.

[14] Mayes, P.E. and Dyson, J.D. (1961) A note on the difference between equiangular and Archimedes spiral antennas. *IRE Transactions on Microwave Theory and Techniques*, **9** (2), 203–204.

6

RF Interoperability

6.1 Introduction

The term RF interoperability is used for operability between systems, whereas the term intra-operability is used for operability within a system, for example between the different LRUs or entities within the same system. Intra-operability will not be addressed in this book. It is a matter for the manufacturer and/or supplier of the system. It is possible for individual LRUs to meet the EMH standards for emissions as well as susceptibility, but then when all the LRUs are connected up together, the total system could fail the standards. This is because the phases of the emissions could result in their combining to give a resultant that is above the amplitude of the individual emissions. These are commonly referred to as 'inter-modulation cross-products'. For further details refer to Chapter 4, where the combination of two waves of the same frequency but of different amplitudes and phases is discussed and the resultants derived for all phase differences and four different relative amplitudes.

For systems to interact, the frequencies have to be above a specified power threshold, and their spatial as well as their time domain characteristics have to be overlapping. When any EM wave transfers from one system to another the systems are said to be 'coupled'. This term is used for systems when this occurs unintentionally.

At the design stage when positions of antennas are continually being changed, a quick and easy means of estimating the coupling between antennas is required. Computer modelling cannot be used at this stage, since the change of position of one antenna is invariably accompanied by the change of positions of a number of adjacent antennas, and a quick method of determining the impact is required. Additionally it is more satisfactory to perform calculations since the coupling has to be estimated over large combinations of antennas at the design stage, and a spreadsheet is more amenable to the frequent changes that occur during the design stages of a new aircraft or complete systems modification of an existing aircraft.

6.2 Coupling between Systems on an Aircraft

The coupling could be due to EM emissions resulting from a combination of conducted and radiated emissions and between a transmitter and a receiver, and could be due to both

Introduction to Antenna Placement & Installation Thereza M. Macnamara

being inside the airframe, both being outside the airframe, or if they are on opposite sides (one inside and one outside) of the airframe.

There is a very fine dividing line between conducted signals that couple capacitatively (or inductively) between physically unconnected circuits, and radiated signals in the near field of the source. However, if the distances involved are very small in terms of wavelengths (of the order of one-thousandth or so) the mechanism of energy transfer can be considered to be induction, whereas for larger distances the mechanism of transfer is through radiation, and the energy is characterized in terms of electric and magnetic fields rather than in terms of voltages and currents.

6.2.1 Conducted Emissions inside the Airframe

The airframe of an aircraft usually consists of a frame made up of hollow rods (called ribs) that run longitudinally from the nose to the tail of the fuselage and are held in place by circular rings called frames as shown in Figure 6.1. The frames usually also have holes in them (as shown in the first part of the figure) to reduce the weight and this also allows cables, pipes, and so on to traverse the length of the aircraft between the outer skin and cabin walls.

Inside the airframe there may be instances where the RF cables from the transmitter and receiver (to the respective antennas) are adjacent to each other as shown in Figure 6.2. This would occur because the cables (RF as well as mains power) are tied together in looms and then run between the cabin wall and outer skin. Thus, for instance, if the transmitter is on the wall of the cabin midway between the floor and ceiling of the fuselage, the RF cable to the antenna could be taken down to floor level and then strapped together in a loom that runs along the circumference of the fuselage to the antenna at the top. The RF cable from the receive antenna of another system could also be strapped into this same loom. Although in theory this should be avoided for systems within the same frequency band, this may not be viable or implemented in practice. Coupling through induced currents would occur and the integration team would then try and minimize the

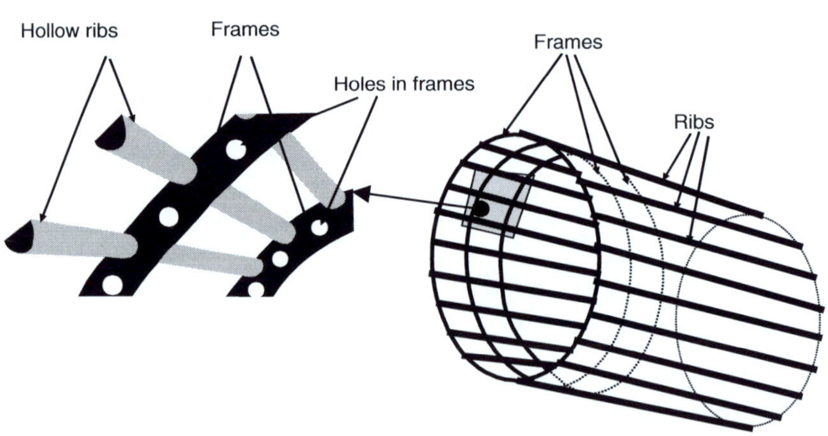

Figure 6.1 Fuselage of an aircraft showing the ribs and frames to which the skin is riveted.

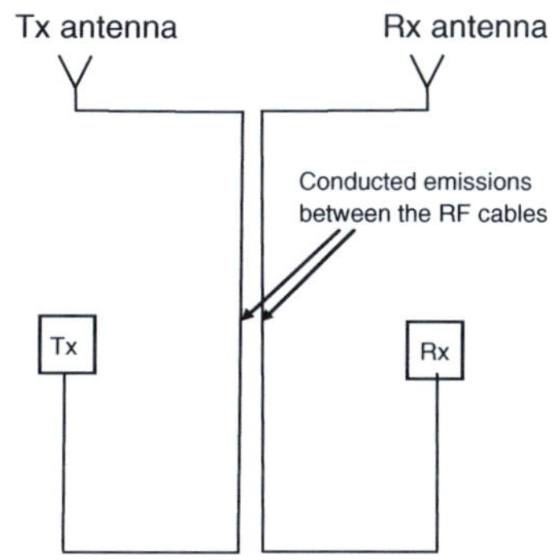

Figure 6.2 Side view of schematic of a receiver system subjected to conducted emissions from a transmitting system.

problem by EM shielding of cables. The term pick-up is used if the interfering source is power (400 Hz for aircraft) and the term cross-talk is used when the actual signal from one cable can be retrieved by another unconnected cable. Double or even triple shielding on cables would be used in preference to rerouting of the cables.

6.2.2 *Radiated Emissions inside the Airframe*

The connectors attached to the RF cables may also have apertures through which radiation is emitted. In order to reduce this the connectors would have 'back shells'. There may also be apertures in the equipment LRUs that would be used for ventilation. If the frequency of the equipment is high even small apertures can result in EM radiation being emitted through them, because the wavelength is short. This radiated emission could then enter the LRU of a 'victim' system and cause interference. The victim system would be said to be 'susceptible' to EM interference.

6.2.3 *Radiated Emissions outside the Airframe*

Radiated emissions outside the airframe would occur between the transmit antenna of one system and the receive antenna of another system. These would also constitute unwanted and unintentional coupling between the systems. Since we are only dealing with emissions with respect to interoperability between on-board systems, we do not need to consider any emissions emanating from off-board systems such as from the ground, satellite or other aircraft.

Radiated emissions outside the airframe form part of the antenna siting process, whereas radiated and conducted emissions within the airframe (as well as conducted emissions through the airframe) are part of the EMH integration process.

6.2.4 *Conducted Emissions outside the Airframe*

The conducted emissions outside the airframe are those resulting from a transmit antenna and a receive antenna sharing the same ground plane. This reduces the isolation between the antennas resulting from their spatial distance apart. However, this coupling due to conducted emissions through the common ground plane is usually very small and a second-order effect. The isolation between two line of sight (LOS) antennas calculated using the formula for radiated emissions gives values fairly close to those measured, indicating that the coupling through the common ground plane does not have any significant effect.

6.2.5 Radiated Emissions through the Airframe

These are a result, for instance, of:

1. emissions emanating from an LRU, RF cable or RF connector inside the airframe being radiated through a window, or other non-metallic part of the airframe and being incident on the antenna of another system, or
2. emissions emanating from an antenna outside the airframe entering the inside of the aircraft through a non-metallic part of the airframe and coupling to the LRU of another system.

Levels of radiated emissions are difficult to measure accurately with respect to specific positions and measurements would normally be performed with a probe connected to a receiver on peak hold so that only the maximum level is obtained. Peak detectors are used in measurements for compliance with military standards such as MIL-STD 462, whereas quasi-peak detectors are recommended by bodies such as Comité International Spécial des Perturbations Radioélectriques (CISPR), for compliance with national standards and legal regulations.

6.2.6 Conducted Emissions through the Airframe

Conducted emissions through the airframe (either from the outside to inside or vice versa) could occur at low frequencies, for instance, if the HF system is transmitting. At these low frequencies, when the HF antenna is radiating, the whole airframe will have currents induced in it and the EM waves would be conducted through the airframe and may couple to RF cables and hence cause interference to systems inside the airframe. Conducted emissions from inside the airframe to the outside are likely to be of a low level and hence are unlikely to be a major problem.

6.2.7 Coupling between Systems due to Radiated Emissions Outside the Airframe

This is the only part of coupling between systems that is within the scope of the antenna integration team. The simplified coupling mechanisms between a transmitting system and a receiving system due to radiated emissions are depicted schematically in the Venn diagram of Figure 6.3. The frequency domain applies to frequencies at a threshold level that is likely to result in malfunction of the receiving system. The problems occur when all three domains (time, frequency and spatial) of a transmitter and receiver overlap.

If they just overlap in two domains, then there should not be a problem. For instance, if they overlap in the spatial and frequency domains, but not in the time domain, that is,

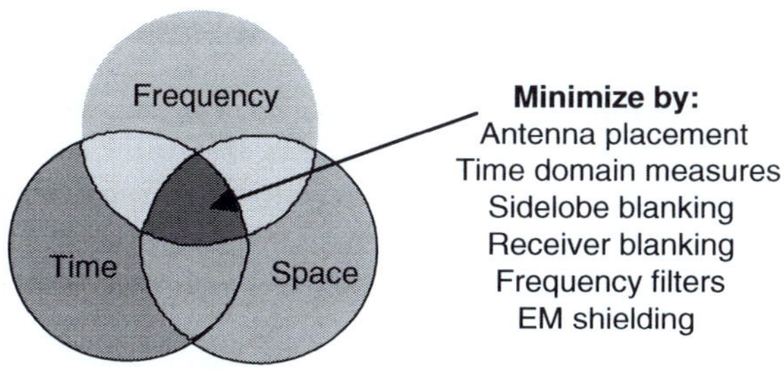

Figure 6.3 Techniques (with respect to radiated emissions) for achieving RF interoperability between a Tx and Rx system.

if the transmitter and receiver are spatially near but are not in the same time domain (i.e. not operating at the same time), no problem exists.

Similarly, if they overlap in the time and frequency domains, but not in the spatial domain, then there will be no coupling between systems. If the signals from one system arrive when the susceptible receiver of a second system is operational, but the two systems are spatially separated or operating in different angular sectors (e.g. upper and lower hemispheres or directional antennas pointing in non-overlapping areas) so that the radiation from the transmit antenna is not incident on the receive antenna, then there will not be any adverse effect on the second system.

If they overlap in the time and spatial domains, but not in the frequency domain, then there will be no coupling between systems, since it means that they are separated in the frequency domain, that is, their frequency bands do not overlap. Of course, what must be borne in mind is that when we refer to the frequency band of a system, we are referring to its entire power frequency spectrum, and not just its carrier or fundamental frequency. Thus we have to consider the actual signal emitted. This signal (in the time domain) is converted to the frequency spectrum by using the Fourier transform. The frequency spectrum (i.e. the amplitude at each frequency) can be measured using a spectrum analyser.

6.3 Techniques for Achieving RF Interoperability

As mentioned earlier, the final antenna layout is determined before all the systems parameters are known and by trade-offs between all the relevant disciplines. Thus measures usually have to be applied when the antennas and other system LRUs are physically installed. Techniques (with respect to radiated emissions) for achieving RF interoperability between systems as depicted in the Venn diagram of Figure 6.3 are:

1. antenna placement
2. time domain measures
3. sidelobe blanking
4. receiver blanking
5. frequency filters
6. EM shielding.

In the case of the frequency domain it should be noted that the amplitude at the different frequencies is important. If the amplitude of the wave at a particular frequency is very low then the impact on the victim/receiving system will be negligible. In the case of electromagnetic interference (EMI) the standards define the level at each frequency or frequency band. In the case of RF interoperability the level will depend on the characteristics of the entire victim system.

6.3.1 Antenna Placement

The isolation between antennas can be increased by increasing their spatial separation. As mentioned in Chapter 3, antenna locations are low in the pecking order of the various disciplines. Thus the positions will be optimized by judicious application of trade-offs based on the other constraints. The initial antenna layout is based on perhaps placing the antennas 3 or 4 wavelengths apart at the lowest frequency in the band. This cannot be achieved at the low end of the VHF band for most aircraft, since at 30 MHz the wavelength is 10 m. In the HF band no attempt would be made to attain this separation. Furthermore, the coupling between systems cannot be calculated since the details required are not available until much later in the system design process. These details include the powers transmitted, losses in the cables, sensitivities of the receivers, and so on. However, the antenna specialist is left with the task of predicting the coupling with the designated antenna locations. Once the positions of the antennas are determined it is very difficult if not impossible to change them if it is found that the isolation between the systems is insufficient from the interoperability point of view. This could be due to the necessary compromises caused by the dearth of real estate available, initially assumed power transmitted by the offending systems being higher and/or the victim system having greater sensitivity, incorrect losses assumed in the RF cables, etc. Higher losses in the RF cables will decrease the coupling between the systems and hence be advantageous (to interoperability), although this is not desirable for the operation of the system.

6.3.2 Time Domain Measures

Time domain measures are used to ensure that the two conflicting systems do not transmit or receive at the same time. This can be accomplished by:

1. interleaving emissions,
2. manually suppressing (the transmitter or receiver), or
3. press to transmit, where a manual switch is depressed so that when the offending system is transmitting the receiver of the victim system is automatically switched off.

Emissions can be interleaved whether they are CW or pulsed transmissions, although the term interleaving is usually only used for pulses. Thus, for instance, the transmissions can be restricted so that the different systems transmit in allocated time frames. Pulsed systems that transmit several pulses per second can stagger the pulses and allow pulses of another system to transmit within the same time frame, but not at the same instant in time. For some systems such as radar the interleaving of pulses can be undertaken easily, but other systems such as transponders (for ATC) that have to operate when interrogated, this may be impossible to achieve.

6.3.3 *Sidelobe Blanking*

This is effectively blanking in the spatial domain. Sidelobe blanking is usually used in the case of high-gain, narrow-beam antennas such as radar antennas. A receiving device called a sidelobe blanker is employed to detect if the received radar signal originates from the sidelobes of the transmitted antenna signal. In this case a broad beam is generated at the level of the blanking required to suppress the sidelobes, by applying the blanking in antiphase with the high-gain antenna beam, effectively 'gating out' the sidelobes from the antenna radiation pattern. Thus the antenna output signal is from the main beam alone or from the main beam and 'near-in' sidelobes. In Figure 6.4 the radar radiation pattern is shown with the blanking applied as a dotted line. This suppresses all the sidelobes apart from the 'near-in' ones. Thus the spatial coverage of the antenna is restricted and radiation from it is not incident on antennas outside the angular sector of the main beam and near-in sidelobes. Note that the overall range of the system is reduced, since the reduction in gain affects the output power radiated as well as the received radar return.

6.3.4 *Receiver Blanking*

In this case the receiver is prevented from functioning by either manually or automatically suppressing it.

The receiver can be blanked by:

1. manual means
2. programming
3. automatic means.

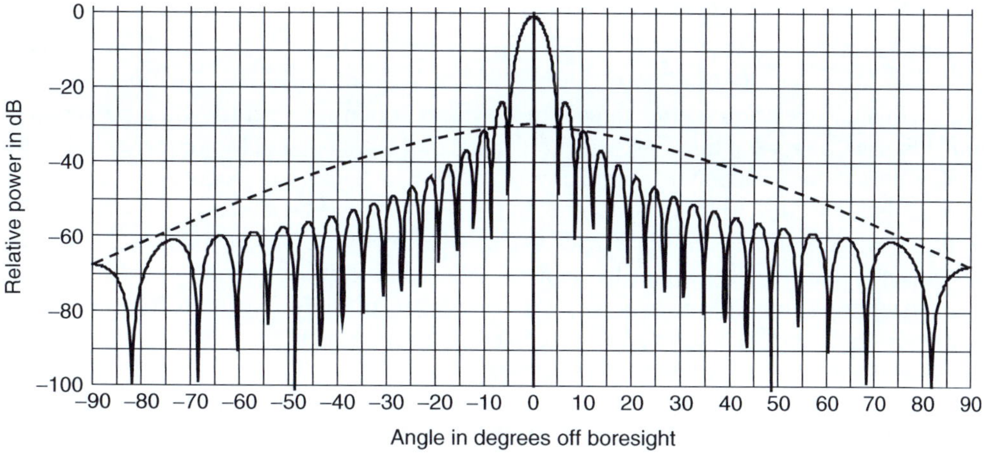

Figure 6.4 Sidelobe blanking.

6.3.4.1 Manual Suppression

Manual suppression is used when there is an unscheduled transmission by an on-board system. This could be the case when, for instance, an aircraft is transmitting to a ground station in the case of an emergency or if the aircraft is trying to communicate with a survivor in the water who has transmitted a mayday signal. In the case of military sorties any departure from the planned mission could result in an unscheduled transmission and would be implemented by the mission commander.

6.3.4.2 Programmed Blanking

Programmed blanking would be employed, when it is known that the on-board systems could be affected by certain transmissions. For instance, if it is known that specific transmissions would interfere with a landing system, then the transmission of the offending system would be programmed to prevent operation during landing. Other cases could occur in the case of military aircraft. These aircraft usually fly on specific missions, and the details are given to the mission commander in the form of a pre-flight message (PFM). This would take the form of electronic media, such as a DVD. A surveillance exercise could, for instance, form a part of the mission, and would necessitate the suppression of all transmitting systems. The mission commander would be in charge of the mission and could override any manoeuvre or detail of the mission as required.

6.3.4.3 Automatic Blanking

Automatic blanking would be in the form of power limiters, which could be inserted in the RF cables or form part of the system LRU. The receiver would be automatically blanked when the received power level got to a pre-specified value. These limiters usually work on the heating effect of the RF power, and thus take a short time to be triggered. However, although this is of order of milliseconds, a time delay of this length could be unacceptable in some cases.

6.3.5 Frequency Filters

The frequency spectrum of the signals of a system (often referred to as the power spectrum) is usually not contiguous over the whole frequency band. The frequency spectrum applies to all the frequencies within the signal and includes the frequencies as a result of modulation as described in Section 6.4. The systems usually operate in fixed channels or fixed spot frequencies in the carrier band. It is possible to insert filters to restrict the systems so as to prevent certain frequencies that would cause problems to other systems from being transmitted. Similarly, systems that are susceptible to certain frequencies can have stop filters inserted before their receivers to prevent these frequencies from causing any adverse effects. Stop band filters prevent a band of frequencies, whereas pass band filters allow a band of frequencies to be propagated through them. If a very small band of frequencies is to be rejected or passed, then a notch filter is used. For instance, notch filters are used at 1030 and 1090 MHz to avoid interference with ATC/SSR interrogating and receiving at these frequencies. Some of these filters are available as through-line

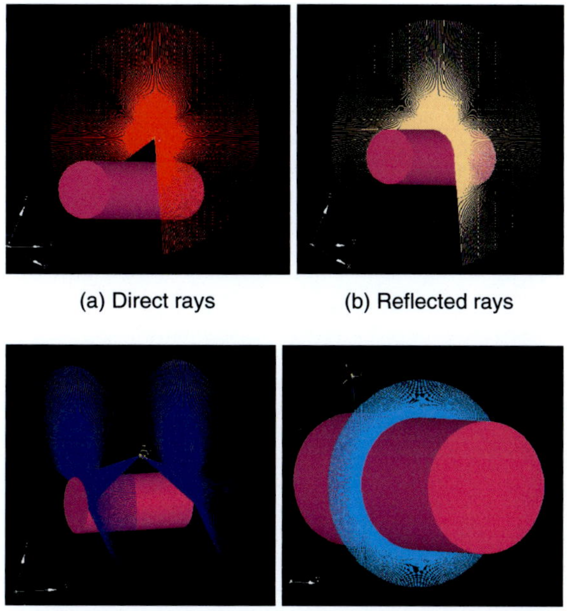

(a) Direct rays (b) Reflected rays

(c) Diffracted rays (d) Creeping rays

Plate 1 Figure 1.2 Graphic illustration of the direct, reflected, diffracted and creeping rays obtained for an antenna located above the surface of a cylinder. Figure 4 of [1]. Reproduced by kind permission of EADS.

(a) Reflected rays (b) Diffracted rays

Plate 2 Figure 1.3 Rays obtained using the EADS ASERIS-HF GTD code. Figure 27 of [2]. Reproduced by kind permission of EADS.

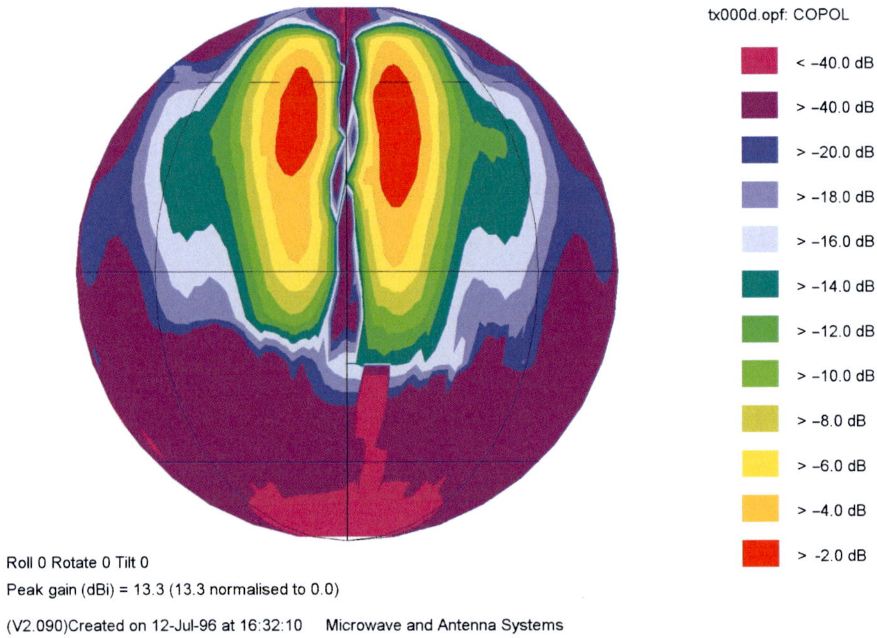

Plate 3 Figure 1.34 Contour plots on a spherical surface. Reproduced by kind permission of ASL [5].

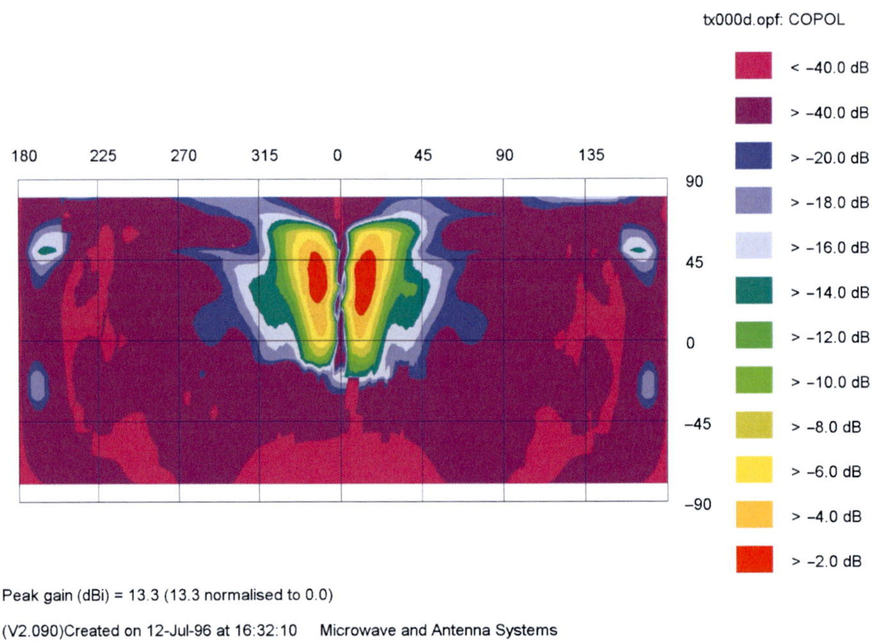

Plate 4 Figure 1.35 Mercator's projection of a contour plot. Reproduced by kind permission of ASL [5].

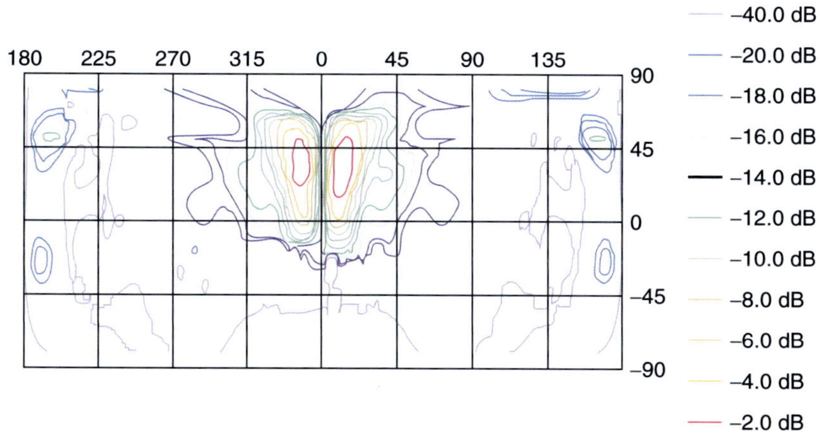

Plate 5 Figure 1.36 Mercator's projection of the radiation pattern, showing lines of equal gain/power levels. Reproduced by kind permission of ASL [5].

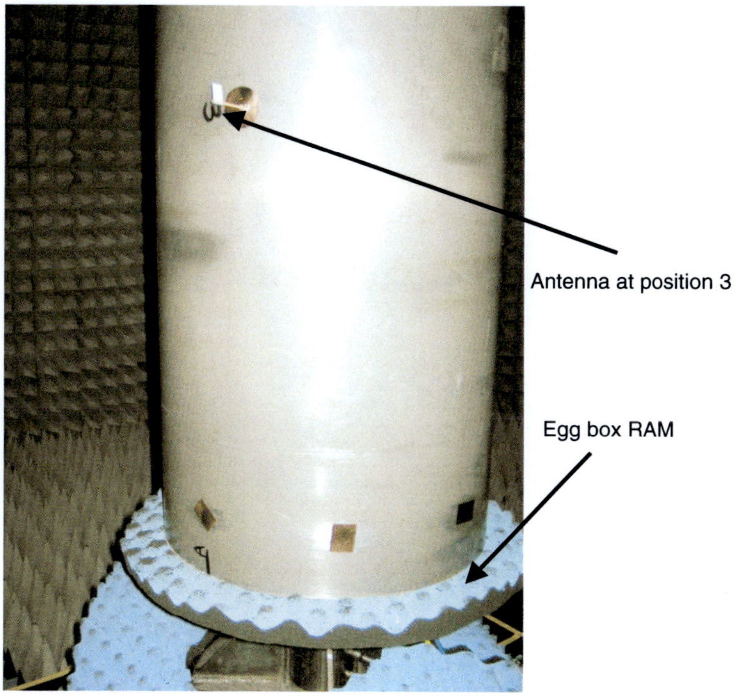

(b) Photograph of the scaled cylinder

Plate 6 Figure 4.17 Diagram and photograph of the scaled cylinder showing the positions of the antennas.

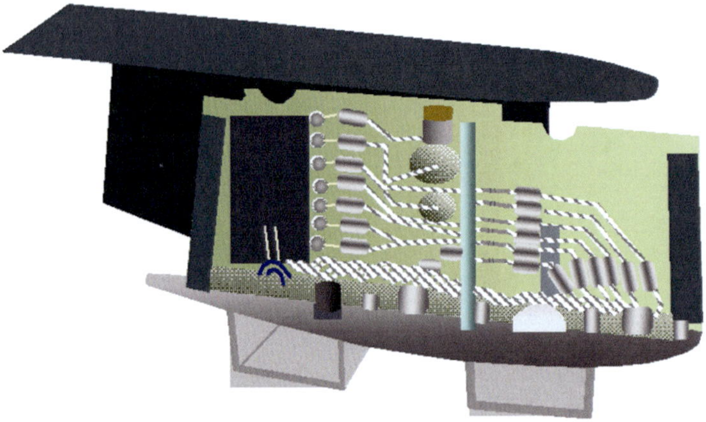

Plate 7 Figure 5.13 A UHF blade antenna showing the complex circuitry on the printed circuit element – manufactured by Chelton – IPAS [8].

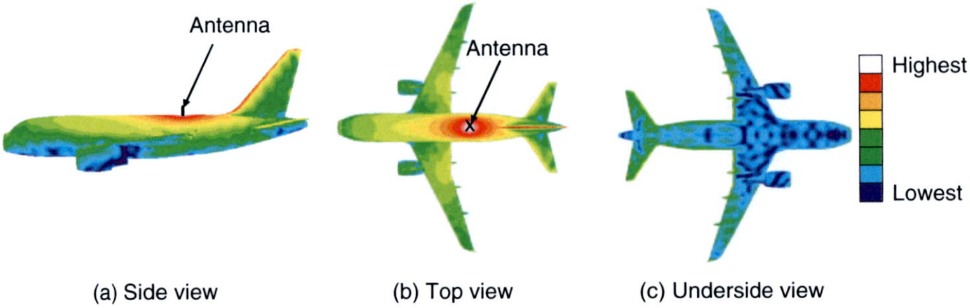

Plate 8 Figure 7.7 Surface currents on IPAS-1 for an antenna on the rear upper fuselage. Reproduced by kind permission of EADS.

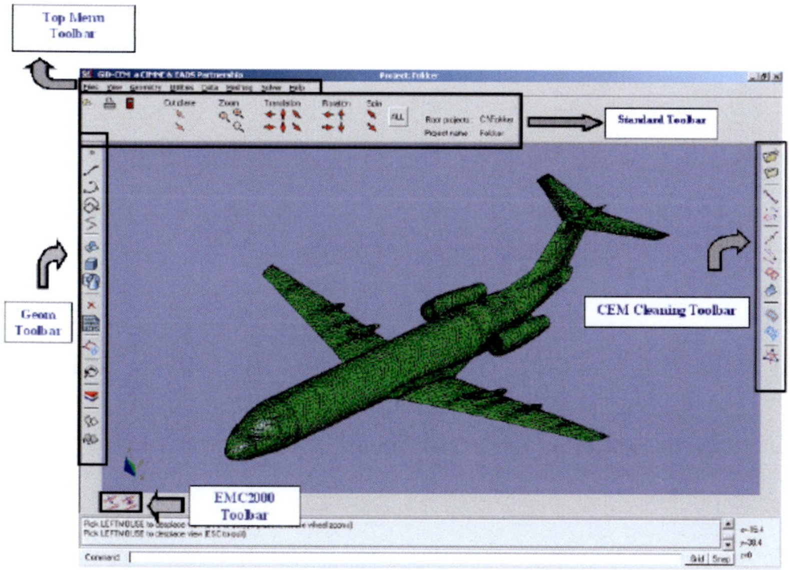

Plate 9 Figure 7.8 A typical screen GiD_CEM screen developed to integrate with EMC 2000. Reproduced with kind permission of GiD_CIMNE.

(a) Surface currents using FMM

(b) Surface currents using hybrid MoM/PO

Plate 10 Figure 7.27 Surface currents calculated with FMM and Hybrid MoM/PO [2]. Reproduced by kind permission of EADS.

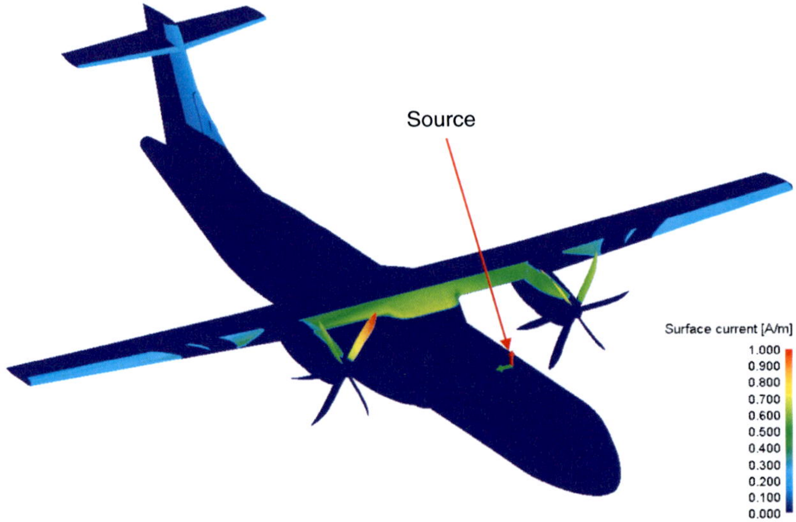

Plate 11 Figure 7.28 Surface currents on an ATR 72 airframe using PO [12]. ©2008 IEEE. Reproduced by kind permission of IEEE.

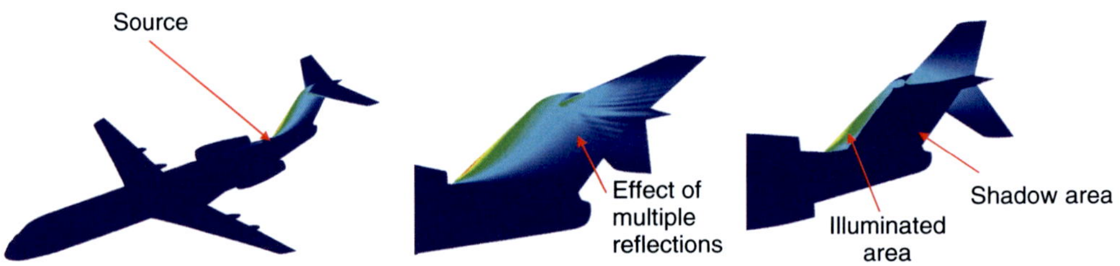

(a) Antenna in front of tailfin (b) Currents using FMM (c) Currents using PO

Plate 12 Figure 7.29 Surface currents on a Fokker 100 airframe using FMM and PO [13]. Reproduced by kind permission of IET.

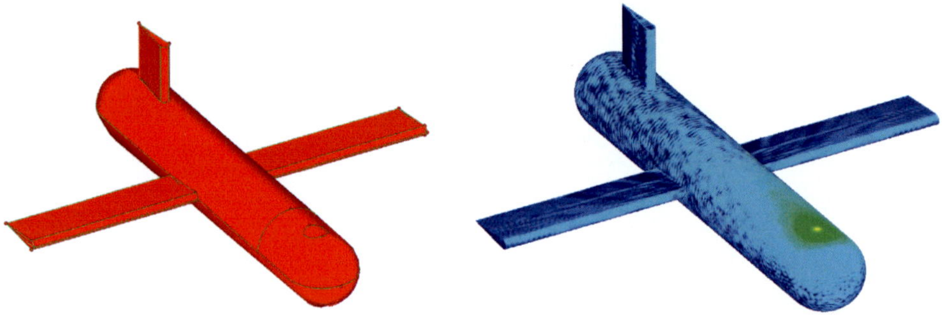

(a) Entire airframe with $\lambda/10$ segments (b) Current distribution

Plate 13 Figure 7.31 Configuration 1 meshing and the corresponding current distribution. Reproduced by kind permission of EADS.

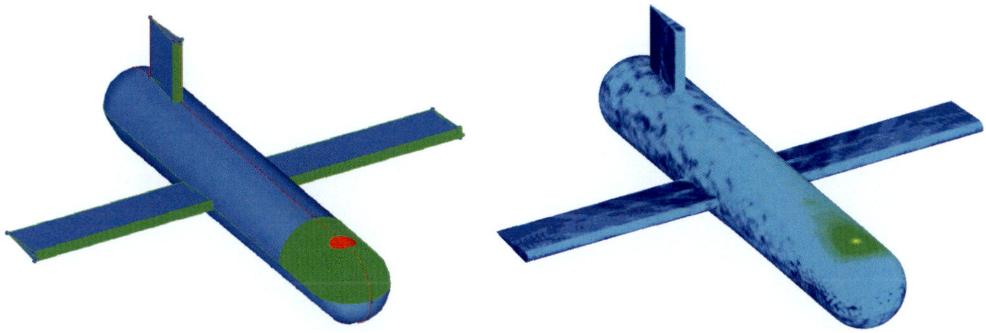

(a) Airframe with $\lambda/10$, $\lambda/6$ and $\lambda/3$ segments (b) Current distribution

Plate 14 Figure 7.32 Configuration 2 meshing (red = $\lambda/10$, green = $\lambda/6$, blue = $\lambda/3$) and the corresponding current distribution. Reproduced by kind permission of EADS.

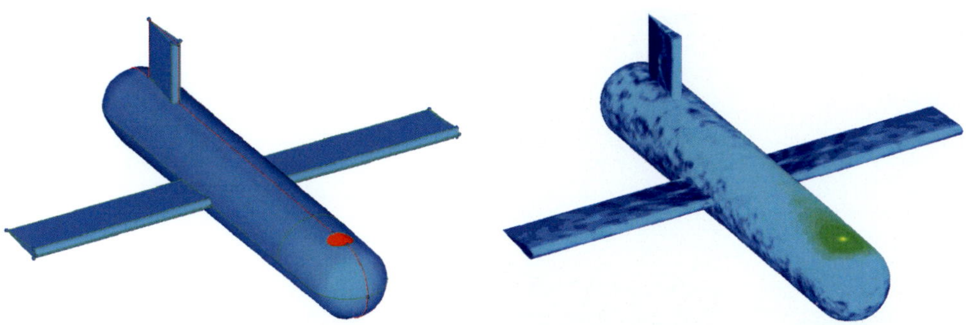

(a) Airframe with $\lambda/10$, $\lambda/6$ and $\lambda/3$ segments (b) Current distribution

Plate 15 Figure 7.33 Configuration 3 meshing (red = $\lambda/10$, blue = $\lambda/3$) and the corresponding current distribution. Reproduced by kind permission of EADS.

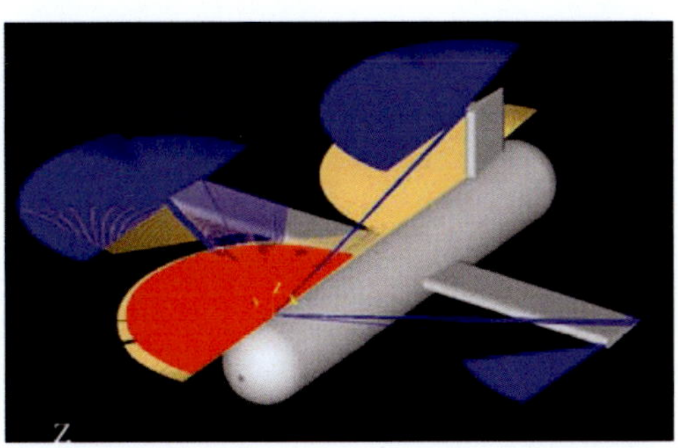

(c) Rays in azimuth plane (red, direct; yellow, reflected; blue, diffracted)

Plate 16 Figure 7.34 Comparison between the azimuth plane plots obtained using FMM, MoM/PO hybrid and UTD at 1 GHz for a simplistic aircraft. Reproduced by kind permission of EADS.

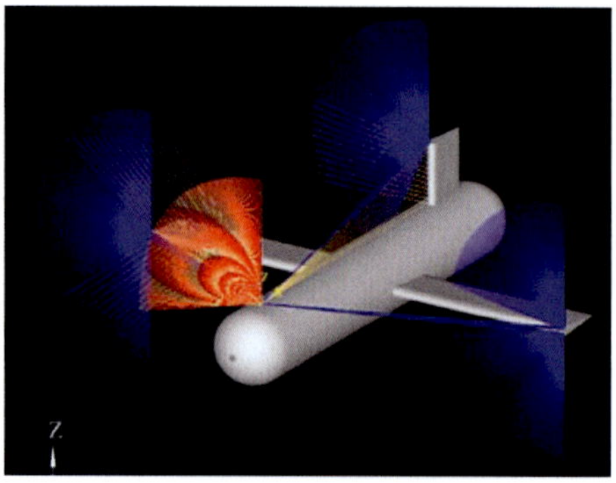

(c) Rays in roll plane (red, direct; yellow, reflected; blue, diffracted)

Plate 17 Figure 7.35 Comparison between the roll plane plots obtained using FMM, MoM/PO hybrid and UTD at 1 GHz for an antenna on the nose on a simplistic aircraft. Reproduced by kind permission of EADS.

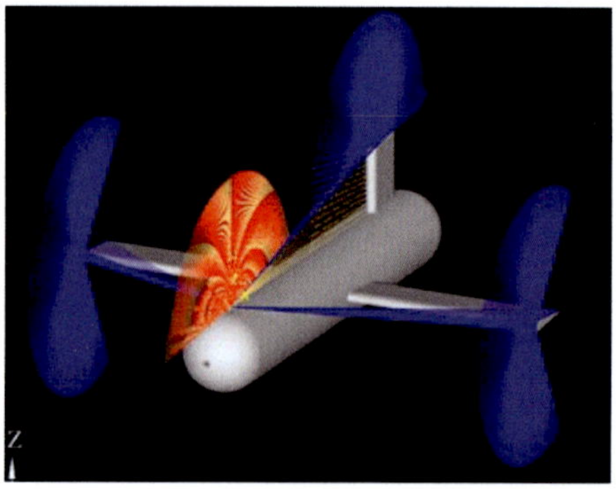

(c) Rays in the pitch plane (red, direct; yellow, reflected; blue, diffracted)

Plate 18 Figure 7.36 Comparisons between FMM in the pitch plane and MoM/PO hybrid as well as UTD at 1 GHz for an antenna on the nose on a simplistic aircraft. Reproduced by kind permission of EADS.

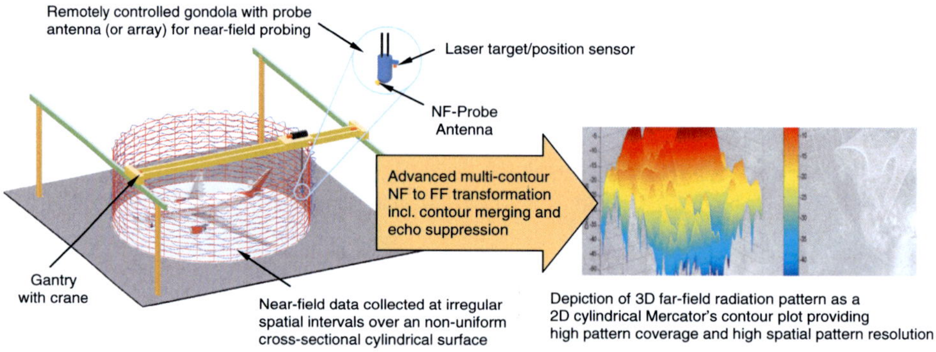

Plate 19 Figure 8.17 A schematic view of the AANTS used as a cylindrical near field facility. Reproduced by kind permission of Astrium EADS.

filters with female connectors (similar to a back-to-back connector) so that they can be inserted in the RF cables connecting the antennas to their systems. Most of these filters take a finite time to react, and in some cases this may result in problems for the victim system.

6.3.6 EM Shielding

EM shielding is used to exclude or confine EM waves. Its increasing importance is in part due to the failure of many manufacturers to consult EM compatibility specialists at the design stage, and then attempting to shield equipment at the production stage to comply with the relevant standards. In the case of radiated emissions outside the airframe, shielding would most probably only be used to shield the HF tuner embedded in the mast of a towel rail mounted on the surface of the airframe.

The extent to which a material performs this function is referred to as its shielding effectiveness (SE). The SE of a material is the attenuation it presents to an electromagnetic wave and is defined as the insertion loss in dB obtained in the presence of the material.

Solid materials prevent the penetration of electric and magnetic fields by three mechanisms:

1. reflection at the air–material interfaces
2. absorption as the fields travel through the material
3. multiple internal reflections (MR) at the material–air interface.

Figure 6.5 shows the path traced by a wave incident at the surface of a shield. If the incident fields are E_1^i and H_1^i then we can see that at the first air–material interface, parts of the fields are reflected (E_1^r and H_1^r) and parts are transmitted (E_1^t and H_1^t). These

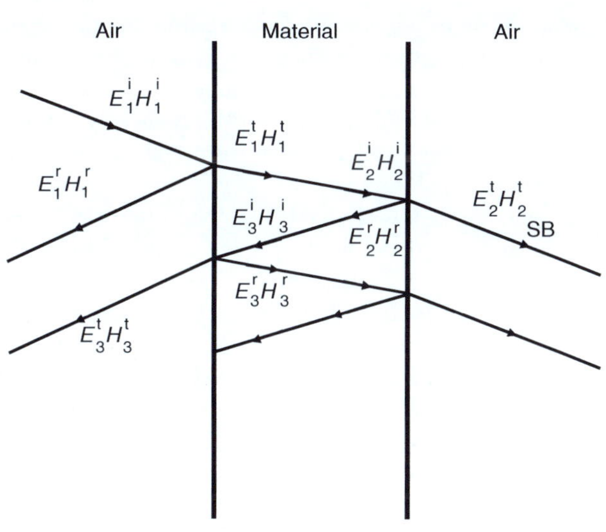

Figure 6.5 Path traced by a wave incident at the surface of a shield.

transmitted fields travel to the second material–air interface and suffer some absorption, so that when they reach this second interface the waves have lower levels (E_2^i and H_2^i). These incident waves at the back surface of the material are also partly transmitted (E_2^t and H_2^t) and partly reflected (E_2^r and H_2^r). The reflected waves are transmitted back to the first material–air interface, suffering some absorption, and the incident fields at this first material–air interface are (E_3^i and H_3^i). At this interface they will again be partly reflected (E_3^r and H_3^r) and partly transmitted (E_3^t and H_3^t). The transmitted waves (E_3^t and H_3^t) due to multiple internal reflection may not be in phase with the first reflected waves (E_1^r and H_1^r) and thus may reduce the resultant reflection loss of the material.

Thus this contribution (MR) to the total SE is also called a correction term. It will only be significant under certain conditions. If the absorption of the shield is greater than 15 dB, or if thickness of the material is greater than the skin depth, then the fields will be greatly attenuated so that the value of MR will be negligible.

The SE depends on:

1. the conductivity of the shield
2. the permeability of the shield
3. the permittivity of the shield
4. the thickness of the shield
5. the frequency of the incident radiation
6. the distance between the source and shield
7. the shape of the shield.

The shield should be grounded, and if made of aluminium it should not be anodized. Anodizing makes the aluminium non-conducting so that it cannot be used as a ground. Electrolytic tin-plated steel on the other hand is a low-cost material that lends itself to the formation of multiple shields that can be soldered together [1]. Shielding materials can also be painted or sprayed onto surfaces. Nickel is often used as paint coating, but copper is almost four times more conductive for the same thickness, and the water based copper is also more environmentally friendly [1]. It is also cheaper and easier to apply.

Steel has a relative magnetic permeability of around 1000, and is a good shield at low frequencies [2]. Copper has good conductivity and is relatively light. It is used in solid and mesh form. Bronze is usually used in mesh form and in cases where magnetic and high frequency performance is not severe.

At low frequencies (i.e. far below 1 GHz) and when the source is in the near field of antennas the electric and magnetic fields are considered separately and the SE can be defined for E-mode and H-mode, respectively. At the higher frequencies when the material can be assumed to be in the far field of the source, the plane wave SE can be used.

For materials of high permeability the attenuation due to reflection at the air–material interface is relatively small compared to the absorption loss. For instance, in the case of a Mumetal plate of 0.1 mm thickness, the plane wave reflection loss at 50 MHz is 26.75 dB whereas the plane wave absorption loss is 4537 dB.

The plane wave absorption loss obtained for different materials of thickness 1 mm is shown in Table 6.1. It can be seen that materials such as permalloy, Hypernick and Mumetal provide the highest levels of shielding. Casings made of these materials are

Table 6.1 Theoretical absorption loss for different materials of thickness 1 mm.

Material	Conductivity relative to copper	Magnetic permeability relative to a vacuum	Theoretical absorption loss in dB of 1 mm at		
			100 kHz	1 MHz	50 MHz
Aluminium	0.61	1	32.45	102.63	725.68
Beryllium	0.1	1	13.14	41.55	293.82
Brass	0.26	1	21.19	67.00	473.77
Cadmium	0.23	1	19.93	63.02	445.60
Copper	1	1	41.55	131.40	929.14
Gold	0.7	1	34.77	109.94	777.37
Hypernick	0.06	80 000	2878.83	9103.66	64372.59
Iron	0.17	1000	541.78	1713.25	12114.48
Lead	0.08	1	11.75	37.17	262.80
Magnesium	0.38	1	25.61	81.00	572.76
Mumetal	0.03	80 000	2035.64	6437.26	45518.30
Nickel	0.2	1	18.58	58.76	415.52
Permalloy	0.03	80 000	2035.64	6437.26	45518.30
Phosphor-bronze	0.18	1	17.63	55.75	394.20
Silver	1.05	1	42.58	134.64	952.08
Stainless steel	0.02	1000	185.83	587.64	4155.23
Tin	0.15	1	16.09	50.89	359.85
Zinc	0.29	1	22.38	70.76	500.36

used around the existing casings of LRUs to provide the shielding and hence reduce the emissions (or susceptibility) to acceptable levels. HF antenna tuning units are often the culprits that radiate unwanted signals. These units have to be placed near the antennas and hence often cannot be relocated to other areas, so that shielding may be the only alternative.

In the case of the cables, single, double or even triple shielding of coaxial cables is used. The shield consists of a mesh, and the higher the operating frequency, the tighter the mesh is required to be. This is because the gaps in the shield must be a small fraction of a wavelength to prevent leakage of the EM fields.

The connectors are also prone to EMI leakage at the periphery of the outer conductor of the coaxial cable and many connectors have back shells for this reason.

6.4 Modulation

When we refer to the frequency of a system, we should not consider just the carrier frequency but all the frequencies in the signal. In order to transmit information, for instance, the carrier frequency is modulated at the transmit end and then demodulated at the receive end in order to extract the information. In Chapter 5, it was explained how the Fourier transform is used convert a waveform or pulse in the time domain to the frequency domain so that the individual frequencies can be obtained together with their amplitudes. There are various forms of modulation, but only a few are discussed below.

A CW signal is unmodulated and has the typical sine wave shape as shown in Figure 6.6a. A radar signal is usually employed to obtain information about a target. This information could be distance (range), angle, shape of target or relative velocity information. A radar signal could be CW but in general it is pulsed. CW signals do not carry information. In order to carry information, the CW signals have to be modulated. There are broadly two types of modulation:

1. analogue methods that use a sine wave as a carrier
2. pulse methods that use a digital or pulse train as the carrier ([3], p. 1).

The most common forms of analogue modulation are amplitude modulation (AM), frequency modulation (FM) and single sideband (SSB). AM systems are narrowband and suffer from distortion due to noise, whereas FM systems give greatly improved signal-to-noise performance, although they require equipment with greater bandwidth.

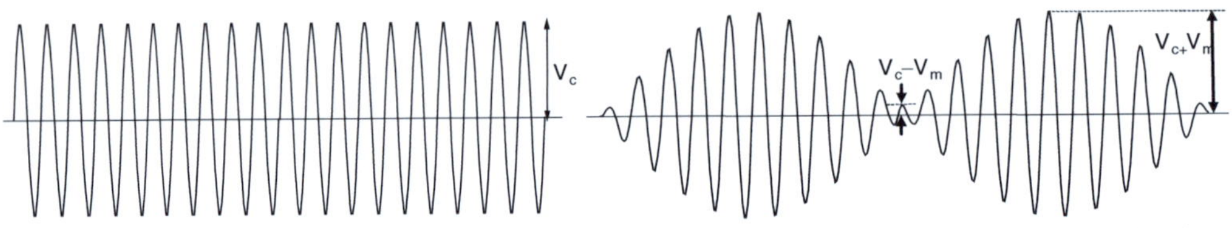

(a) Unmodulated carrier signal (b) Single frequency amplitude modulated signal

Figure 6.6 Amplitude modulation.

The modulation affects the power available at a specific location and must be taken into account when calculating the electric field strength. The British Standard [4] specifies that the power for a transmitter has to multiplied by the square of the modulation factor m to allow for the effect of this modulation.

6.4.1 Amplitude Modulation

AM was one of the earliest forms of modulation. It is a means of transmitting information at the desired carrier frequency. A CW carrier signal of amplitude V_c, as shown in Figure 6.6a, is modulated by a single frequency signal of amplitude V_m, and the modulated signal produced by the superposition of the modulating signal on the carrier signal has amplitudes that vary sinusoidally between $V_c + V_m$ and $V_c - V_m$, as shown in Figure 6.6b. The modulation index (or index of modulation), m_a, is the ratio of the amplitudes of the modulating signal (V_m) to that of the (unmodulated) carrier V_c, and it can take any value from 0 to 1 for percentage modulations from 0 to 100%. A modulation index of 0 represents no modulation, whereas $m_a = 1$ represents the maximum modulation.

The ratio of the amplitude of the carrier signal V_c to the maximum amplitude of the modulated signal, $V_c + V_m$ is given by

$$\frac{\mathrm{V_c}}{V_c + V_m} = \frac{1}{1 + V_m/V_c} = \frac{1}{1 + m_a}. \tag{6.1}$$

Thus for a modulation index m_a of 0.8 (i.e. for 80% modulation) the ratio of the amplitude of the unmodulated CW to that of the AM modulated wave is 1 : 1.8 or 0.56.

The instantaneous voltage of the modulated signal V_{md} is given by

$$V_{md} = V_c \sin \omega_c t + \left(\frac{m_a V_c}{2}\right) \cos(\omega_c - \omega_m) t - \left(\frac{m_a V_c}{2}\right) \cos(\omega_c + \omega_m) t \tag{6.2}$$

where

V_c is the amplitude of the carrier

ω_c is the angular frequency of the carrier, $\omega_c = 2\pi f_c$, where f_c is the carrier frequency

V_m is the amplitude of the modulating signal

ω_m is the angular frequency of the modulating signal, $\omega_c = 2\pi f_m$, where f_m is the modulating frequency.

The carrier power is proportional to V_c^2, while the power in each sideband is proportional to $m_a^2 V_c^2/4$, and the power in both sidebands is proportional to $m_a^2 V_c^2/2$. Thus the total power is $V_c^2 + m_a^2 V_c^2/2$.

The ratio of the power in both sidebands to the total power is

$$\frac{\text{sideband power}}{\text{total power}} = \frac{m_a^2 V_c^2/2}{V_c^2\left(1 + m_a^2/2\right)} = \frac{m_a^2}{2 + m_a^2} \tag{6.3}$$

The modulation index m_a usually varies between 0.3 and 0.5, thus for $m_a = 0.5$, the ratio of the power in both sidebands to the total power is 0.25/2.25 or 1/9.

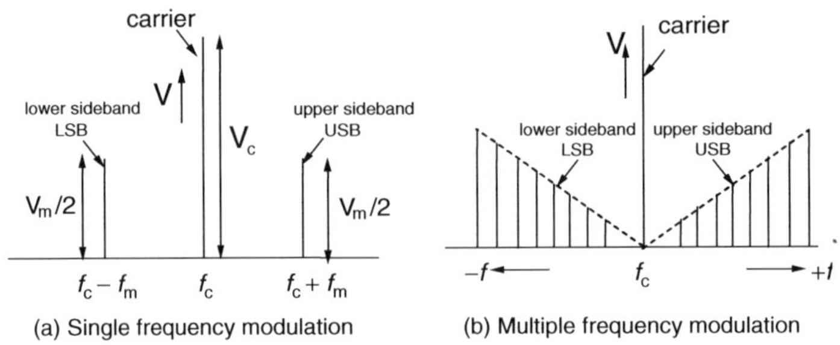

Figure 6.7 Frequency spectrum for single and multiple frequency amplitude modulation.

If we look at the frequency spectrum of the modulated signal, we get the three frequencies at f_c, $f_c - f_m$ and $f_c + f_m$ as shown in Figure 6.7a. The lower frequency $f_c - f_m$ is known as the lower sideband (LSB), and $f_c + f_m$ is known as the upper sideband (USB). This is the frequency spectrum for a single frequency modulating signal. In general, speech and music modulating signals will have many frequencies and the frequency spectrum of the modulated signal will cover a wide range of frequencies in the LSB and USB as shown in Figure 6.7b.

The value quoted for the radiated power is that of the unmodulated power. For speech and music the instantaneous electric field at the modulation peaks may be two or three times greater when the music or speech is at its loudest volume ([4], p. 4).

Another version of AM known as modulated continuous wave is used for Morse and other coded transmissions.

6.4.2 Single Sideband Modulation

In SSB modulation only one sideband is transmitted, whilst the other sideband and the carrier are suppressed. The information can be retrieved from the sideband. In this system there is a considerable saving of power and bandwidth, and in addition distortion due to carrier fading is avoided ([3], p. 19). In the case of HF the USB is used in the aviation industry, whereas the LSB is used for amateur radio.

6.4.3 Frequency Modulation

In FM the frequency of transmission is varied to carry the information. When the instantaneous modulating voltage is increasing (has a positive gradient), the frequency of the modulated signal increases, and when the gradient is negative the frequency decreases, as shown in Figure 6.8. The modulated carrier signal has a frequency deviation of Δf_c. The modulation index m_f is defined as the ratio of the frequency deviation Δf_c of the modulated carrier, to the frequency of the modulating signal f_m ($m_f = \Delta f_c/f_m$). When m_f is small (~0.2) there are few sideband frequencies but they have high amplitudes, whereas when m_f is large (~5) there are more sideband frequencies but they have smaller amplitudes ([3], p. 28). The frequency deviation of FM systems is determined by the frequency bandwidth available.

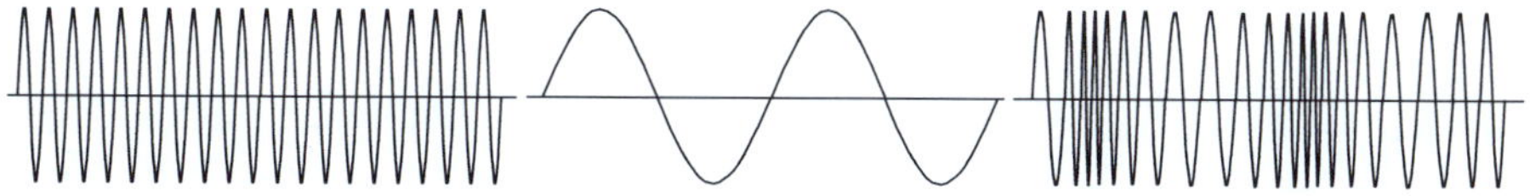

Figure 6.8 Frequency modulation.

The output power remains constant so that the modulation factor is unity, and no allowance has to be made to the output power. Frequency shift keying, phase shift keying and phase modulation are all forms of FM.

6.4.4 Pulsed Radar

Pulsed radar can also be regarded as a form of modulation. In this case the pulses are generated at the pulse repetition frequency (PRF) n in s^{-1}. This value multiplied by the pulse width (PW) or duration t_p in seconds gives the duty cycle. This duty cycle is the square of the modulation factor m, (i.e. $m^2 = nt_p$) and will always be less than one. Both the PRF and the PW have to be known in order to calculate the duty cycle. The peak power P_{pk} is given by

$$P_{pk} = \frac{P_m}{nt_p} \tag{6.4}$$

where

P_{pk} is the peak power in watts
P_m is the mean power in watts
n is the PRF in s^{-1} and
t_p is the PW in s.

The PRF is the number of pulses in one second and the period T is the time for one pulse. Thus $T = 1/\text{PRF}$. It is assumed that the pulses are rectangular; however, most pulses are not.

The frequency to be considered for the pulsed radar is not just the carrier frequency but the frequencies generated by the pulse. If the duration of the pulse is (say) 1 μs, the bandwidth can be assumed to be approximately 2/1 μs, which equates to 2 MHz. The bandwidth is inversely proportional to the width of the pulse, so that if we have a very narrow pulse the bandwidth of the receiver has to be much larger than is the case for wide pulses. The radar transmits a pulse, and the time taken for the pulse to be reflected back by the target gives the distance/range of the target. By decreasing the width of the pulse, more pulses can be sent in the same time period. If the returned pulse is wider than the transmitted one, it can be deduced that it has been returned from a single target that is long (in the dimension parallel to the transmitted beam, but the furthest end is not obstructed by the front end) or that there is more than one target in that direction that is returning the pulse. By having very narrow pulses, the returns will be separated, and thus the narrower the pulse the greater is the resolution of the radar. However, if a target is very far away its return may be received after the second pulse is transmitted and thus may be assumed to be the return from the second pulse instead of the first. Thus increasing the number of the pulses per second (the PRF) could also result in ambiguity.

6.5 Coupling due to Radiated Emissions through the Antennas

When an antenna on an airframe radiates, the EM signals emitted can be received by any other antenna and its associated receiver (of another system). The two systems are then said to be coupled. The amount of coupling between the two systems is calculated in order to anticipate any problems caused by the offending system and instigate relevant preventative measures.

The coupling between two systems depends on:

1. the power of the transmitter
2. the losses due to the cables and other units of both systems
3. the coupling between the antennas due to radiation
4. the sensitivity of the victim receiver.

The terms ‘isolation’ and ‘coupling’ are used synonymously to denote the power from an emitter that is received by a sensor. However, if the isolation is increased this is meant to indicate that the two are not so tightly coupled, that is, the coupling is reduced. Strong or tight coupling means that there is less isolation.

The power of the transmitter, losses and the sensitivity of the receiver can easily be obtained for the systems under consideration. However, the coupling between antennas is a more complex problem. The coupling between two antennas depends on:

1. their spatial separation – this is not the shortest distance between them, in the case of antennas on opposite sides of the fuselage
2. the absolute gains of the antennas at the angles at which the direct, reflected creeping and diffracted rays leave the transmit antenna and are incident on the receive antenna
3. the conductivities of the ground planes
4. the electrical resistance between their ground planes. In the case of a single structure like the airframe, the antennas share the same ground plane, which effectively reduces this resistance to zero for a perfect conductor, which in turn reduces the isolation between the antennas.

If we have the two antennas in free space without any obstacles the coupling between the two antennas would be via the direct ray alone as shown in Figure 6.9.

In the case of antennas on structures or in the vicinity of other obstacles, the second-order effects that we have to consider are the paths from antenna 1 to antenna 2, with each ray undergoing one specular reflection (first-order reflection), in the case shown in Figure 6.10.

The path length of one ray is $AR_1 + R_1B$ and the path length of the other ray is $AR_2 + R_2B$.

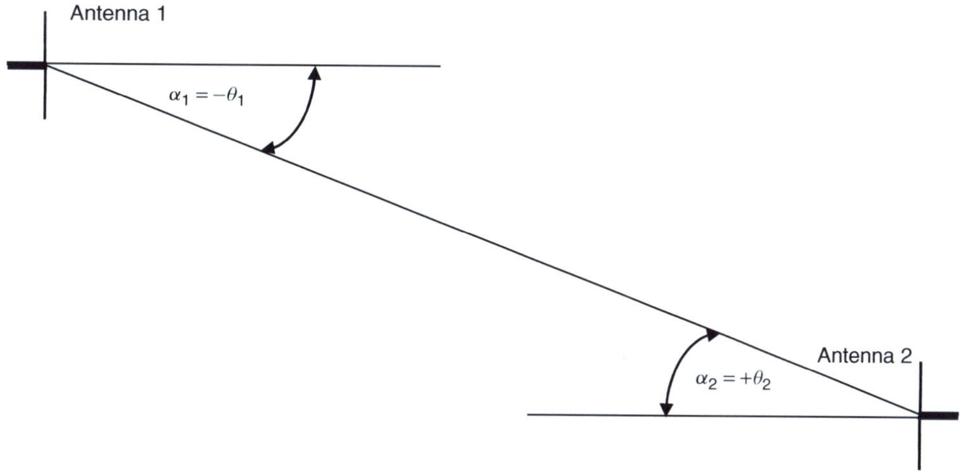

Figure 6.9 Antennas in free space.

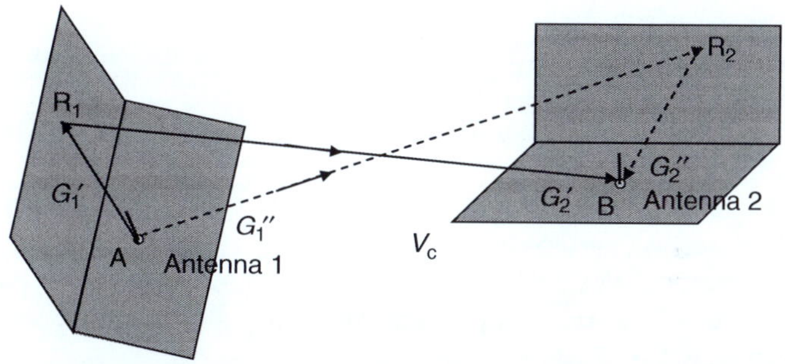

Figure 6.10 Antennas showing first-order reflections.

The level of the coupling from antenna 1 to antenna 2 depends on the gains of the individual antennas in the particular directions and the loss over the total path length, as well as the reflectivity of the surfaces at which the reflections occur.

For the ray reflected at the first structure we have to consider the gain of antenna A for ray AR_1 and the loss over the total path length $AR_1 + R_1B$, as well as the reflectivity of the surfaces at which the reflections occur. Thus for the AR_1 part of the ray we have to consider the gain of antenna 1 in the direction AR_1 (i.e. G_1'), and for the R_1B part of the ray we have to consider the reflectivity of the surface at point R_1 as well as the gain of antenna 2 in the direction R_1B (i.e. G_2'), as well as the path losses. Similarly, for the second ray AR_2 and R_2B we have to consider the path losses, different reflectivities and different gains G_1'', G_2''.

The second-order reflections that we have to consider are the paths from antenna 1 to antenna 2 with each ray undergoing two reflections (third-order effects), in the case shown in Figure 6.11.

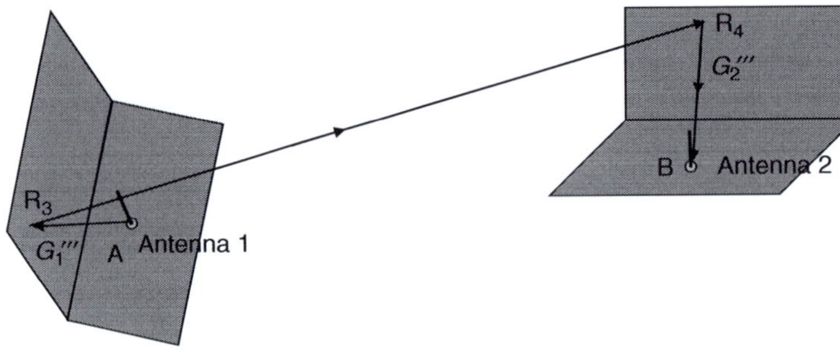

Figure 6.11 Antennas showing second-order reflections.

The level of the coupling from antenna 1 to antenna 2 depends on the gains of the individual antennas in the particular directions and the total path lengths, as well as the reflectivity of the surfaces at which the reflections occur. Thus we would have to

1. consider the gain of the antenna G_1''' in the direction AR_3,
2. allow for the reduced level of the ray (due to the path loss between A and R_3) when it strikes the first surface,
3. calculate the level of the ray (due to the reflectivity of the first surface) when it is reflected off the first surface
4. allow for the reduced level of the ray (due to the path loss between R_3 and R_4) when it strikes the second surface
5. calculate the level of the ray (due to the reflectivity of the second surface) when it is reflected off the second surface
6. allow for the reduced level of the ray (due to the path loss between R_4 and B)
7. calculate the level of the ray entering antenna 2 depending on the gain of antenna 2 (G_2''') in the direction of R_4B.

In addition to the specular reflections there is also the coupling due to diffraction, creeping waves and diffuse reflections. The levels of these higher-order interactions are given in Table 7.2 in Chapter 7. It can be seen that the specular reflections at flat surfaces are the highest levels (same as the direct wave), with creeping waves being −6 dB (compared with the direct wave 0 dB) and diffraction at straight edges being −10 dB (compared with the direct wave 0 dB).

For LOS antennas on the same side of the fuselage (i.e. both on the upper or both on the lower surface) the specular reflected wave from one antenna is unlikely to be incident on the second antenna. This is borne out by the fact that the calculated values of coupling (which do not allow for any reflections) are very close to the measured ones. In certain cases where parts of the airframe are at the same height as the fuselage (upper or lower surfaces) there could be specular reflections of waves from the first antenna that are incident on the second antenna.

The coupling is conveniently expressed in dB because it is a ratio, and thus the actual output power does not have to be known at the outset. The power transmitted by the transmit antenna can be added, when known, in dBm to get the actual power in dBm at the receive antenna. The same applies to the antenna gains (in dBi), which can be added to or subtracted from the coupling to obtain the received power (in dBm) at the terminals of the receive antenna.

The coupling is usually calculated (using separable variables) in three different ways, depending on whether the antennas are in LOS of each other or not.

6.6 Coupling between Systems with LOS Antennas

When we refer to the coupling between systems, we really mean the power from the transmitter of one system that enters the receiver of the other system. It is undesirable for the signals transmitted by one system to be received by a second system. This could result in malfunction of the second system, because of the interference with the desired signal that it should be receiving or transmitting.

We have seen in the previous section that in the case of LOS antennas on the same side of the fuselage, we usually only have to consider the direct wave from one antenna that is incident on the second antenna. No allowance is made for second-order effects such as any specular reflection. To obtain the coupling between an offending transmitter and a victim receiver we also have to consider the loss due to the RF cables and connectors, the transmitter power and receiver sensitivity. We have to consider each pair of systems in the same frequency band. In each band one would normally work out the coupling at the low, middle and high ends of the band. There is also what is known as out-of-band coupling (as described in Section 6.6.3) since the antennas can transmit/receive (albeit at a lower level) frequencies outside their specified operating band.

Additionally if any characteristic/parameter is not known, several values are used and the calculation carried out in each case. In order to accommodate all the calculations, a spreadsheet is usually set up.

If there are n possible systems for which the coupling is being calculated, then the number of combinations between each pair of systems, is ${}^{n}\mathrm{C}_2$ (pronounced as 'n C two', or 'n take two'), which is defined as:

$$ {}^{n}\mathrm{C}_2 = \frac{n!}{(n-2)!2!} \tag{6.5} $$

The exclamation mark '!' used in this way stands for factorial, and $5! = 5 \times 4 \times 3 \times 2 \times 1$.

If we have five systems, for example, the number of combinations is given by

$$ {}^{5}\mathrm{C}_2 = \frac{5!}{(5-2)!2!} = \frac{5 \times 4 \times 3 \times 2 \times 1}{(3 \times 2 \times 1) \times 2 \times 1} = 10. \tag{6.6} $$

For three frequencies in each band, the number of calculations increases to 30. Additionally three values of gain may be used for each antenna. This increases the number of calculations to 90.

The spreadsheet could be linked to a database, so that as the definitive values of the variables become available the calculations are updated.

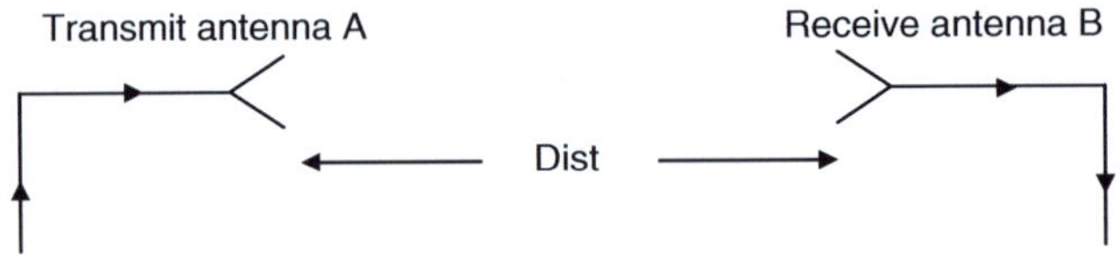

Figure 6.12 Coupling between two LOS antennas.

6.6.1 *Spatial Isolation*

The spatial isolation can be calculated using Friis' formula, which can be derived from first principles. Let us consider the two antennas depicted in Figure 6.12 at a distance R apart. Antenna A is transmitting and antenna B is receiving. It can be shown that the effective area A_r presented by a receiving antenna to a plane wave ([5], Chapter 6) is given by

$$A_r = \frac{G_r \lambda^2}{4\pi} \tag{6.7}$$

where

A_r is the effective area of the antenna in m^2
G_r is the linear (or numeric) gain of the antenna
λ is the wavelength of the radiation in m.

If the transmitting antenna has a gain of G_t and is radiating a power of P_t watts, then the power density P_d (the power per unit area) at a distance R from the receiving antenna, is given by

$$P_d = \frac{P_t G_t}{4\pi R^2}. \tag{6.8}$$

The power P_r received by antenna B, is the power density times the area A_r that it presents to the incident wave, and by combining Equations 6.7 and 6.8 we get

$$P_r = P_d A_r = \frac{P_t G_t}{4\pi R^2} \frac{G_r \lambda^2}{4\pi}, \tag{6.9}$$

This equation can be rewritten as

$$P_r = P_t G_t G_r \left(\frac{\lambda}{4\pi R} \right)^2. \tag{6.10}$$

It can also be written in dB as

$$10 \log_{10} P_r = 10 \log_{10} P_t + 10 \log_{10} G_t + 10 \log_{10} G_r + 20 \log_{10} \left(\frac{\lambda}{4\pi R} \right). \tag{6.11}$$

Equation 6.11 is known as Friis' transmission formula. The last term in Equation 6.11 is dependent on the wavelength of the propagating frequency and on the distance between the antennas. It is known as the space attenuation or loss factor L_e, since it is only dependent on the distance and wavelength. It is always negative in the far field ($R > \lambda/\pi$) for wire antennas or $> 2D^2/\lambda$ for aperture antennas (where D is the largest dimension of the antenna), but is usually written as a positive quantity since it is taken to represent a positive loss rather than a negative gain. If we just want to calculate the loss due to space, without taking the gains of the antennas, the power transmitted or the power received, then we just need to use the last term in Equation 6.11. This term can be split into two parts

$$20\log_{10}\left(\frac{\lambda}{4\pi x R}\right) = 20\log_{10}\left(\frac{\lambda}{R}\right) + 20\log_{10}\left(\frac{1}{4\pi}\right). \quad (6.12)$$

The last term on the right-hand side is equal to -21.98 dB and is usually approximated to -22 dB. If $R = 1\lambda$, then the first term on the right-hand side becomes zero. Thus the space loss between antennas placed λ apart is approximately 22 dB. If the antennas are two wavelengths apart (i.e. $\lambda/D = 1/2$), the first term on the right-hand side is 20 $\log_{10}(1/2)$ which equates to -6 dB. Thus the isolation between two antennas 2λ apart is 28 dB. Each time the distance apart is doubled the isolation is increased by 6 dB, so if the antennas are 4λ apart, the isolation is 34 dB. This formula can be easily used in the initial stages of the placement of antennas on the fuselage.

Figure 6.13 shows the variation of this space loss or attenuation L_s in dB, versus distance, with frequency as a parameter. If we look at the horizontal dotted line which represents a loss of 22 dB, we can see on the 100 MHz plot that the distance that gives us 22 dB loss is 3 m. This is the wavelength λ at 100 MHz. If the frequency is increased to 1 GHz, then to obtain a space loss of 22 dB the distance is reduced to 0.3 m or 30 cm, since this is the wavelength at 1 GHz. Similarly, at lower frequencies of 10 and 1 MHz, a space loss of 22 dB is obtained at the larger distances of 30 and 300 m, respectively.

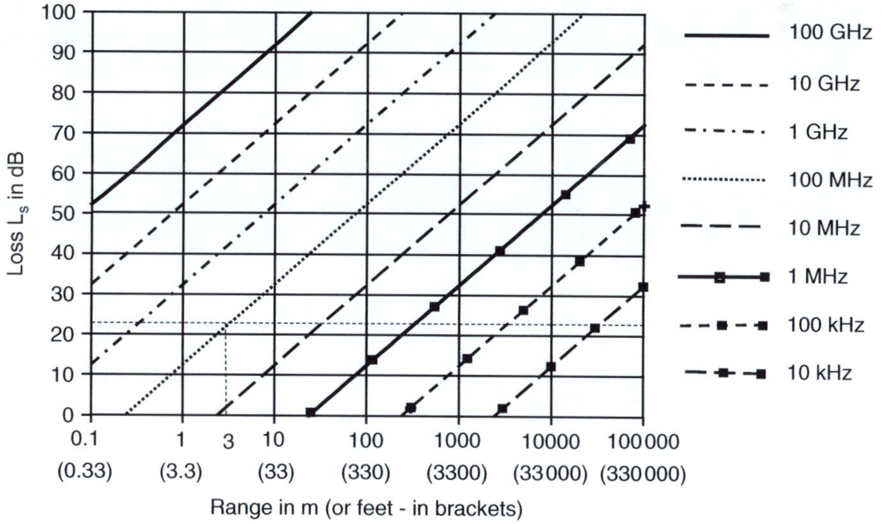

Figure 6.13 Variation of space attenuation with distance, with frequency as a parameter.

6.6.2 *Calculation of Coupling between Systems with LOS Antennas*

Instead of using the power from the transmitter in watts or milliwatts, it is more convenient to convert the power into dBm since this can then be added to the losses which are in dB. The sensitivity of the receiver is also usually given in dBm, so again it is a simple matter of adding this value to the losses. We must be very careful to use the correct signs, so that powers below 1 mW would give us negative dBm, whereas powers above 1 mW would give us positive dBm. Similarly, losses should be given negative signs and gains positive signs.

Consider two antennas on the upper fuselage of an aircraft as shown in Figure 6.14. The rear antenna is connected to the transmitting system and the forward antenna is connected to the receiving system. Suppose that the power transmitted by the first system is 10 mW (which is +10 dBm) and the receiver sensitivity of the receiving system is −84 dBm. The latter means that the receiver will not detect a signal from the transmitting system if it is below this level. The gain of the transmit antenna has to be known at the angle pointing towards the receive antenna. In Figure 6.14 the transmit antenna is aft of the receive antenna, so the absolute gain (dBi) has to be known at 0°.

The radiation patterns shown in Figure 6.15 were measured on a 1/15th scaled model of a Fokker 100 airframe. This work was undertaken as part of the EU IPAS research project Deliverable 16 [6]. From the radiation pattern of Figure 6.15a the gain in the forward (0°) direction is +5.5 dBi. In the case of the receive antenna the gain that we have to consider is that at 180°. The value obtained from the Figure 6.15b is −0.36 dBi.

A typical example of a schematic to aid the calculation of the coupling is shown in Figure 6.16. The cable losses are usually specified as losses per unit metre at each frequency. Thus suppose the cable loss at 120 MHz is stated as 0.4 dB/m and the first cable length to the break/join in the cable is 12.5 m, the total loss over this length is 0.4 dB/m times 12.5 m, which gives 5 dB. The back-to-back connector results in a loss of around 0.5 dB, and then a second cable length of 12.5 m to the antenna will result in a further 5 dB loss. Similarly for the receiving system, the cable length to the break is 7.5 m, resulting in a loss of 3 dB. Assuming that there are two breaks in the receiving system cables, the loss of two back-to-back connectors amounts to 1 dB, and there is a total further cable loss of 6 dB for both the other two 15 m lengths. The space loss at 120 MHz between the two antennas is 24.03 dB.

Note that although the power and sensitivity are given in dBm, these can be algebraically added to the losses that are in dB. Similarly, antenna gain in dBi can also be added. Thus the calculation of the coupling between the two systems shown in Figure 6.16 would be $+10 - 5 - 0.5 - 5 + 5.5 - 24.03 - 0.36 - 3 - 0.5 - 3 - 0.5 - 3 = -29.39$ dBm.

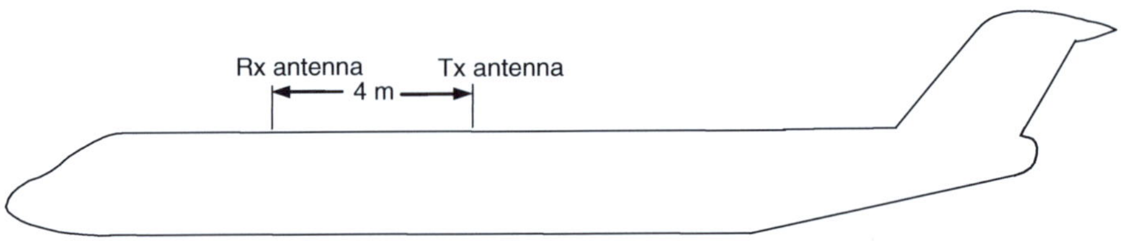

Figure 6.14 Antennas on the upper fuselage of an aircraft.

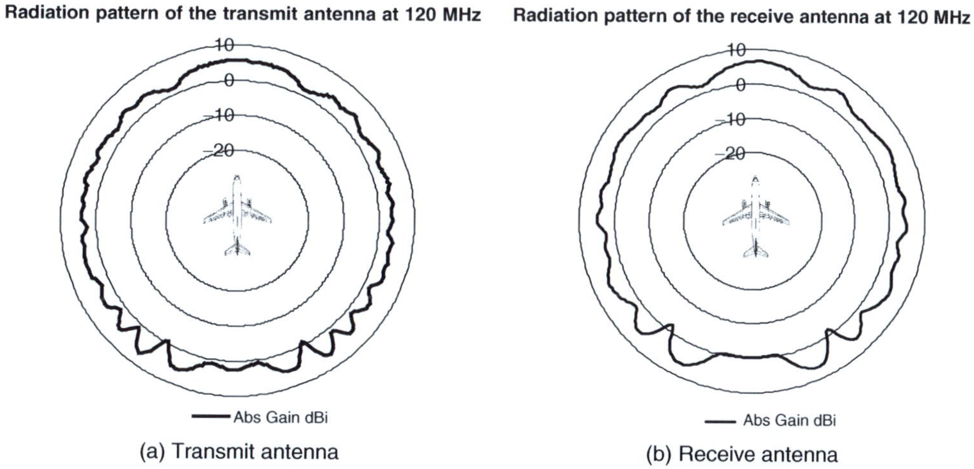

Figure 6.15 Azimuth radiation patterns of the transmit and receive antennas at 120 MHz.

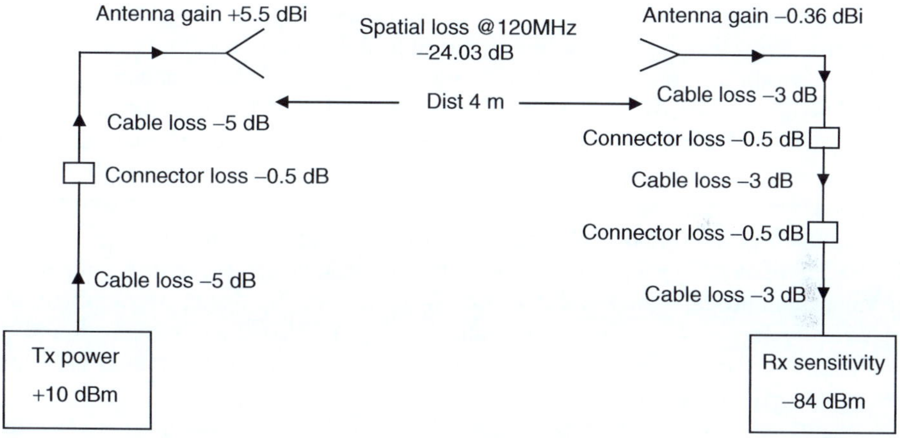

Figure 6.16 Losses between a transmitter of one system and the receiver of another system.

Since the sensitivity of the receiver is −84 dBm, the level of −29.39 dBm is above this level and so the receiving system would detect the transmission from the transmitting system. Thus some measures have to be taken to prevent interference between these two systems. These measures would most probably be in the tine domain, so that the receive system would be turned off when the transmitter of the first system is operating.

6.6.2.1 Comparison with Measurements

The coupling between antennas were obtained by measurements on antennas mounted on a 1/12th scaled cylinder with a diameter of 0.406 m, details of which are given in Chapter 4.

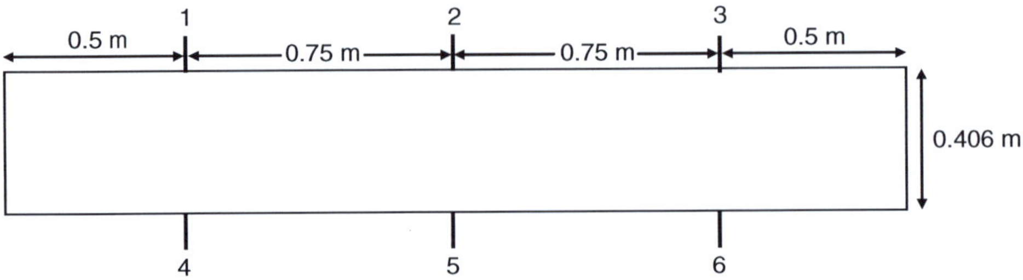

Figure 6.17 Schematic of the cylinder used for coupling measurements.

The frequency also needs to be scaled by the same factor, so the measurements were performed at 1458, 2916 and 13080 MHz, which are 12 times the respective equivalent aircraft frequencies of 121.5, 243 and 1090 MHz. Tuned scaled monopole antennas were installed in the positions shown in Figure 6.17. The coupling was measured between pairs of antennas. For the LOS antennas the coupling was measured between each combination of the antennas 1, 2 and 3 and between each combination of the antennas 4, 5 and 6.

The gains of each antenna in the direction of the other antenna was taken as 3.1 dBi at the lower two frequencies of 121.5 and 243 MHz, and 3.5 dBi at 1090 MHz. However, since the size of the ground plane affects the tilting of the beam, it is probable that antenna 1 looking towards antennas 2 and 3 has its gain at the horizon at a higher value than antenna 2 looking towards antenna 1. The same applies to the antennas on the lower surface. An increased gain would result in less isolation (i.e. greater coupling) between the two antennas and correlate more closely with the measured values especially in the case of the coupling between antennas 1 and 3. In addition, no allowance has been made for the common ground plane, which would also decrease the isolation.

The values of coupling calculated using Friis' formula are compared with the measured values and are tabulated in Table 6.2. At the highest frequency where the antennas are electrically further away from each other, it appears that the common ground plane does not have a marked effect, which is manifest in the smaller differences between the measured and calculated values.

The results for the 121.5 MHz measured and calculated coupling are plotted in Figure 6.18. It can be seen that the antennas that are 1.5 m apart (i.e. 1 and 3, and 4 and 6) show the greatest variations from the measured values with the maximum excursion being about 2.5 dB.

The same trend as for the 121.5 MHz plots is followed for the 243 MHz plots as shown in Figure 6.19, with the maximum difference being about 3 dB.

For the highest frequency of 1090 MHz, the calculated values correlate much better with the measured ones, as shown in Figure 6.20. As explained above, this is possibly caused by the reduced influence of the common ground plane due to the larger electrical distance as well as the fact that the increase in gain due to the larger ground plane is not as marked as it at the lower frequencies.

Table 6.2 Comparison between the measured and calculated values of coupling between LOS antennas.

Between antennas	121.5 MHz			243 MHz			1090 MHz		
	Measured values	Friis coupling	Difference	Measured values	Friis coupling	Difference	Measured values	Friis coupling	Difference
1 and 2	−26.28	−27.03	−0.75	−31.30	−33.05	−1.75	−46.66	−45.29	1.37
1 and 3	−30.57	−33.05	−2.49	−36.52	−39.08	−2.56	−51.55	−51.31	0.23
2 and 3	−26.85	−27.03	−0.18	−31.89	−33.05	−1.16	−46.30	−45.29	1.01
4 and 5	−26.18	−27.03	−0.86	−30.89	−33.05	−2.16	−45.53	−45.29	0.24
4 and 6	−31.28	−33.05	−1.78	−35.93	−39.08	−3.15	−49.86	−51.31	−1.45
5 and 6	−26.76	−27.03	−0.27	−31.67	−33.05	−1.38	−45.04	−45.29	−0.25

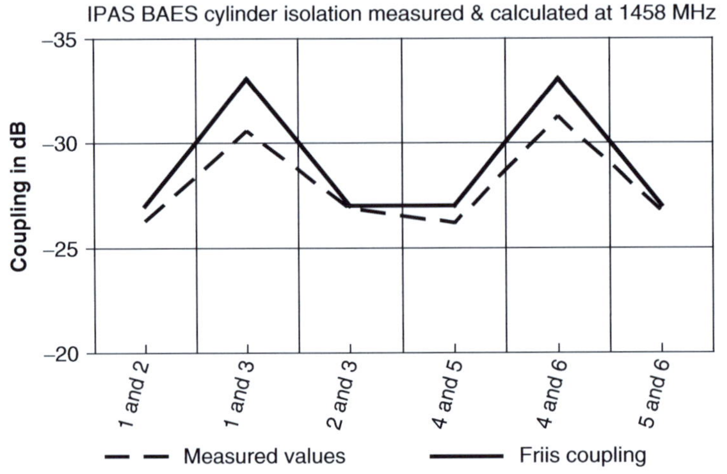

Figure 6.18 Comparison between the measured and calculated coupling between LOS antennas at the aircraft/scaled frequency of 121.5 MHz.

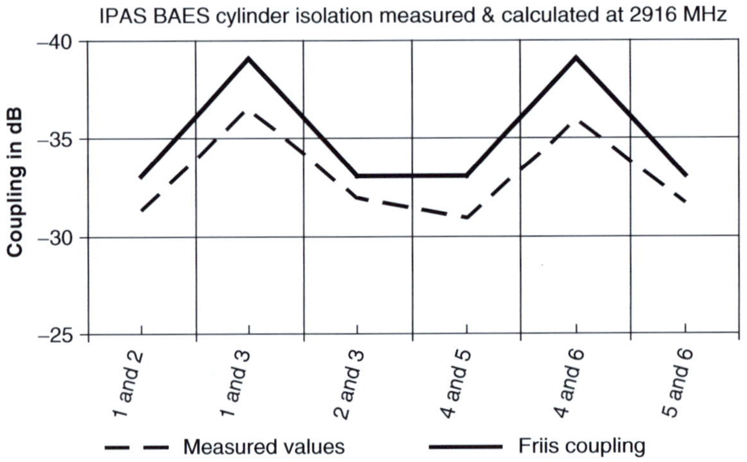

Figure 6.19 Comparison between the measured and calculated coupling between LOS antennas at the aircraft/scaled frequency of 243 MHz.

6.6.3 Out-of-Band Coupling

This is a misnomer since if the frequencies of the two systems are not in the same frequency band, there would be no coupling between the systems. Out-of-band coupling refers to coupling outside the frequency band of the main carrier or fundamental band of the transmitter. In Section 6.4 we have seen how a carrier wave that is unmodulated does not carry any information. When a wave is modulated it has a frequency spectrum outside its carrier frequency band. These are the out-of-band frequencies that are referred

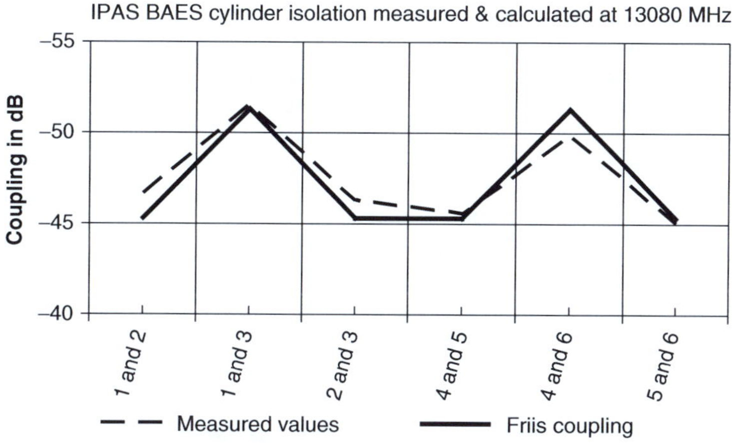

Figure 6.20 Comparison between the measured and calculated coupling between LOS antennas at the aircraft/scaled frequency of 1090 MHz.

to. However, when measuring the out-of-band coupling the transmitted frequencies are in band (they are the carrier frequency band) for the whole of the transmitter system, whereas the receive antenna is outside this band of frequencies.

In the IPAS Deliverable 8 [7] out-of-band coupling measurements were performed on antennas on the same surface of the cylinder, because the coupling between antennas on opposite sides and out of band were expected to be below the noise floor of the receiver in the measurement set-up.

The frequency ranges are referred to as low (centred around 1458 MHz), mid (centred around 2.916 GHz) and upper (centred around 13.08 GHz). These refer to aircraft centre frequencies of 121.5, 243 and 1090 MHz, respectively.

The seven coupling runs are shown in Table 6.3. The comparison between the in-band coupling (low band to low band) and out-of-band coupling for runs 1 and 2 (low to mid band antennas 1 to 2 and 1 to 3) is shown in Figure 6.21. It can be seen that the differential coupling between antennas 1 and 2 and between antennas 1 and 3 is maintained in both the in-band and out-of-band plots. In the case of the out-of-band isolation the coupling is greater (isolation less) as the frequency approaches the tuned frequency (2916 MHz) of the mid-band antenna. This is to be expected since the gain of the mid-band antenna is greater near its tuned frequency and hence the isolation is reduced. This is also apparent in the in-band plots, where there is a dip around the tuned frequency (1458 MHz) of the low-band antennas.

As expected, the coupling in the in-band as well as the out-of-band plots between antennas 1 and 3 is 6 dB less (higher isolation) than that between 1 and 2, because the separation between 1 and 3 is double that between antennas 1 and 2.

The comparison between the low in-band and the low to high out-of-band coupling is shown in Figure 6.22. It can be seen that the out-of-band provides isolation that is around 35 dB more than the in-band figures. This is to be expected since the gains of the antennas are greatly reduced at these high out-of-band frequencies. Note also the greater number of ripples due to the higher frequencies as expected.

Table 6.3 Out-of-band coupling runs.

	Transmit antenna number	Tuned to frequency range	Transmit frequency range	Receive antenna number	Tuned to frequency range
Run No. 1	1	Low	Low	2	Mid
Run No. 2	1	Low	Low	3	Mid
Run No. 3	2	Low	Low	3	Mid
Run No. 4	1	Low	Low	2	Upper
Run No. 5	1	Low	Low	3	Upper
Run No. 6	1	Mid	Mid	2	Upper
Run No. 7	1	Mid	Mid	3	Upper

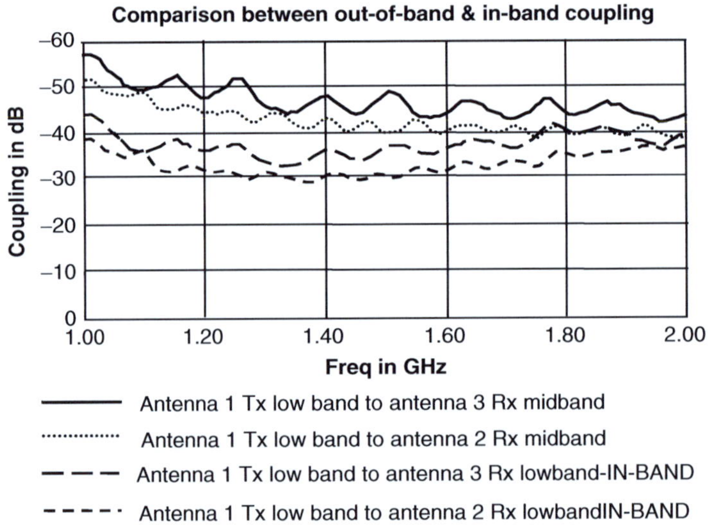

Figure 6.21 Out-of-band coupling between the antennas on the cylinder.

6.7 Coupling between Systems for Antennas on Opposite Surfaces of the Fuselage

Measurements on aircraft have shown that the calculated values are very erratic at the higher frequencies, sometimes lower than the measured frequencies and sometimes higher. The difference can be as much as 23 dB and could be either positive or negative, that is, sometimes the measured values are 23 dB worse and sometimes they are 23 dB better. This clearly means that the variation is up to 46 dB – that is, approximately 40 000 in linear terms.

The coupling between antennas on opposite sides of the fuselage (i.e. one on top and one on bottom) is determined by:

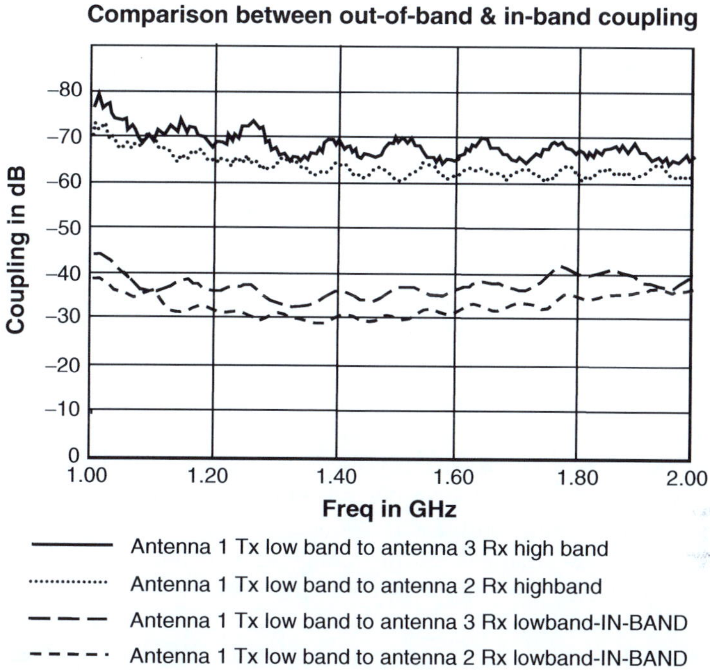

Figure 6.22 Comparison between the in-band and out-of-band (low to high band) coupling between the antennas on the cylinder.

1. the direct wave entering the receive antenna from the transmit antenna – this would not occur for antennas that are not in LOS
2. the reflected power(s) of one or more orders of reflections, entering the receive antenna from the transmit antenna
3. the power from the creeping wave(s) entering the receive antenna from the transmit antenna
4. the power from direct or reflected wave(s) diffracted at edges, and so on, entering the receive antenna from the transmit antenna
5. the power from diffracted wave(s) that is/are subsequently diffracted at edges, and so on, entering the receive antenna from the transmit antenna.

In the initial stages of design of an antenna layout, the positions of antennas are continually changed and it is not economically viable to perform any computational modelling to determine the coupling. Thus the coupling is estimated by using simple calculations that do not take into account any second- or higher-order effects such as reflections and diffraction. This could increase or decrease the coupling between antennas depending on the vectorial combinations of the interactions.

The formulas used also assume that the aircraft is a right circular cylinder of infinite length without wings, tailfin, tailplane, etc. In general, the exclusion of sections like the wings will cause the calculations to err on the side of caution, since the wings provide additional shading. However, it is recognized that there may be instances of vectorial combination that cause enhancement, and hence could increase the coupling (decrease the isolation).

6.7.1 Problems with Estimating Coupling at the Higher Frequencies

In order to verify correlation between measurements and calculations it is important to measure the coupling on the correct diameter of fuselage. For the IPAS project instead of using a fuselage with a large diameter, the measurements were performed on a cylinder which, when scaled up by a factor of 12, would approximate that of a single-aisle aircraft. Thus the number and spacing of the ripples obtained in the roll plane are nearly the same as could be expected on a fuselage of this diameter. The antennas were mounted on a cylinder with a diameter of 0.406 m. Details of the cylinder are given in Chapter 4 and the antenna positions are shown in Figure 6.23 together with the angles subtended by the antennas.

The coupling between antennas was measured and compared with the calculated values using two different formulas that were derived empirically but based on measurements performed on cylinders, with rudimentary 'wings' in the case of the Bull and Smithers derivation. At the lower frequencies the coupling calculations correlate fairly well with measurements but at the higher frequencies (around 1 GHz) the correlation is very poor.

The problems with estimating coupling at the higher frequencies arise from at least the following:

1. The change of levels with frequency is more rapid,
2. The change of levels with position is more rapid.

6.7.1.1 Change in Coupling with Frequency

The rapid changes with frequency (as the frequency is increased) can be seen by comparing graphs shown in Figure 6.23 of the measured coupling over two frequency bands. The low frequency band of 1–2 GHz corresponds to aircraft frequencies of approximately 83–166 MHz, and the high band of 12.5–13.5 GHz corresponds to aircraft frequencies of approximately 1.04–2.08 GHz. The low band coupling is smoother with less variation and

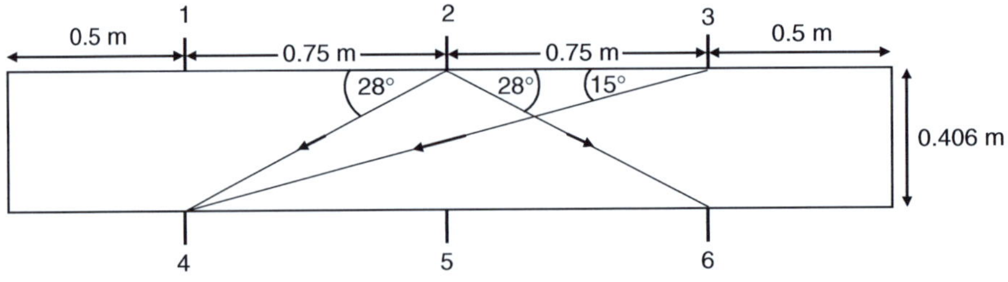

Figure 6.23 The schematic of the cylinder used for coupling measurements showing the angles subtended by the antennas.

the difference between a trough and peak is much smaller, resulting in less pronounced 'spikes'. Thus a change in frequency over the low band will not result in a large change in coupling. Since most signals transmitted are modulated there are a number of frequencies (in addition to the fundamental frequency) being transmitted at any instant in time. This means that the coupling between the antennas is also changing more rapidly over the high band than it does over the low band.

6.7.1.2 Change in Coupling with Position

The change of gain (level) with position (as the frequency is increased) can be seen by comparing the left-hand graph of Figure 6.24 with the right-hand one. The measured roll plane radiation plots were obtained on a aircraft for an antenna placed on the upper fuselage, at the equivalent aircraft frequencies of 120 and 1093 MHz. It should also be noted that the number of ripples depends on the distance traversed by the waves. Thus if the fuselage diameter (circumference) is large there are more ripples, as explained in Chapter 4. We can see that at the low frequency (where the electrical circumference of the fuselage is small) the pattern is much smoother than that obtained at the higher frequency. There is a single null at the top and bottom where the antennas are likely to be placed (diametrically opposite each other). Thus a change in position at the low frequency will not result in a large change in coupling. In the case of the higher frequency plot, there are rapid variations with small positional changes. Thus computational modelling is not adequate, unless it is performed at a number at positions and the arithmetic mean taken. The exact position of the antenna when it is installed on the airframe cannot be guaranteed to be the same as the modelled position, and furthermore the computations assume that the antenna is a point source and that the phase centre of the radiation coincides with the mechanical centre of the antenna footprint, that is, at the centre of the antenna baseplate in the case of monopoles/blades. Airframes also have ribs and frames (see Figure 6.1) that may necessitate changes in the actual locations of the antennas since it is preferable to avoid puncturing a frame or rib. The actual positions of the ribs and frames may also vary across the fleet of a particular aircraft, resulting in different positions relative to the datum. For these reasons predicting the coupling and ensuring that the antennas will be installed at the specified locations is a problem.

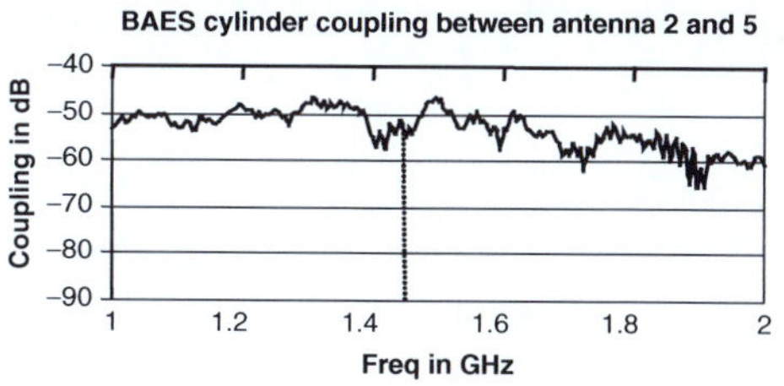

(a) Coupling at low band A/C 83 to 166 MHz

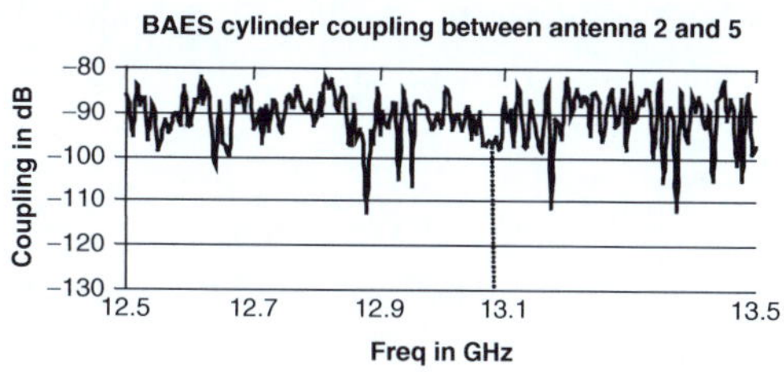

(b) Coupling at high band A/C 1.04 to 2.08 GHz

Figure 6.24 Comparison of the variation with frequency of the coupling at low and high frequencies on a cylinder. Note that the frequencies used are 12× scaled, so the aircraft centre frequencies are 121.5 MHz and 1090 MHz.

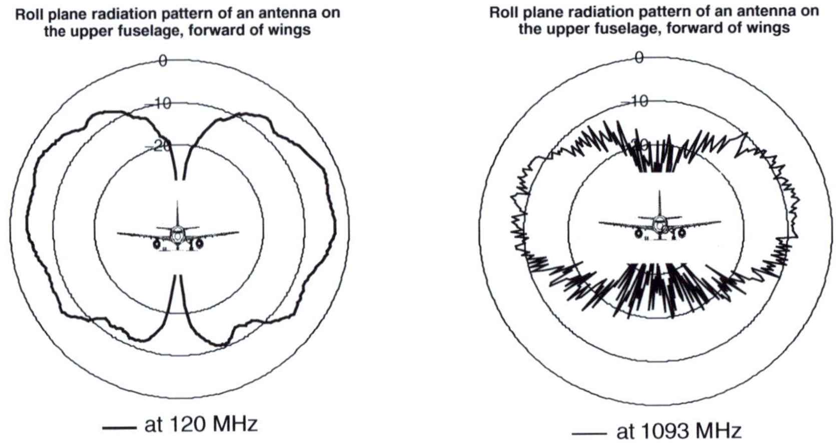

Figure 6.25 Comparison of the variation with position of the gain at low and high frequencies in the roll plane. Note that the frequencies used are 15x scaled, so the aircraft frequencies are 120 MHz and 1093 MHz.

6.8 Existing Formulas Used for Calculating Coupling between Two Antennas on Opposite Surfaces

There are two commonly used formulas in the calculation of coupling between two antennas on opposite surfaces of the fuselage:

1. Bull and Smithers
2. simple diffraction.

For antennas on opposite sides (upper and lower fuselage) these formulas are applied to antennas on a right circular cylinder of infinite length. The nose, tail, wings, tailplane, tail fin, and so on, are not included. Both Bull and Smithers [8] and the simple diffraction formulas are adaptations of Friis' formula with an empirical element which takes into account measured data.

In both formulas no allowance is made for the angle subtended at one antenna by the other. Since the gain of a monopole antenna is dependent on the angle, this an important consideration when calculating the coupling.

6.8.1 Bull and Smithers

Bull and Smithers [8] measured the coupling between two antennas on opposite sides of the circumference of a right circular cylinder as shown in Figure 6.25, and then derived an empirical formula based on Friis' free space formula. The coupling between two antennas on opposite sides of the cylinder is given by

$$-20\log_{10} kF^{1.75} - 20\log_{10}(D+L) - 28 + \mathrm{SA}, \tag{6.13}$$

where
$k = r/15$ and r is the radius of the cylinder in m
F is the frequency in MHz
D is the distance along the circumference of the cylinder in m
L is the distance along the length of the cylinder in m
SA is the Siarkiewicz and Adams factor that allows for the gains of the antennas.

When the upper and lower antennas are both on the centreline, the distance D is half of the circumference.

It should be noted that $20\log_{10}(D + L)$ is $(D + L)^2$ in linear terms, and is not the same as $D^2 + L^2$ which is used in the simple diffraction formula in Section 6.8.2.

The gains used for the antennas in this formula depend on the length on the antennas compared with the lengths of tuned antennas (i.e. quarter-wave monopoles) at the fundamental frequency.

Since the antennas used on the scaled model are the scaled quarter-wave tuned antennas, no (Siarkiewicz and Adams) correction factors were used in the calculations.

6.8.1.1 Comparison with Measurements

The values of coupling calculated using the Bull and Smithers formula are compared with the measured values and are tabulated in Table 6.4. It can be seen that the calculated values do not follow the same trend as the measured values, in that they are sometimes less and sometimes higher than the measured values. The differences between the measured and calculated values vary between +11 and −10.89 dB, a total variation of nearly 22 dB.

The results are plotted in Figure 6.26 for the 121.5 MHz coupling measured and calculated using the Bull and Smithers formula. The fact that the trend of the measured values is not followed is clearly demonstrated in the plots.

When the frequency is doubled to 243 MHz the coupling calculated using the Bull and Smithers formula shows a steady increase of around 10 dB for all the relative positions of the antenna pairs (Figure 6.27). Again the coupling values do not follow the same trend as the measured ones.

At the highest frequency of 1090 MHz the coupling calculated using the Bull and Smithers formula shows a steady increase of around 23 dB for all the relative positions of the antenna pairs (Figure 6.28), and again these values do not follow the same trend as the measured ones.

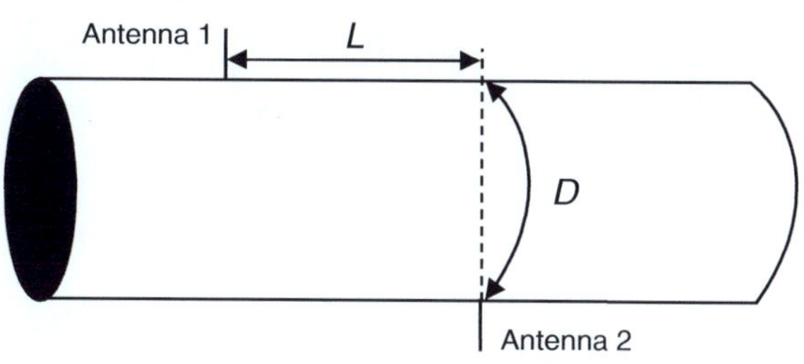

Figure 6.26 Bull and Smithers coupling parameters for antennas on opposite sides of a cylinder.

Table 6.4 Comparison between the measured and calculated values of coupling between antennas on opposite sides of the cylinder using the Bull and Smithers formula.

Between antennas	121.5 MHz			243 MHz			1090 MHz		
	Measured values	Bull and Smithers	Difference	Measured values	Bull and Smithers	Difference	Measured values	Bull and Smithers	Difference
1 and 4	−46.24	−41.45	4.79	−60.14	−51.99	8.15	−74.06	−74.80	−0.74
1 and 5	−46.64	−48.21	−1.56	−57.74	−58.74	−1.00	−78.36	−81.56	−3.20
1 and 6	−42.36	−51.96	−9.60	−55.11	−62.49	−7.38	−77.08	−85.31	−8.23
2 and 4	−46.16	−48.21	−2.04	−56.74	−58.74	−2.00	−73.16	−81.56	−8.40
2 and 5	−49.62	−41.45	8.17	−57.32	−51.99	5.33	−85.80	−74.80	11.00
2 and 6	−46.03	−48.21	−2.18	−54.59	−58.74	−4.15	−82.50	−81.56	0.94
3 and 4	−42.55	−51.96	−9.41	−57.18	−62.49	−5.31	−74.42	−85.31	−10.89
3 and 5	−46.39	−48.21	−1.81	−56.33	−58.74	−2.41	−83.20	−81.56	1.64
3 and 6	−42.46	−41.45	1.01	−58.22	−51.99	6.24	−79.67	−74.80	4.87

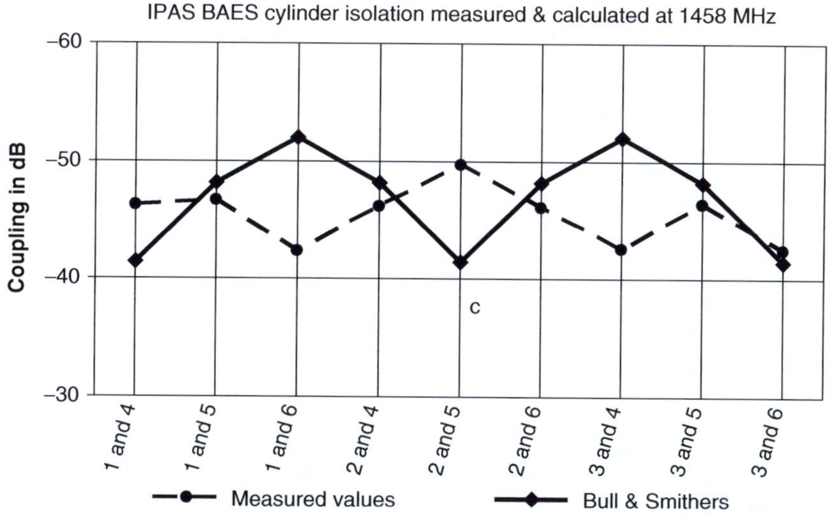

Figure 6.27 Coupling measured between the pairs of antennas on the opposite sides of the cylinder and that calculated using the Bull and Smithers formula at the aircraft/scaled frequency of 121.5 MHz.

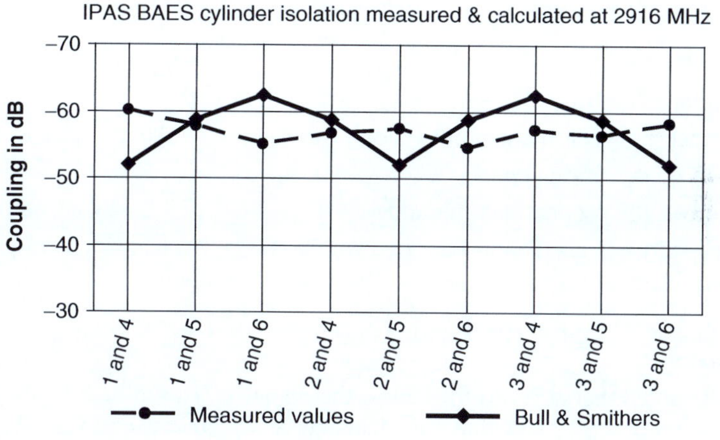

Figure 6.28 Coupling measured between the pairs of antennas on the opposite sides of the cylinder and that calculated using the Bull and Smithers formula at the aircraft/scaled frequency of 243 MHz.

6.8.2 Simple Diffraction

This formula considers the antennas on opposite sides of the curved surface of a right circular cylinder of infinite length as shown in Figure 6.29a. The other parts of the aircraft, such as the wings, tail, tail fin, and so on, are not taken into account.

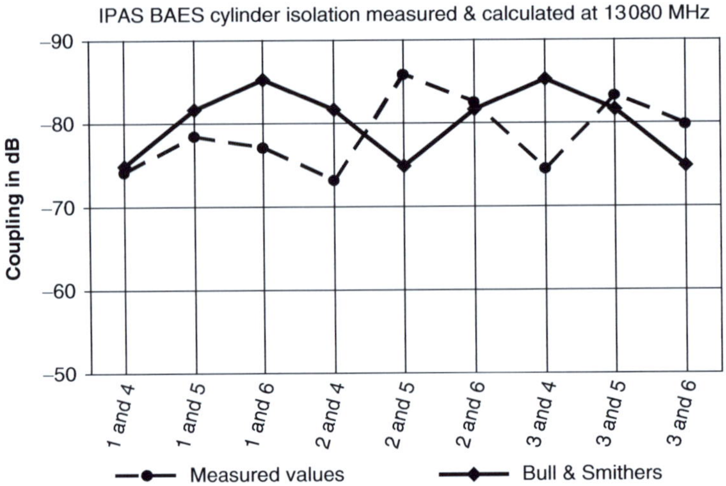

Figure 6.29 Coupling measured between the pairs of antennas on the opposite sides of the cylinder and that calculated using the Bull and Smithers formula at the aircraft/scaled frequency of 1090 MHz.

The coupling in dB is given by

$$20\log_{10}\left(\frac{\lambda}{4\pi R}\right) + 10\log_{10} G_1 + 10\log_{10} G_2 - \text{SF}, \tag{6.14}$$

where

λ is the wavelength in m

R is the distance between the antennas along the surface of the cylinder in m

G_1 is the gain of the first antenna

G_2 is the gain of the second antenna and

$\text{SF} = A/(\eta A + \xi)$ is a correction factor in which $A = r\theta^2\sqrt{(2\pi/\lambda)}$, with r the radius of the cylinder in m and θ the angular separation between antennas in radians.

If $A \geq 26$, then $\eta = 3.34 \times 10{-}3$ and $\xi = 0.5621$. If $A < 26$, then $\eta = 5.478 \times 10^{-3}$, and $\xi = 0.5083$.

It should be noted that $r\theta$ is equivalent to the distance D used in the Bull and Smithers formula in Section 6.8.1, and that the distance R is equivalent to the square root of $D^2 + L^2$ as opposed to the distance $(D + L)$ used in the Bull and Smithers formula.

In the case of simple diffraction the gains to be used are not specified, that is, it is not clear if they should be the arithmetic mean of the gains over the whole sphere, the peak gains, the gains at the horizon or the estimated gains at the angles subtended by each antenna at the other, as shown in Figure 6.30b. The gain values used in the calculations were 3.1 dBi for each antenna.

6.8.2.1 Comparison with Measurements

The values of coupling calculated using the simple diffraction formula are compared with the measured values and displayed in Table 6.5. At the lowest frequency the calculated

Table 6.5 Comparison between the measured and calculated values of coupling between antennas on opposite sides of the cylinder using the simple diffraction formula.

	121.5 MHz			243 MHz			1090 MHz		
	Measured values	Simple diffraction	Difference	Measured values	Simple diffraction	Difference	Measured values	Simple diffraction	Difference
1 and 4	−46.24	−49.32	−3.07	−60.14	−63.45	−3.31	−74.06	−103.87	−29.81
1 and 5	−46.64	−48.96	−2.32	−57.74	−61.91	−4.17	−78.36	−98.00	−22.75
1 and 6	−42.36	−49.35	−6.99	−55.11	−61.07	−5.96	−77.08	−92.81	−15.73
2 and 4	−46.16	−48.96	−2.80	−56.74	−61.91	−5.17	−73.16	−98.01	−24.85
2 and 5	−49.62	−49.32	0.30	−57.32	−63.45	−6.13	−85.80	−103.87	−18.07
2 and 6	−46.03	−48.96	−2.93	−54.59	−61.91	−7.32	−82.50	−98.00	−15.50
3 and 4	−42.55	−49.35	−6.80	−57.18	−61.07	−3.89	−74.42	−92.81	−18.40
3 and 5	−46.39	−48.96	−2.57	−56.33	−61.91	−5.58	−83.20	−98.00	−14.80
3 and 6	−42.462	−49.32	−6.86	−58.22	−63.45	−5.22	−79.67	−103.87	−24.20

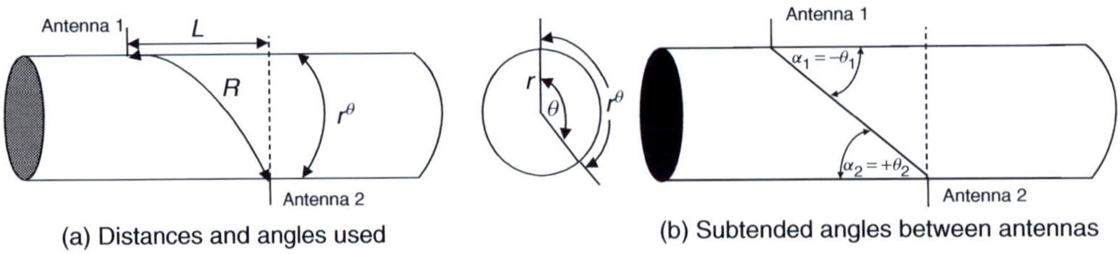

Figure 6.30 Simple diffraction coupling parameters for antennas on opposite sides of a cylinder.

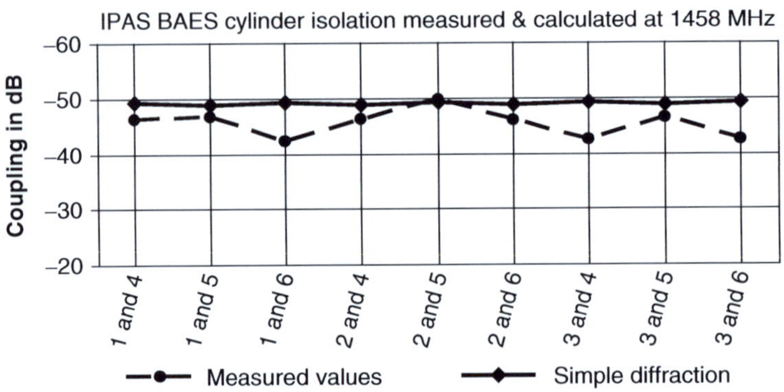

Figure 6.31 Coupling measured between the pairs of antennas on the opposite sides of the cylinder and those calculated using the simple diffraction formula at the aircraft/scaled frequency of 121.5 MHz.

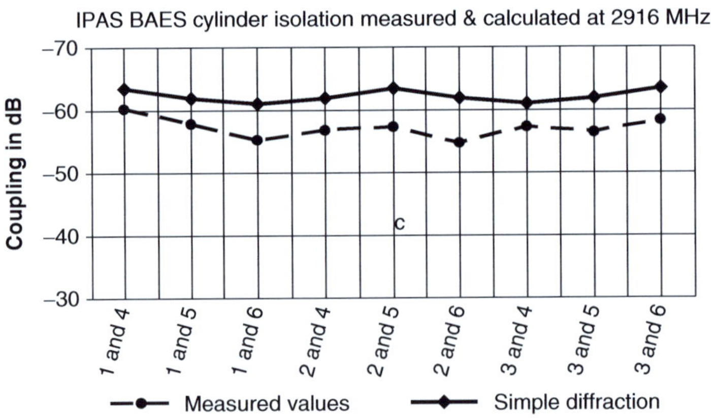

Figure 6.32 Coupling measured between the pairs of antennas on the opposite sides of the cylinder and those calculated using the simple diffraction formula at the aircraft/scaled frequency of 243 MHz.

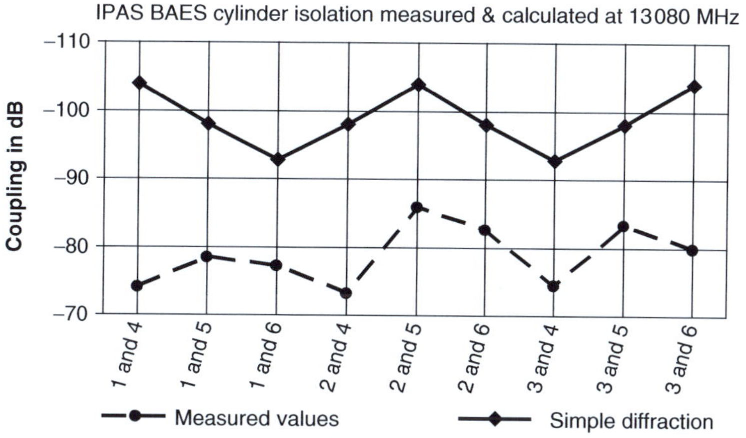

Figure 6.33 Coupling measured between the pairs of antennas on the opposite sides of the cylinder and those calculated using the simple diffraction formula at the aircraft/scaled frequency of 1090 MHz.

values are within 7 dB of the measured ones although they show better isolation than the measured ones. It can be seen that at the two higher frequencies, although the calculated values follow the same trend as the measured values, the differences between the measured and calculated values vary by as much as −29.91 dB, and again they are very optimistic in that they show a much greater isolation than the measured ones.

The graphs of Figures 6.30–6.32 show the comparison between the measured and calculated coupling.

6.9 Derivation of an Empirical Formula that Correlates with the Measured Data

It can be seen that the Bull and Smithers formula gives values nearer the measurements, but the simple diffraction formula follows the measured trend more closely. Neither of these formulas take account of the variation of gain with the subtended angle between the antennas or the variation of gain with frequency. The former is more likely to result in a larger variation than the latter.

Under the IPAS project Deliverable 8 [7] a number of adaptations of the Bull and Smithers, simple diffraction and Friis formulas were used to obtain a better fit to the measurements on the cylinder as well as measurements of coupling between antennas on the Fokker 100 scaled model. The formulas were used with different distances between the antennas such as shortest distance (through the cylinder), Bull and Smithers distance and the simple diffraction distance. Different values of gain were also used depending on the frequency as well as on the angles subtended between the antennas. It can be seen from Figure 6.33 that the angle subtended between antennas 1 and 6, as well as between antennas 3 and 4, is 15°, between antennas 1 and 5, 2 and 4, 3 and 5, or 2 and 6, is 28°, and between the diametrically opposite pairs is 90°.

Table 6.6 Values of gain (for the best fit) at the different frequencies and angles subtended between the pairs of antennas.

Subtended angle in degrees	90			28			15		
Frequency in GHz	1.458	2.916	13.08	1.458	2.916	13.08	1.458	2.916	13.08
Gain in dBi	−9	−12.5	−15	−6	−8.5	−12	−2	−5	−9

The best fit was obtained by using the Friis formula with shortest distance (through the cylinder), and the gain values shown in Table 6.6. These values of gain are more realistic since they decrease with frequency as would be expected. The gains were also selected to give the best fit for the different subtended angles.

The calculated values obtained using the gains (shown in Table 6.6) are compared to the measured ones on the cylinder in Table 6.7. It can be seen that at the lowest frequency the difference between the measured and calculated values is very small with the largest difference being +3.72 dB, that is, it shows that the calculated value has less isolation than the measured one. The next largest deviation is −3.44 dB. Both of these occur for diametrically opposite antennas – that is, coupling between 2 and 5, and between 3 and 6.

At the middle frequency of 243 MHz the calculated value has 5.82 dB less isolation than the measured one.

At the highest frequency of 1090 MHz the calculated value has 8.84 dB less isolation than the measured one, and again this occurs for diametrically opposite antennas – that is, coupling between 2 and 5.

At all three frequencies the largest difference is positive, that is, the calculated values err on the side of caution.

The graphs showing the comparison between measured and calculated coupling for the three frequencies are shown in Figures 6.34–6.36.

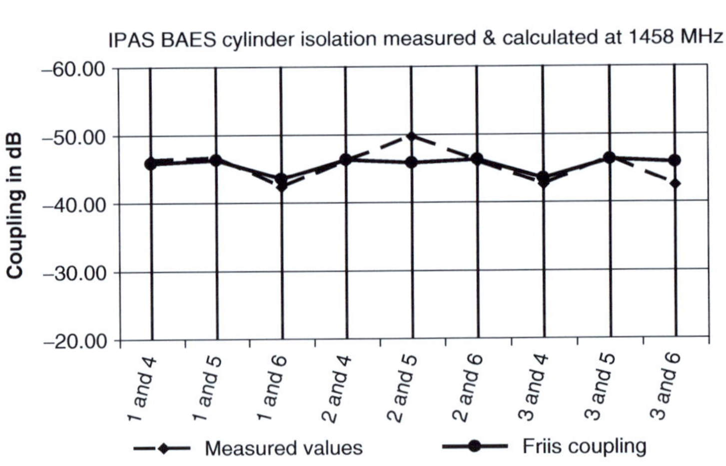

Figure 6.34 Coupling measured between the pairs of antennas on the opposite sides of the cylinder and those calculated using the Friis formula and the gains shown in Table 6.6 at the aircraft/scaled frequency of 121.5 MHz.

Table 6.7 Comparison between the coupling obtained using Friis' formula with gains that vary with frequency and angles between the antennas.

	121.5 MHz			243 MHz			1090 MHz		
Antennas	Measured values	Friis coupling	Difference	Measured values	Friis coupling	Difference	Measured values	Friis coupling	Difference
1 and 4	−46.24	−45.90	0.34	−60.14	−58.92	1.21	−74.06	−76.96	−2.90
1 and 5	−46.64	−46.35	0.29	−57.74	−57.37	0.37	−78.36	−77.41	0.95
1 and 6	−42.36	−43.56	−1.20	−55.11	−55.58	−0.47	−77.08	−76.62	0.46
2 and 4	−46.16	−46.35	−0.19	−56.74	−50.92	5.82	−73.16	−76.29	−3.13
2 and 5	−49.62	−45.90	3.72	−57.32	−58.92	−1.61	−85.80	−76.96	8.84
2 and 6	−46.03	−46.35	−0.32	−54.59	−57.37	−2.78	−82.50	−77.41	5.09
3 and 4	−42.55	−43.56	−1.01	−57.18	−55.58	1.60	−74.42	−76.62	−2.20
3 and 5	−46.39	−46.35	0.04	−56.33	−57.37	−1.04	−83.20	−77.41	5.79
3 and 6	−42.46	−45.90	−3.44	−58.22	−58.92	−0.70	−79.67	−76.96	2.71

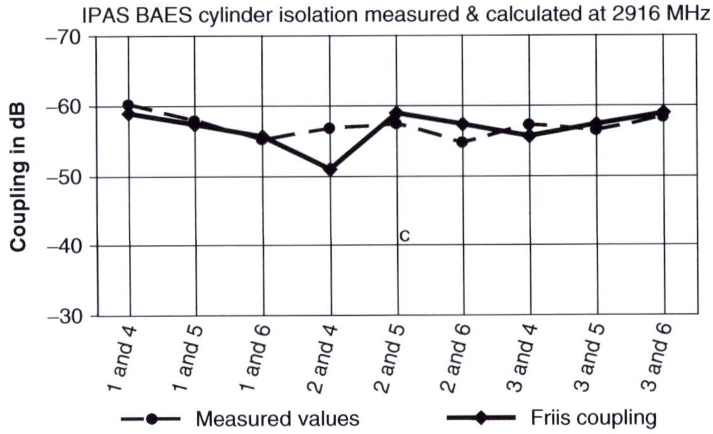

Figure 6.35 Coupling measured between the pairs of antennas on the opposite sides of the cylinder and those calculated using the Friis formula and the gains shown in Table 6.6 at the aircraft/scaled frequency of 243 MHz.

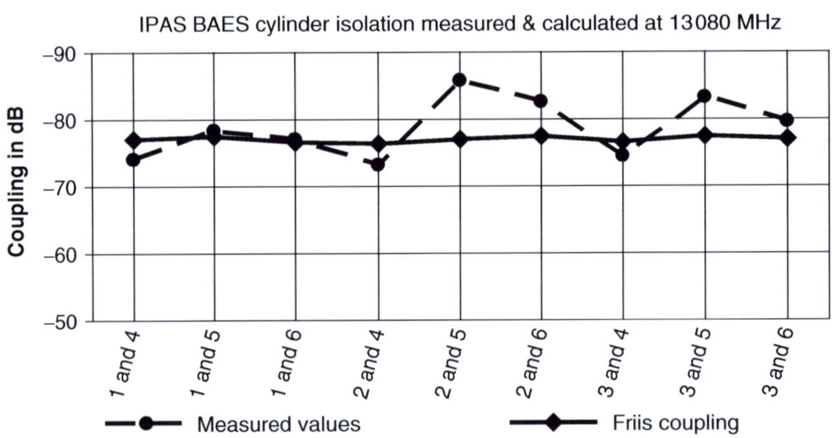

Figure 6.36 Coupling measured between the pairs of antennas on the opposite sides of the cylinder and those calculated using the Friis formula and the gains shown in Table 6.6 at the aircraft/scaled frequency of 1090 MHz.

These values were also used for calculating the coupling between pairs of antennas on the Fokker 100 as shown in Figure 6.37. Only the couplings between antennas on the top and bottom of the aircraft are compared with calculated values using Friis' formula for the distance between antennas and Friis' formula for the spatial isolation. The NLR scaled antenna as described in Section 7.1.5 was also installed in position 4 and the coupling measured between it and scaled monopoles in positions 1 and 2.

The angle between the antennas in positions 2 and 4 on the Fokker 100 is 24.9°, so this is very near the angle of 28° between positions 2 and 4 on the cylinder, and the

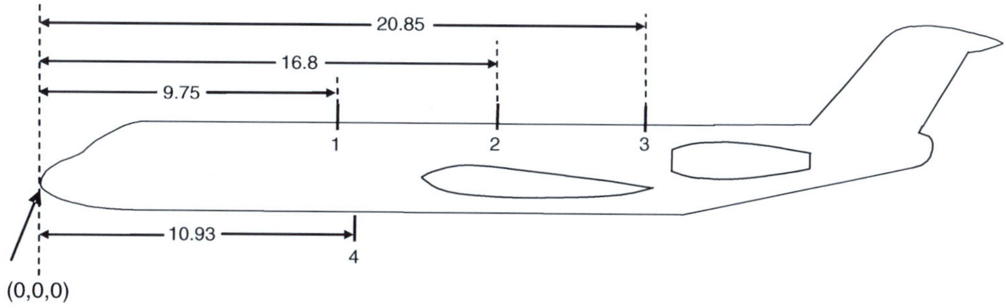

Figure 6.37 Antenna positions on the Fokker 100 scaled model.

scaled-down (from 1.8 to 3.6 GHz) frequencies of 120 and 240 MHz on the Fokker 100 are also very similar to the scaled-down frequencies of 121.5 and 243 MHz used on the cylinder. Thus gains of −6 and −8.5 dBi respectively were used.

The angle between the antennas in positions 1 and 4 on the Fokker 100 is 88°, so this is very near the angle of 90° between positions 2 and 5 on the cylinder. Gains of −9 and −12.5 dBi were used for 120 and 240 MHz respectively.

The coupling was measured for the tuned antennas in positions 1, 2 and 4 with scaled monopole antennas, as well as with the NLR scaled antenna in position 4 only (the last two values). These are shown in Figure 6.38.

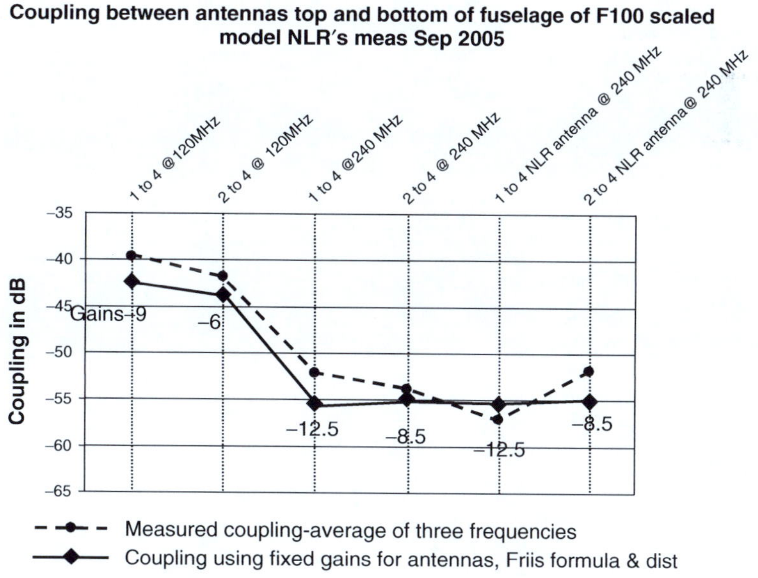

Figure 6.38 Coupling measured between the pairs of antennas on the opposite sides of the Fokker 100 scaled model and those calculated using the Friis formula and the gains shown in Table 6.6 at the aircraft/scaled frequencies of 120 and 240 MHz. Reproduced by kind permission of NLR.

The correlation is very good and follows the same trend as the measured values for all the scaled monopoles, showing around a 3 dB worse coupling (i.e. less isolation) than the measured values. It is always preferable to err on the side of caution.

However, in the case of the NLR scaled antenna, the correlation is not so good for the coupling between it and antenna 1. The antenna radiation pattern of the NLR antenna is slightly different from the scaled monopoles, and this could account for the difference compared with a scaled monopole at this position.

There were no measurements performed at the highest frequency, so this formula could not be checked at frequencies above 240 MHz.

References

[1] Duncan, D. (1984) A low cost method of RFI/EMI shielding. *EMC Technology*, October/November, pp. 75–76.

[2] Hardy, S.M. (1992) A less expensive Tempest alternative. *Journal of Electronic Defense*, **15** (6), 54–61.

[3] Connor, F.R. (1982) *Modulation*, 2nd edn, Edward Arnold.

[4] British Standard BS6656. Guide to prevention of inadvertent ignition of flammable atmospheres by radio-frequency radiation.

[5] Slater, J.C. (1942) *Microwave Transmission*, Chapter 6, McGraw-Hill, New York.

[6] Verpoorte, J. (2006) Verification of modelling tools by measurements and computation on scaled aircraft, Deliverable D16 Version 3, IPAS-TR_3/NLR/JV 061020-3, 20 October. Installed Performance of Antennas on AeroStructures (IPAS), an EU Specific Targeted Research Project, November 2003 to January 2007, Contract No AST3-CT-2003-503611.

[7] Macnamara, T.M. (2006) Canonical test bodies, Deliverable D8, IPAS/F_8/BAES/TM061013, 13 October. Installed Performance of Antennas on AeroStructures (IPAS), an EU Specific Targeted Research Project, November 2003 to January 2007, Contract No AST3-CT-2003-503611.

[8] Bull, D. and Smithers, B.W. (1979) Aerial isolation: a study of the interaction between co-sited aerials. *The Radio and Electronic Engineer*, **49** (6), 289–296.

7

Computer Modelling Techniques

7.1 Introduction

This chapter addresses the steps involved in computational modelling of installed antennas, to enable engineers to judge the best way forward as well as the requirements and limitations of generic modelling tools. It does not go into any detail of the mathematics behind the computation. However, it addresses the logistic problems of modelling antennas on large structures and qualitatively explains the differences in the various types of modelling software.

Comparison between predicted antenna radiation patterns and those measured on full-scale aircraft as well as on scaled models of the airframe is shown to demonstrate verification of the computational tools used. The term validation is often used when referring to computational software. However, this usually applies to checking the software code for bugs. The term verification, on the other hand, is usually used to indicate that the software has been checked for accuracy in the predicted results obtained. This verification is undertaken by comparing the predicted results with measured ones. If good correlation is obtained then it can be assumed that both the measured results and the predicted ones are accurate. However, when the correlation is not good, then it is usually assumed that the measured results are more accurate than the predicted ones.

Note that the correlation cannot be done in a mechanical manner, for instance by looking at a two-dimensional array of the magnitudes of the predicted and measured power levels or electric fields. The measured radiation pattern may have to be shifted along the angular scale (in the case of rectangular/Cartesian plots) or rotated (in the case of polar plots) to get good correlation. The two-dimensional array would show that there was no correlation between the predicted and measured gains.

Often the phase centre error correction defined in Chapter 8 has to be applied to get good correlation. This correction varies sinusoidally and is difficult to apply to polar patterns, since when using spreadsheets the angular values are plotted at equal intervals. In polar plots the angles are plotted over the circumference of a circle, so that the values are plotted at equal intervals over 360°, regardless of the number of angles in the array. However, in the case of rectangular plots the phase centre error correction can be applied relatively easily, since non-integral angles can be plotted. Details of this correction are given in Chapter 8.

7.2 Overview of Computer Modelling

Since the antenna pattern, when installed on the airframe, is very different from that on a standard ground plane or in free space, knowledge of this installed radiation pattern is often required at an early stage of the design process, sometimes before the airframe physically exists and often before physical installation. Computer modelling is one way in which installed patterns can be examined.

This requires that the exterior surface of the airframe be modelled, and then the antenna placed at the proposed position on the CAD model of the airframe. The radiation pattern is then computed at the required frequency and cut. The performance of the antenna has a major impact on the performance of the entire system to which it is connected.

7.2.1 Reasons for Computer Modelling of Antennas on Structures

Measurements of radiation patterns are usually undertaken either as far-field measurements (on outdoor sites, compact ranges or anechoic chambers) or as near-field measurements, usually indoors, from which the far-field patterns are derived.

In the case of far-field measurements only one pattern cut is obtained from each measurement, whereas in the case of near-field measurements, although each measurement run takes longer, other cuts can be derived, depending on the type of facility used. In addition, measurement costs can be very high. Computational modelling can be used to obtain any desired cut in the post-processing stage, and it is easy to change the frequency, although a new CAD model may be required to cover different frequencies. Some software can compute the patterns over a whole range of frequencies in a single run.

The modelling of the antennas on structures using computational techniques is undertaken for one or more of the following reasons:

1. The geometric surface model of the airframe is easy and quick to generate as a canonical surface, that is, a structure made up of canonical objects (planes, cylinders, ellipsoids, etc.). It may take longer and be more difficult in the case of generation of a CAD model which is based on the true surface of the airframe.
2. It is easier to change antenna position to accommodate changes in the early stages of the design phase.
3. It is easier to run the software than to perform measurements on scaled models or mock-ups.
4. The frequencies can be changed very easily in most cases for re-runs.
5. Modifications to the airframe can be accommodated more easily than in the case of the physical scaled models.
6. Different radiation cuts can be viewed without re-running the software.
7. Contour plots can be produced in most cases to view the spherical spatial coverage.
8. It can be used in the case single antenna retrofits, where the construction of a scale model (if one is not readily available) or a programme of flight trials would not be justified.
9. With some software packages the radiation patterns over a range of frequencies can be obtained in a single run.
10. User-friendly software can be used by a less skilled workforce than would be required for measurements.

7.2.2 Systems Performance Using Antenna Modelling

When an aircraft is moving the angle subtended by an off-board receiver, for instance, is continually changing. Thus systems engineers are usually interested in the probability of maintaining the link between on-board and off-board receivers/transmitters, rather than the individual gain of the antenna at a particular angle. In these cases it is more useful to look at the radiation pattern of the antenna and derive the percentage probability that the gain will be above a given level in a specified angular sector.

For instance, consider the case of investigating the likelihood of detecting the distress signal of a survivor of a wreck at sea. If we work out the space loss to a survivor at a certain distance, and we know the power output of the distress beacon or SOS transmitter, then we could be looking at the following scenario.

Suppose that the output power of a distress signal at 243 MHz is +10 dBm (10 mW), and the aircraft has a specification to pick up distress signals from a distance of up to 40 miles (64 km). Space attenuation/loss for this distance at 243 MHz is 69.13 dB. Assuming that the gain of the survivor's antenna is 0 dBi in the direction of the aircraft, the power level at the aircraft antenna would be −59.13 dBm.

Assuming the system losses (cables, connectors, etc.) are 12 dB, the power at the aircraft receiver would be −71.13 dBm. This assumes that the aircraft antenna gain is 0 dBi. If the sensitivity of the receiver is −80 dBm then an aircraft antenna gain down to approximately −9 dBi can be tolerated. However, if the survivor's antenna gain in the direction of the aircraft is less than 0 dBi then the tolerance of aircraft antenna gain is reduced by this amount. These link (power) budgets have to be worked out for each type of scenario being considered.

Some antenna computation programs can give the probability of obtaining this gain over an angular sector by dividing the angular sector where the gain is above −9 dBi by the total angular sector (e.g. 360° for a line radiation pattern). If we look at the total number of angles in a 360° line radiation pattern that have a particular gain value and plot this distribution, we could get a plot similar to that shown in Figure 7.1. We can convert this to a probability distribution by dividing the area down to a particular value of gain by the total area under the curve. Multiplying this fraction by 100 gives this probability as a percentage. For instance, if we want to derive the probability of getting a gain value above −9 dB we would divide the shaded area under the curve between +3 and −9 dB, by the total area under the curve (shown in Figure 7.1). If this gives a figure of 55%, this tells us that the probability of getting a gain above −9 dB is about 55%. This can be done for different levels of gain values, resulting in the probability distribution shown in Figure 7.2.

It can also be seen from Figure 7.2 that if we had a more sensitive receiver then we could tolerate a lower gain from the antenna. For instance, a receiver sensitivity of −91 dBm would require an antenna gain of −20 dB, and thus the probability would increase to about 68%.

The same process can be used to obtain a probability distribution for 3D coverage. In practice only the angular coverage actually required to be covered by the system has to be considered. For instance, if we assume that the aircraft has a maximum altitude of 40 000 ft (12.192 km) and the range of the system is 64 km, the angle of depression is

$$\sin^{-1}(12.192/64) \approx 11^\circ$$

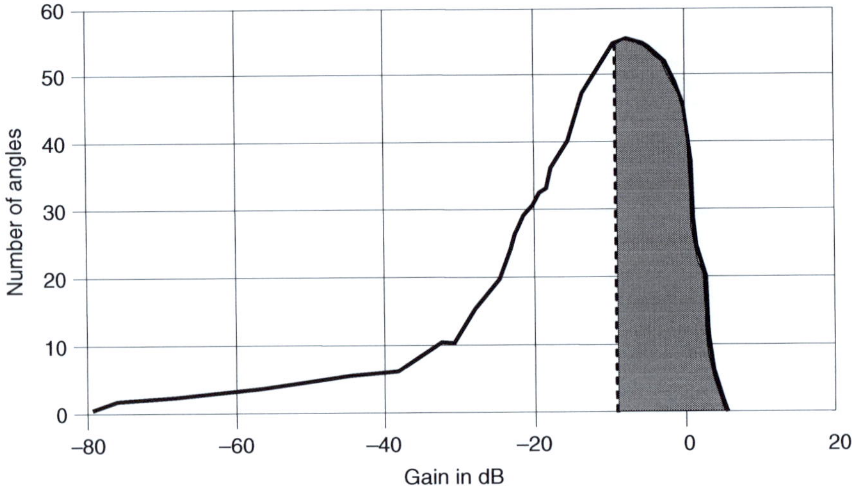

Figure 7.1 Distribution of gain values over the line radiation pattern.

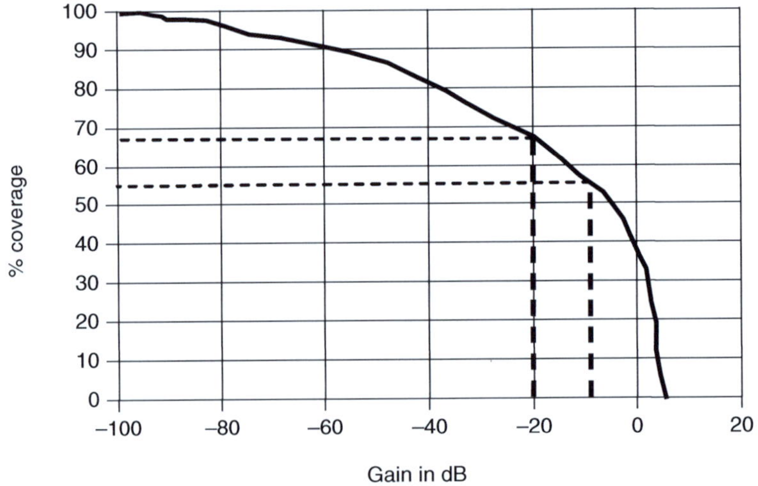

Figure 7.2 Percentage coverage at different levels of gain.

(since the aircraft only has to consider survivors on the ground or sea). This assumes that the aircraft is flying straight and level. Most aircraft (excluding fighter aircraft) fly in a nose-up attitude of (say) 2 to 4°, so we have to add this value to the angle of depression. The angle of depression reduces as the aircraft flies towards the survivor, if the aircraft maintains the same altitude, but in most cases the aircraft would descend and hence the angle of depression is likely to be the same or smaller. If the aircraft is flying directly above the survivor it is likely to be at a low altitude, and hence the actual distance to the survivor would be relatively short resulting in a lower space loss and thus accommodating a lower antenna gain.

Thus the angular sector over which the probability of detection has to be considered would be 360° in azimuth, and the angular sector in elevation would have to be worked out depending on the ranges and altitudes covered by the operating spatial envelope of the particular aircraft. The space loss would also have to be worked out for each case.

7.2.3 Postdiction

Postdiction can also be used in cases of major modifications to an airframe. In this case the measurements are available for the airframe prior to the modifications. These could be scaled model or in-flight measurements and the computer modelling is performed to obtain correlation with these measurements – hence the term 'postdiction', as opposed to prediction. This is useful in obviating the requirement for the fabrication of a new scale model. In this case CAD models are produced of the old as well as the new airframe, and the correlation shown between the computations obtained with the old CAD model and the measurements on the unmodified airframe/aircraft. It is then accepted that the computations obtained with the new CAD model of the modified airframe would also correlate with measurements without the need to undertake the measurements on the modified airframe.

7.2.4 Modelling the Exterior Surface

The exterior surface of the airframe and any significant excrescence is also modelled, but all the obstacles are modelled as passive ones. They cannot be modelled as parasitic ones. On a real aircraft there may be some excrescences that would be resonant and re-radiate in the same way as chaff used to confuse radar or other electronic measures that irradiate military aircraft. This would occur, for instance, in the case of excrescences that are a resonant length but are not grounded, and antennas that are not matched to the transmission lines (to which they are connected) and hence re-radiate the EM wave incident on them. Another cause of error is resonant slots in the airframe at doors and other apertures in the airframe. This becomes very important at high frequencies, say, above NATO I-band.

The airframe surface is modelled as a perfect electric conductor (PEC), even though airframes are made of aluminium and usually consist of several plates riveted to the ribs and frames that form the shell of the airframe. The frames are the circular rings and the ribs run longitudinally from the front to the rear of the airframe as shown in Figure 6.1. There may also be doublers and triplers on the aircraft skin. These are additional plates riveted to the original skin to repair damage such as holes in the skin.

Non-conducting surfaces such as radomes cannot be modelled by some of the modelling tools. Furthermore, there are difficulties in modelling complex shaped radomes such as those used on the nose cones and sandwich radomes.

When the airframe is modelled as a PEC, we can assume that the interior of the aircraft is shielded from the EM wave on the surface of the airframe and thus does not need to be included in the model.

However, in those cases where the radome is modelled, the EM wave passes through the radome into the interior of the aircraft, which is not modelled. The computational electromagnetics (CEM) engineer has to decide whether it is better to model the radome as:

1. a dielectric without including the metallic objects inside the radome, or
2. a metallic surface that would obviate the requirement for modelling the interior.

The decision is taken by examining, in the first instance, the angle at which the direct wave from the antenna strikes the radome.

If the angle of incidence at which the wave strikes the radome is large, then there is a distinct possibility that this angle is greater than the critical angle and the wave is subjected to total internal reflection. In this case the wave will be reflected at the inner surface of a sandwich radome and thus modelling the radome as a metallic surface might be more appropriate. Although it should be appreciated that the reflected waves could be of different amplitudes and phases (as well as being at different angles) compared with the case where the radome is a dielectric.

If it can be shown, from line of sight (LOS) examination, that the incident wave would not be incident on any object inside the radome (or be subjected to total internal reflection) and would emerge from the other side of the radome, then it might be better not to model the radome at all and thus effectively treat it as free space. We know from the laws of refraction that the emergent ray from a parallel sided block is displaced from the incident ray. Additionally it undergoes some attenuation. In the case of a shaped radome it is more difficult to predict the exact effect of the radome, but replacing the radome with a metallic section causes more errors since it results in LOS blockage of the EM radiation.

The same arguments are applied to the cockpit of the aircraft. Most commercial airliners have a metal cabin with windows that are near vertical and not in LOS of most of the antennas. However, in the case of fighter aircraft the cockpit has a canopy which the pilot lifts to board the aircraft. This canopy is also constructed so that it can be broken to enable the pilot to eject in an emergency. For this reason the canopy is usually made of Plexiglas or of a similar material and has a very similar dielectric constant to that of single skin radomes.

In this case, the pilot is often in LOS of the antennas and thus ideally should be included in the modelling. The human head is very complex to model, but in some cases it is modelled as a sphere of water. In most cases the pilot is not modelled at all. An alternative approach is to model the dielectric first as a PEC, and then omit the entire radome but model the structure inside the dielectric surface. This will provide bounds to the radiation pattern.

7.2.5 Modelling the Antenna

Blade antennas can be modelled as simple monopoles. Under the auspices of the IPAS project [1] a scaled antenna was fabricated for NLR to represent the radiating element and mimic the performance of a full-scale UHF Sensor Systems blade ([2], p. 47). This scaled antenna (of height 19 mm) was effectively a defeatured antenna, in that it does not have the exact radiating elements of the full-scale antenna. The scaled antennas normally used for scale model measurements are simple monopoles tuned to a fixed frequency. One such example attached to the back of a flanged sub-miniature amphenol (SMA) connector is shown in Figure 7.3a, together with the NLR scaled antenna in Figure 7.3b,c.

Measurements were performed on the Fokker 100 scaled model with this scaled antenna positioned on the lower fuselage behind the nose landing gear (NLG) in position 4 as shown in Figure 7.4.

These radiation patterns were compared with those obtained with a simple monopole antenna at the same position. Very little difference was found between the antenna modelled as a simple tuned monopole and the NLR scaled antenna, in the case of the azimuth patterns as shown in Figure 7.5a. The differences in the forward sector could be attributable to the metallic strip at the front of the NLR antenna. There are some differences between the tuned monopole and the NLR scaled antenna, in the forward sector of the pitch plane, whereas in the rear sector the differences are much less marked, as shown in Figure 7.5b. However, these differences are relatively minor. Although it cannot be assumed that all blades have the same performance as simple monopoles, it may be safe to assume that the differences in the radiation patterns are likely to be a second-order effect. When using the method of moments (MoM) the radiating element can be modelled to a limited extent, but this cannot be done in the case of the geometric/unified theory of diffraction (GTD/UTD).

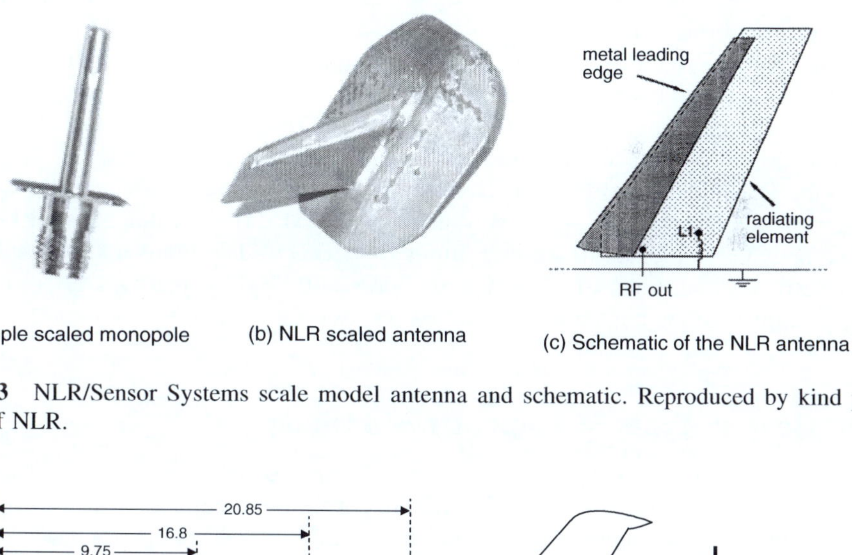

(a) Simple scaled monopole (b) NLR scaled antenna (c) Schematic of the NLR antenna

Figure 7.3 NLR/Sensor Systems scale model antenna and schematic. Reproduced by kind permission of NLR.

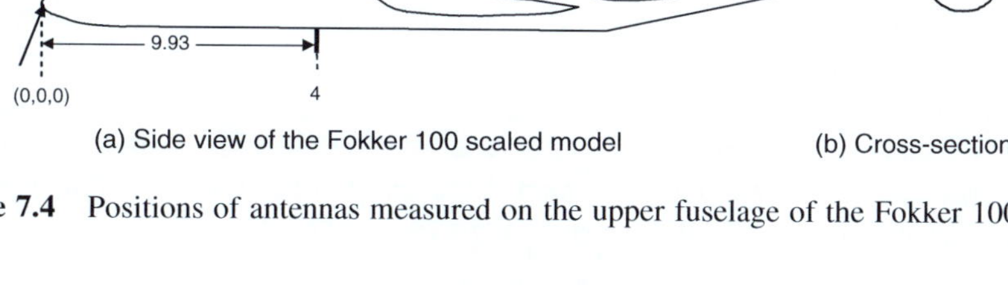

(a) Side view of the Fokker 100 scaled model (b) Cross-sectional view

Figure 7.4 Positions of antennas measured on the upper fuselage of the Fokker 100 scale model.

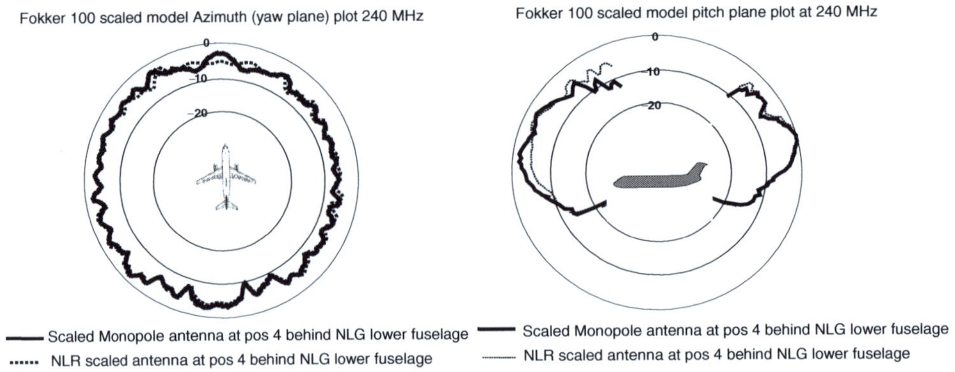

Figure 7.5 Azimuth and pitch plane radiation patterns of the NLR scaled antenna and a simple tuned monopole on the lower fuselage of a Fokker 100 scaled model behind the nose landing gear.

7.2.6 The Frequency Gap

At the low frequencies the MoM and finite difference time domain (FDTD) are used, but as the frequency increases the wavelength decreases and since the wire segments/cells have to be fractions of the wavelength, there is a corresponding two- or threefold increase in their number. This results in the hardware requirements and computation time rising to unrealistic and unmanageable levels.

At the higher frequencies, GTD/UTD is used, but this has a low frequency limit, since the dimensions of objects have to be greater than 0.8 wavelengths and they have to be in the far field of the antenna. Between the low and high frequency methods there was what was known as a 'frequency gap', where MoM could not be used because of the computation times and hardware limits, and GTD/UTD could not be used because the electrical dimensions of obstacles are too small. Hybrid methods and advances in computational techniques have been developed to bridge this frequency gap.

7.3 Generic Types of Computer Modelling

The different types of computer modelling software can be categorized by the methods used to model the structure on which the antennas are installed. In this book only the generic antenna modelling software is considered, and not proprietary software marketed by the different institutions. Many companies use their own proprietary software that could be commercially adapted packages or code written by their own personnel that is not marketed or available for external use. This chapter only discusses the radiation patterns of the antennas on airframes.

Aircraft engineers require knowledge of the spatial as well as the frequency performance of the systems connected to the antenna. The spatial performance is characterized by the radiation pattern of the antennas. In general the coupling between antennas is best calculated or measured for the reasons described in Chapter 6.

7.3.1 Classification of Generic Forms of Computer Packages

Most CEM specialists classify the different modelling software depending on the solution of the mathematical equations used. Thus, generic codes can be divided into the following basic types:

1. boundary element method (BEM)
2. FDTD
3. asymptotic methods such as GTD, UTD
4. physical optics (PO)
5. multi-domain (MD).

For the non-specialist systems engineer these classifications may not be very meaningful. The different types of software could be re-classified depending on the method of modelling the structures and qualitatively explaining the physics behind the modelling. Thus, the most common forms of modelling the radiation patterns of antennas on structures could be classified depending on how the structure is modelled and the physics underlying the production of the radiation pattern. The structure can be modelled as a volume, a surface or using canonical shapes, and the EM radiation can be considered as waves or rays.

In the BEM (which is also sometimes known as the boundary integral equation method or boundary integral method), boundary conditions are given and the method fits these boundary values into the integral equation, rather than values throughout the space defined by a partial differential equation. The BEM is often more efficient than other methods, including finite elements, in terms of computational resources for problems where there is a small surface to volume ratio. The BEM applies the finite element method to an integral formulation of Maxwell's equations. This reduces the computation to solving for currents on the interface between two (or more) media. Usually the media are PEC and free space. The method can be extended to the treatment of certain homogeneous dielectrics using surface equivalence methods, and to inhomogeneous dielectrics by using volume elements in addition to surface patches.

The MoM is a special case of BEM with a direct solver, although this definition is not universally accepted. In the case of MoM the structure is modelled as a surface, and the feed of the antenna is modelled as a voltage across a wire. This induces currents on the surface of the airframe that radiate into free space. In the case of a monopole, for instance, one of the main differences between MoM modelling and real physics is that the radiation occurs across a gap at the base of a monopole, whereas in MoM the voltage is applied across a wire.

In the case of FDTD the spatial domain around the structure is modelled as a volume made up of cuboid cells that consist of PEC, dielectric or free space, and the EM wave is transmitted through these cells. Each cell is homogeneous (i.e. has the same characteristics throughout its volume) with its own EM properties, and the size of each cuboid is chosen so that the linear dimension of the cuboid, in the direction in which the wave is travelling, is between about $\lambda/10$ and $\lambda/5$ at the highest operating frequency. A pulse is applied as the source and thus the radiation patterns over a number of frequencies can be obtained in a single computational run.

The fast multipole method (FMM) is not considered as a distinct class of generic code, since it is basically a MoM code as far as the governing mathematical model for the physics is considered. It is an accelerated method for solving the boundary integral equations and hence decreasing the computation times.

In the case of GTD/UTD the structure is modelled using canonical shapes (cylinders, ellipsoids, cones and plates) and the waves that are radiated into free space are subjected to the laws of reflection and diffraction, including creeping waves. One definition of a canonical shape is a shape that can be defined according to a general rule or formula.

In the case of PO the structure is modelled using a CAD or canonical model and the waves are considered as rays that are radiated into free space and subjected to the laws of reflection, with only limited consideration being given to diffraction.

MD, as its name suggests, involves the use of a combination of methods and it is thus sometimes referred to as a hybrid method.

7.3.2 Verification of Modelling by Showing Correlation with Measured Data

Modelling enables verification of the antenna coverage at

1. a larger number of frequencies
2. a larger number of angles and pattern cuts.

Usually measurements are done at a few frequencies and one or two pattern cuts and then these are used to provide verification of the modelling software. If good correlation is obtained between the measured and the predicted data then it can be assumed that all the modelled data is valid.

The measured radiation patterns are considered to be the benchmark against which the computed patterns are verified. Most of the measured results are undertaken on subscale models of airframes.

At HF frequencies (2–30 MHz) there are no measured results on scaled models because the measurements are usually undertaken on a mock-up of part of the full scaled aircraft. One of the main reasons for this is that it is very difficult to physically scale down the HF antenna and tuner. In many cases the HF antenna is integrated into part of the airframe, for instance into the dorsal fin or tail fin.

7.4 Method of Moments

This method is based on the solution of the Stratton–Chu integral equations that are based on Maxwell's equations. The equations are stated in terms of magnetic and electric fields. MoM uses both types of fields and requires the solution of magnetic field integral equations and electric field integral equations.

This was one of the first types of modelling software programs that was written for antennas on structures and had the proprietary title of NEC (Numerical Electromagnetic Code). It was a product of the Lawrence Livermore National Laboratory (LLNL), which was managed from its inception in 1952 until September 2007 by the University of California for the US government. LLNL is currently managed by Lawrence Livermore National Security. NEC was written in Fortran and was available to research and defence institutions. It has since been modified and rewritten in other languages, various versions being available commercially, some of which can be run on standard desktop computers.

In NEC the surface of the structure is modelled as surface patches or wire segments. In simplistic terms the surface on which the antenna is placed, along with any other surface that is illuminated by the radiated wave, is modelled either as a closed surface using patches or any other surface using a wire grid. This is because patches cannot have two sides and must include a volume, whereas this restriction does not apply to wire grids.

In the case of surface patches, the structure should either be flat or have a curvature whose radii vary in a continuous manner. In the case of the airframe of an aircraft, there are sharp discontinuities, for instance, where the fuselage and wings meet. Patches, which are usually square, must have dimensions of less than $\lambda/5$ (i.e, a maximum area of $\lambda^2/25$), or there must be a minimum of 25 patches per square wavelength.

In the case of the wire grid, however, a closed surface is not required and thus the wire grid can be used to represent both sides of a plate, depending on the electrical dimensions of the mesh.

Thus, in general, the airframe is modelled using wire grids rather than patches. The electrical length of the wire segments has to be short because the current has to be assumed to be constant (or of a simple polynomial form) over the wire. The length of each segment has to be a fraction of a wavelength, resulting in a large number of wires at high frequencies. This results in a massive amount of central processing unit (CPU) time as well as stringent hardware requirements, restricting NEC to low frequencies. However, with the advent of greater computing power, coupled with hardware miniaturization, desktop machines can now be as powerful as the large Cray computers of a few years ago.

The grid spacing, like the diameter of the wires used for the grid, depends on the wavelength of the operating radiation. If we have a long wavelength (i.e. low frequency) a very open mesh can be used. Meshes are usually square, rectangular or triangular, the latter being the most common for aircraft surfaces. They can also be a mixture of these shapes. The length l of each wire segment has to be less than a quarter of a wavelength, and $\lambda/10$ is usually taken as a good compromise between accuracy and CPU runtime.

The diameter d must be less than a quarter of the segment length l. Thus if the wire segment length is $\lambda/10$, the wire diameter must be less than $\lambda/40$.

At HF, for instance, where the lowest frequency is 2 MHz, the wavelength is 150 m, and small aircraft that are about 10 m long can be modelled as 'matchsticks' – that is, a single segment $\lambda/15$ long can be used to represent the entire fuselage as long as the fuselage diameter is less than $\lambda/40(= 3.75\,\text{m})$, which is the case for most small aircraft. No other detail of the fuselage is required and the wings can also be modelled as wire segments, but they would be orientated at the correct angle to allow for the sweep and angle that the actual wing makes with the horizontal. If the wings are tilted upwards this angle is called the dihedral, whereas if they are tilted downwards the angle is called the anhedral. Most commercial aircraft have dihedrals, whereas fighter and combat aircraft have anhedrals.

Because of the limitations on the electrical dimensions of both patches and segments, a wire mesh model used at low frequencies cannot be used at a high frequency if the electrical dimensions do not meet the criteria. Thus the airframe may require several wire mesh models to cover a wide range of frequencies. The wire mesh model may take several months to construct manually, but many packages are available to convert the surface model of the airframe into a wire grid model for CEM. The CAD surface model could be one used by aerodynamicists and may have to be adapted before conversion into the wire grid format required for the EM computational modelling. The segments are joined together to form grids or meshes, usually triangular or rectangular. The runtime of the software can range from a few minutes at HF to several days for a large airframe at high frequencies.

The number of segments increases rapidly with the size of an object. For instance, if we consider a cube of four wavelengths and we use a segment length of $\lambda/10$ then we would have 40 segments for each edge of the cube. The number of segments for each face of the cube would be 1600 and since there are six sides to a cube, this would be a total of 9600 for the surface area of the cube. For a cube of eight wavelengths each face would require 6400 segments and the total number for all six faces would be 38,400 segments. Each doubling of the length results in a quadrupling of the number of segments. In the case of aircraft and other structures it is more difficult to calculate the total number of segments that would result from the meshing of the CAD model. The processing time is directly proportional to $N^{2.5}$ where N is the number of segments.

For example, a wire-grid model of a P-3/CP-140 aircraft had 327 segments for 2–30 MHz [3]. This corresponds to a segment length of 0.2λ. A model for a BAC-111 aircraft at HF had 7736 segments and took 30 000 seconds to run on a Cray-2 computer. Using the new commercial tools available can result in large runtime reductions and hence reduced CPU costs.

Smaller segment lengths are used near the antenna or in areas of high electric fields and where the angles of incidence are small.

For the IPAS project, a scaled model called IPAS-1 that was a single aisle aircraft was constructed from timber and conductively coated. A computer CAD model was produced from the surface of the scaled model by scanning it with a laser. The CAD surface was then meshed for modelling the performance of antennas on its surface. The surface was meshed at 1 GHz for ATCantennas mounted on the lower fuselage as shown in Figure 7.6.

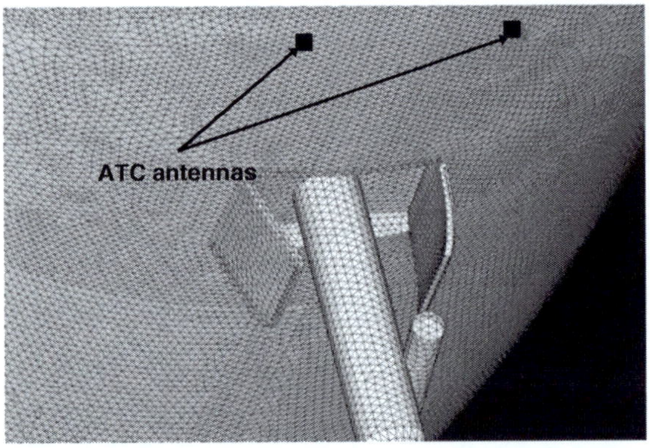

Figure 7.6 Meshed IPAS-1 with landing gear deployed, for two ATC antennas on the lower fuselage ([2], Figure 152). Reproduced by kind permission of EADS.

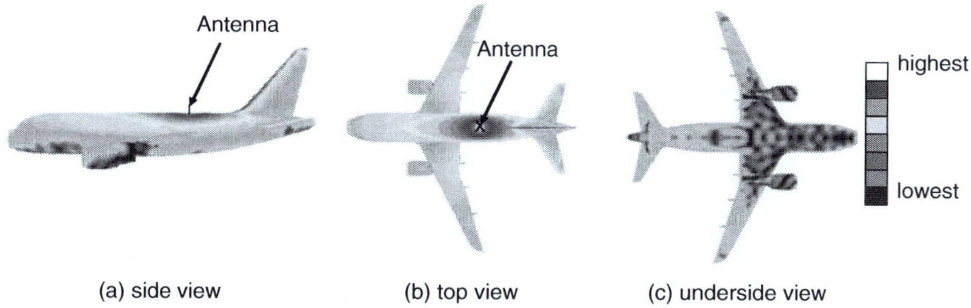

Figure 7.7 Surface currents on IPAS-1 for an antenna on the rear upper fuselage. Reproduced by kind permission of EADS. See Plate 8 for the colour figure.

The surface currents can be seen on the surface of the CAD model in Figure 7.7, and this provides a useful insight into the analysis of the far-field radiation patterns obtained [2]. For instance, for an antenna on the upper fuselage aft of the wings, it can be seen that the surface currents spread to the lower fuselage to the aft as well as right down to the front of the lower fuselage.

7.4.1 Enhancements

Computer modelling using MoM has been improved and extended to larger problems by:

1. improvements in creating the wire grid surface model
2. improvements in processing
3. extensions to the mathematics to enable its application to materials other than perfect conductors.

7.4.1.1 Improvements in the Creation the Wire Grid Surface Model

Software packages are now available to convert the surface CAD models to wire grid models. The conversion as been simplified using a graphical user interface (GUI) which enables the user to manipulate the wire grid by dragging and dropping the nodes (joins between two wires). Apart from a reduction in the time scales, this also allows less experienced personnel to undertake the modelling.

Additionally, areas can be selected to increase or decrease the wire grid or mesh density. Thus areas where the variation of surface currents is more rapid can be changed to a finer grid spacing (i.e. a tighter mesh).

Areas that would affect the far-field radiation pattern could also warrant the use of a finer mesh. For instance, if a wing is directly illuminated by the antenna, the specular reflection from it could provide a large contribution to the far-field radiation pattern and a finer grid spacing should be used in this area. Similarly, near the antenna, where the fields are of greater magnitude and are changing more rapidly with distance, a finer grid should be used.

Some packages also have the ability to clean up and reduce the number of wire segments by the automatic removal of duplicated and unnecessary wires, as well as to increase or

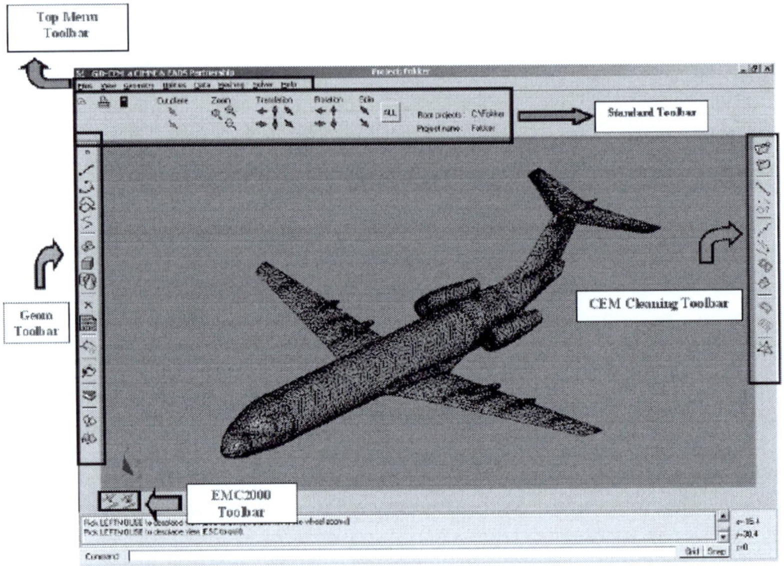

Figure 7.8 A typical screen GiD_CEM screen developed to integrate with EMC 2000. Reproduced with kind permission of CIMNE. See Plate 9 for the colour figure.

decrease the grid spacing and segment length in areas that are near to or far from the source of the radiation. Compass provide one such package that has the proprietary name of GiD_CEM. It was developed by CIMNE (International Centre for Numerical Methods) which is part of the University of Barcelona. Under the IPAS project [1], CIMNE worked closely with the French arm of EADS to integrate GiD_CEM with EADS's proprietary software EMC 2000 so that the wire grid models produced by EADS could be input to GiD_CEM, modified and re-output to EMC 2000 for the computation modelling to be undertaken. A typical screen is shown in Figure 7.8.

Among the features of the GiD_CEM software is the ability to:

1. import and export files of standard format, such as IGES, ACIS, Parasolid, DXF, Rhino, VDA
2. create and edit the geometry, in order to automatically repair bugs in the CAD model
3. collapse the grid to coalesce parts or to remove small details
4. create missing entities, such as wires and nodes
5. attach data to geometrical/CAD entities (physical properties, boundary conditions, etc.) and automatically transfer this data to mesh entities
6. automatically generate meshes of different specified sizes and shapes
7. edit the mesh, by collapsing points, splitting wires and smoothing the surfaces
8. post-process the results to allow graphical visualization of scalar or vector fields.

The collapsing function stated in point 3, is used to collapse or to merge two points that are near each other. The separation between the two points can be selected by increasing or decreasing the 'tolerance' – the maximum distance between the two points. This is

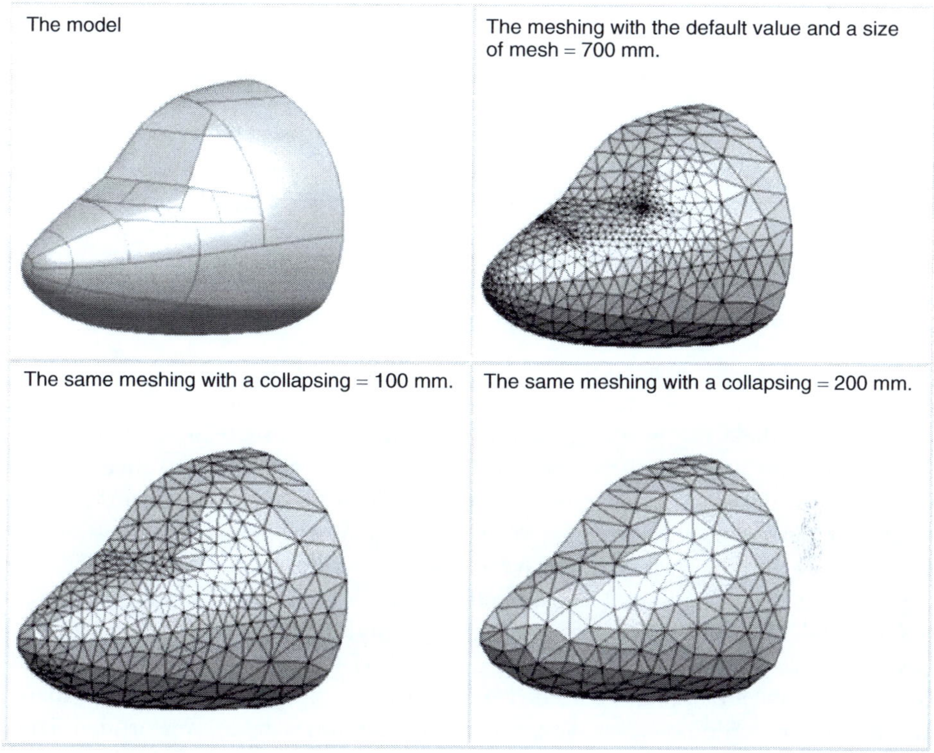

Figure 7.9 A CAD model meshed using different tolerance options in GiD_CEM. Reproduced with kind permission of CIMNE.

illustrated in Figure 7.9. The CAD model of the nose cone is shown in the first part of the figure (top left), and the second part of the figure (top right) shows the model meshed with a user-defined mesh of size 700 mm. However, because there are small surfaces in the geometry, the real elements are much smaller than 700 mm. Thus the surface is meshed with a much smaller segment size.

The third (lower left) and fourth (lower right) parts of the figure show the meshes when the collapsing function is used to collapse points that are separated by 100 mm and then to collapse points that are separated by 200 mm, respectively. It can be seen that when the tolerance is increased to 200 mm, the details in the shape of the nose cone are lost in some areas. This would not be critical for low frequencies or for antennas far from the nose cone or on the lower fuselage. However, in cases such as the MLS antenna (frequency 5.2 GHz), which is often mounted on the nose cone, misrepresentation of the installed radiation pattern would result.

7.4.1.2 Improvements in Processing

The increase in the computing power of desktop computers, accompanied by the reduction in size and cost, has resulted in larger problems being undertaken on standalone computers that a few years ago could only be handled by larger Cray computers with parallel processors. However, with the larger airframes and higher frequencies used for aircraft systems, the problems have also increased.

Newer methods have been developed for combining the CPU resources of desktop machines and reducing the total computation time required. GridSystems have successfully linked standard desktop computers and by using the idle resources have distributed the computational problems across the network. This work was developed under the IPAS project to solve different types of electromagnetic problems.

7.4.1.3 Extension to Non-Perfect Conductors

The BEM has also been applied to MoM to model the interface between the two media, such as PECs and homogeneous dielectrics. In the case of aircraft, the nose cone is usually made of dielectric materials such as fibreglass, which is relatively transparent to EM radiation. This nose is called a radar dome or radome and houses the weather/search radar. The distortion and attenuation of the EM radiation is dependent on the electrical thickness of the radome. In the case of a radar signal, the transmitted as well as the received signals are subject to the adverse effects of the radome. The weather radar operates in the I band (around 9.5 GHz) and thus the wavelength is around 3 cm. Ideally the radome thickness should be less than 0.1λ (around 3 mm). This thickness would not be able to withstand the environments encountered by the aircraft or birdstrike. Thus in most cases sandwich radomes are used. These consist of three or more layers of materials with different dielectric constants. In some cases there might be a layer of low dielectric constant sandwiched between two layers of higher dielectric constants, or vice versa. In this way the adverse effect of the thickness is reduced, and the sandwich radome is more acceptable than a radome made from a single layer of the equivalent thickness. However, the sandwich radome is far more difficult to model than the single-layer radome. The shape of the radome, especially at the pointed front end, results in the shifting of the radar beam, known as boresight pointing error (BSE). Most of the MoM modelling software available is suitable for modelling the performance of antennas on perfectly conducting surfaces. The incorporation of dielectric materials is in its infancy, but further developments are expected in the not-too-distant future.

7.4.2 Comparison with Scaled Model Measurements

The radiation patterns predicted using MoM were compared with the measurements performed on a scaled model of a single aisle airframe at 2 GHz, and since it was a 1/18th scaled model, the equivalent aircraft frequency is 111 MHz. The three principal plane cuts are shown in Figure 7.10. The correlation between the predictions using MoM and the measurements is very good in the azimuth and roll plane cuts. In the case of the pitch plane plot, although the correlation is very good in the upper hemisphere that in the lower hemisphere shows more beating than the measured pattern. However, this increase errs on the side of caution, in that the predicted pattern is worse (lower levels) than the measured pattern. In the case of a moving platform such as an aircraft, the angular change in the radiation pattern will result in a “filling-in” of the nulls (since the angular position of a

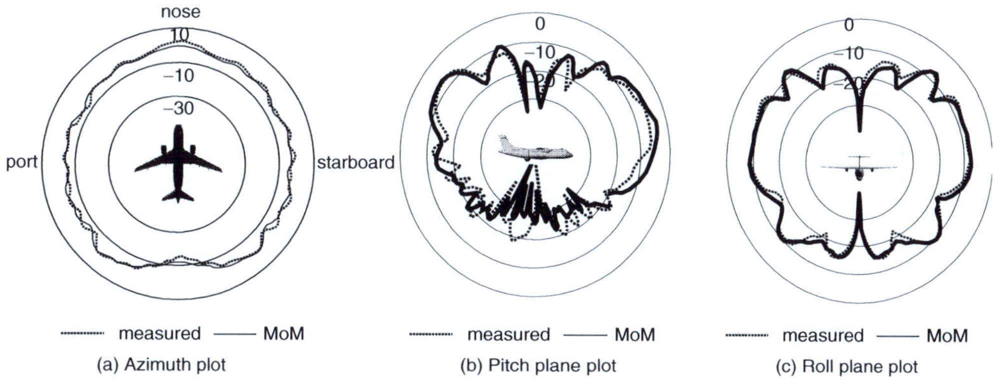

Figure 7.10 Comparison between the predicted radiation patterns using MoM and measurements on a scaled model of IPAS-1 (a single aisle airframe). Reproduced by kind permission of EADS.

fixed transmitter/receiver at a ground station is continuously changing) and may therefore not pose a problem.

7.4.3 Fast Multipole Method – BEM with Accelerated Solver

The FMM accelerates MoM by grouping together interactions between parts of the structure on which the antenna is installed as well as other obstacles. This leads to a significant reduction is memory resources. Since the aircraft is a closed body the combined field integral equation can then be applied since this provides a far more rapid convergence than the electric field integral equation.

MoM requires considerable CPU resources. However, using the hybridization provided by the multi-level fast multipole method (MLFMM) technique, the time was reduced by a factor of 8.5. The DLR modelled a monopole on an ATR 42. In the immediate vicinity of the antenna the mesh size was $\lambda_0/16$, whereas the rest of the airframe was modelled using $\lambda_0/8$ segments. At 126 MHz the number of segments would be 4245 triangles for half the airframe, instead of 8490 triangles for the full airframe. This is because reflection in the plane of symmetry (the vertical longitudinal plane through the centre of the fuselage) provides the second half.

The MLFMM calculations have been performed using the full airframe (8490 triangles). At higher frequencies (e.g. 1 GHz) the mesh size in the immediate vicinity of the antenna was $\lambda_0/10$, whereas the rest of the airframe was modelled using $\lambda_0/5$ segments. At 1 GHz a mesh of 193,212 triangles is required.

7.5 Finite Difference Time Domain

The electric and magnetic fields are dependent on each other. Their variation in time and space is defined by Maxwell's equations. These are in differential and integral forms. In simplified qualitative terms, the change with time of the electric field depends on the change in space of the magnetic field. The change in time is the differential/derivative

dE/dt or the partial derivative $\partial E/\partial t$ and the change in space of the magnetic field is the curl of the vector H. The curl is a term used to define the change in all three dimensions in space of the magnetic field.

Thus Maxwell's first equation is given by

$$\operatorname{curl} H = \sigma E + \varepsilon \frac{\partial E}{\partial t},$$

where

σ in the conductivity in siemens per m

ε is the permittivity in farads per m.

For a simplified explanation of Maxwell's equations and the curl of a vector, refer to [4], Chapter 3.

The name FDTD refers to the finite difference in the electric field in the time domain. Maxwell's equations are solved for the electric field at an instant in time, then the magnetic field is solved at the second instant in time, again followed by the electric field solutions, and so on, in what is commonly known as 'leap-frogging'.

The FDTD is a volume element method, in which the structure is divided into small cells. The cells are usually cuboid, although different shapes and mixtures of shapes have been used in extensions to the theory. The number of cells depends on the frequency of the EM radiation. The higher the frequency the smaller is the wavelength and hence the greater the number of cells required.

The excitation is applied to the antenna and the electric and magnetic fields in the neighbouring cells are computed at time steps after that – see [5]. The time step is a function of the smallest cell size in the structure. An important factor in the FDTD method is that the electric and magnetic fields are computed at points separated by half a cell. The field should be nearly constant over a cell. The structure and the surrounding media have to be modelled. In the case of antennas on aircraft, the surrounding medium is air.

The main advantage in the case of FDTD is that dielectrics can be modelled because each cell can be given different properties. Thus individual cells can have their own dielectric constants, permitting the modelling of radomes juxtaposed to the PEC airframe. Another important feature is that the dimensions of some cells can be reduced so that higher resolution can be achieved in areas of large curvature, for instance.

In the near field of an antenna, the field varies rapidly and the cell dimensions may be required to be as small as $\lambda/30$ or even $\lambda/50$. Problems occur in these cases, where there are cells of different sizes, since the requirement to have the approximately the same field variation across adjacent cells cannot be maintained.

The other problem is what is known as 'staircasing'. The cells that are used to model curved surfaces are like a Lego™ brick model, and so the curved surface is modelled like steps. Where there are small/tight radii of curvature, the cells have to be smaller in order to follow the exterior profile more closely.

Because FDTD uses a pulse excitation, a wide range of frequencies can be computed in one run. This is because a pulse contains a whole range of frequencies, and thus using a pulse excitation is equivalent to having a range of frequencies. The narrower the pulse the greater is the bandwidth of the frequencies. This offers considerable savings in time over the other methods, where each frequency has to be computed separately.

Because the whole volume up to the far field has to be modelled to get a far-field radiation pattern, theoretically the problem volume is infinite. To truncate the volume to be modelled, absorbing boundary conditions (ABC) are applied. This effectively absorbs the wave like a perfect match, in the same way as we would terminate the antenna terminals to prevent an unconnected antenna from re-radiating an EM wave that is incident on it.

Because the volume to be modelled is large, this method is not often used for large aircraft since the CPU times can be very long. However, it has been used for modelling part of the airframe, when the rest of the airframe is not in LOS of the antenna and hence it was not necessary to include the rest of the airframe in the modelling. Thus it can be used, for instance, to model a TCAS antenna on the nose of the front fuselage of a small aircraft. Only the nose and the cabin/cockpit would need to be modelled in this case.

7.5.1 Comparison with Measurements

Measurements on a 1/15th scaled model of a Fokker 100 scaled model aircraft were compared with the predicted radiation patterns for positions 1 and 2 as depicted in Figure 7.4. Modelling was undertaken using FDTD software at the highest frequency of 16.4 GHz (corresponding to an aircraft frequency of 1093 MHz) for positions 1 and 2 only. The correlation between the relative measured and computed data for the azimuth plots is very good, as manifested in the very good comparison in the shape of the radiation patterns (Figure 7.11). There are differences in the absolute levels, which is quite common, although the correlation with the absolute data is much better for position 2.

In the case of the roll plane plots the correlation in the upper hemisphere is very good for both positions. However, the correlation in the lower hemisphere for position 1 is not as good as it is for position 2. For position 1 the predicted pattern does not have the same number and depth of nulls as the measured patterns and the levels are also slightly higher. This indicates that the predictions for creeping waves are not as accurate. These higher levels compared with the measured plots are not desirable since this gives a more optimistic view of the coverage that can be achieved in practice. For roll plane plots of position 2 where the antenna is above the wings the lack of creeping waves into the lower hemisphere results in the predicted patterns being more accurate, and there is a remarkable correlation with the measured pattern.

7.6 GTD/UTD

In mathematics the term asymptote is used to denote that fact that a variable approaches a certain value, but only attains that value at infinity. For instance, the decay of a radioactive material asymptotically approaches zero in infinite time. The term asymptotic is commonly used in GTD and UTD for a similar reason, and the computations are based on approximations of Maxwell's equations.

In these cases the waves are considered as rays that are reflected (subject to Snell's laws of reflection) and diffracted at edges. Creeping waves around curved surfaces are also included in these computations. In both cases the rays are assumed to be plane waves and thus the nearest interaction point from the source has to be in the far field of the antenna.

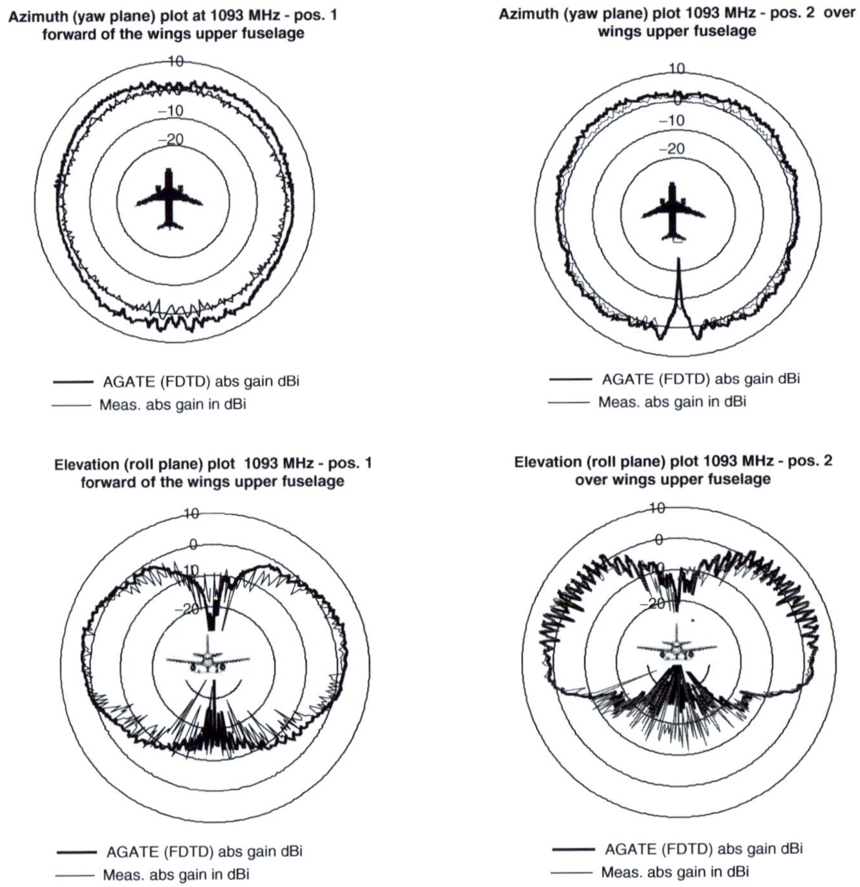

Figure 7.11 Fokker 100 Scaled model measured radiation patterns compared with computed patterns modelled using FDTD software at positions 1 (forward of the wings) and 2 (between/above the wings).

These methods are used when the structure size is greater than about 10 wavelengths. Reflection and diffraction coefficients are available for canonical surfaces, such as cylinders, ellipsoids, cones and plates.

7.6.1 Geometric Theory of Diffraction

This appears to be a contradiction in terms, since diffraction is a phenomenon of physical optics, as opposed to geometric optics that deals with the rectilinear propagation of light, that is, light travelling in straight lines. This theory is based on geometric optics and the physical theory of diffraction.

Using ray-tracing results in there being no dependence on the size of the structure. If the frequency is increased, there is no resulting increase in the runtime which is a major drawback in the case of MoM and FDTD methods.

The main drawback of GTD and UTD is the limitation on the electrical dimensions of obstacles/excrescences that can be modelled. No linear dimension can be less than λ, although the diameter of cylinders can be 0.8λ. There is a restriction on the relationship between the length and its radius, in that the length of a cylinder must be greater than its radius.

The low frequency limit is set by the diameter of the cylinder that is required to be included in the modelling. Thus, for instance, if the diameter of an engine is 1 m this must equate to 0.8λ, giving a value of 1.25 m (1/0.8) for λ. This corresponds to a frequency of 240 MHz and thus GTD cannot be used for frequencies below this.

7.6.2 Uniform/Unified Theory of Diffraction

UTD attempts to deal with the semi-lit region (sometimes known as the penumbra). It is used for antennas on curved surfaces and smoothes out the boundaries between the fully lit and semi-lit regions, as well as between the semi-lit and full shadow regions.

7.6.3 Creeping Waves Around Cylinders

If a monopole is placed on a cylinder of circular cross-section, the creeping waves are propagated around the circumference, shedding power tangentially as they travel over the curved surface. The angular distance that the wave traverses before having negligible power left depends on the electrical radius of the circular cross-section. This is depicted in Figure 7.12a, which shows the angular extent that the waves traverse for different cylinder radii. The smaller the radius of curvature the more the wave travels around the circumference. For instance, if the radius of curvature is 0.2λ the wave goes around about

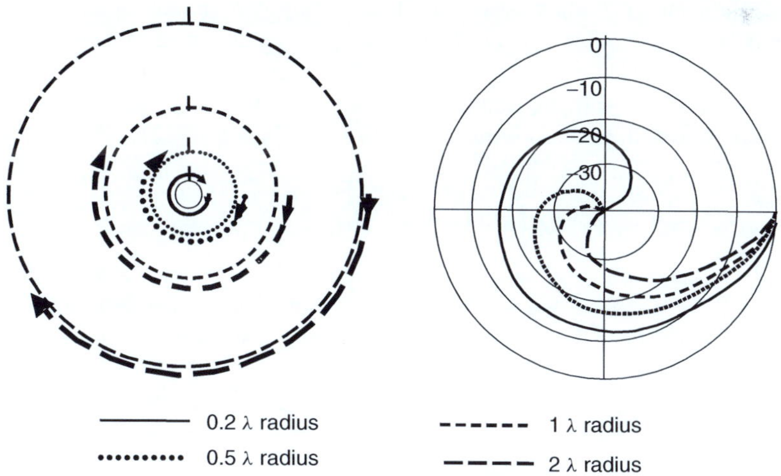

(a) The angular distance travelled
(b) The relative amplitude of the wave

Figure 7.12 The creeping waves around a cylinder.

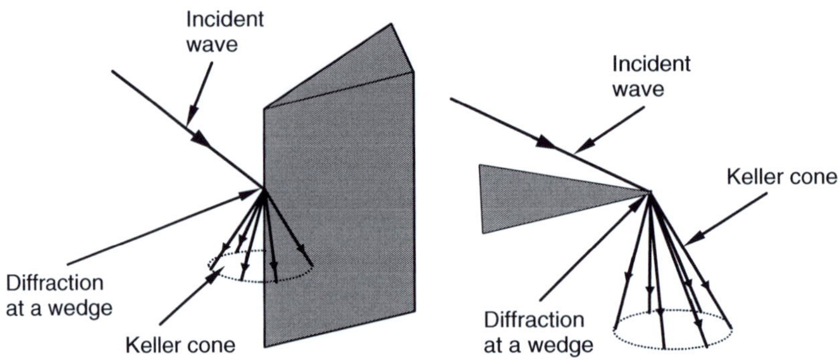

Figure 7.13 The Keller cone showing diffraction at the edge of a wedge.

290°. Figure 7.12b shows the relative amplitude of the wave as it moves around cylinders of different radii. The ray will travel round the cylinder several times but radiates at a low level which will produce a low-level fast ripple in the far-field radiation pattern.

In PO a ray of light striking an edge acts as a secondary source (sometimes called a Huygens source) and the secondary source radiates spherically like the diamond rings seen in the total eclipse of the sun.

In GTD the secondary source radiates into a cone of light (known as the Keller cone) as shown in Figure 7.13, with the half angle of the cone, as well as the angle at which the cone radiates, dependent on a number of factors such as the angle the edge makes with the incident ray, the frequency, the polarization, the distance of the source from the edge, etc., – see [6]. Thus there is a sharp cut-off of the angular distribution of the diffracted ray, unlike the case in the real world where a sphere of diffracted rays is present. Furthermore, if the cone half angle and/or the angle at which the cone is formed is incorrect, then the secondary (diffracted) rays may not appear in the final radiation pattern. Thus any subsequent reflection or diffraction of these secondary rays will also not appear in the final radiation pattern. This can be seen in some of the radiation patterns where there is sharp step showing the presence of the direct (LOS) wave but the diffracted rays and any rays resulting from other interactions are not present.

7.6.4 Limitations of GTD/UTD Modelling

For GTD/UTD modelling, the obstacle modelled has to be several wavelengths away (about 10 to 12 for aperture antennas) and each dimension of the structure has to be at least one wavelength. Thus details in the structure cannot be modelled and low frequencies cannot be used.

Diffraction coefficients are used to compute the diffracted waves from different types of surfaces and edges. For curved surfaces only lower-order terms exist, but for a curved edge higher-order terms may be required [7].

In the case of curved surfaces the reflecting surface should be greater than one wavelength square, and the smallest radius of curvature should be greater than half a wavelength.

In the case of a curved edge the total length of the diffraction edge should be greater than one wavelength and the smallest radius of curvature should be greater than half a wavelength.

The diffraction coefficients used depend on several factors.

1. The wave incident on the point where diffraction occurs must be plane over a square of side 0.5λ or 1λ.
2. The distance from the source of all parts of the square must be relatively equal.
3. The amplitude of the wave must not vary appreciably over this square area.

The wave striking an obstacle is assumed to have a plane wavefront locally. By 'locally' is meant over 0.5λ to 1λ. The distances of diffraction points from the source are assumed to be constant over a planar 1λ × 1λ square of the wavefront. It is also assumed that the field does not vary appreciably over this wavefront. If we consider a portion of the spherical wavefront, we can see from Figure 7.14 that as the wave moves further away from the source, the wavefront is less curved and the curvature matches a planar square more closely.

There is an error in the amplitude as well as the phase that increases with the radius of curvature as well as the length of the side of the square.

If we calculate the difference in the space loss between the centre of the planar square and the edge, we can get the variation of the amplitude and phase due to this difference in the distance the wave has travelled, compared with the case that matches the actual wavefront of the wave at different distances/radii.

The variation of the amplitude and phase is calculated as shown in Table 7.1.We can see that for a wavefront of radius 0.5λ, if the length of the planar edge is doubled from 0.5λ to 1λ the maximum amplitude variation increases more than threefold (from 0.48 to 1.51) and the maximum phase variation is approximately 3.5 times as great (75° instead of 21.2°).

Figure 7.15 shows the maximum amplitude and phase variation as a function of the radius of the wavefront for two planar squares of edge lengths 0.5λ and 1λ.

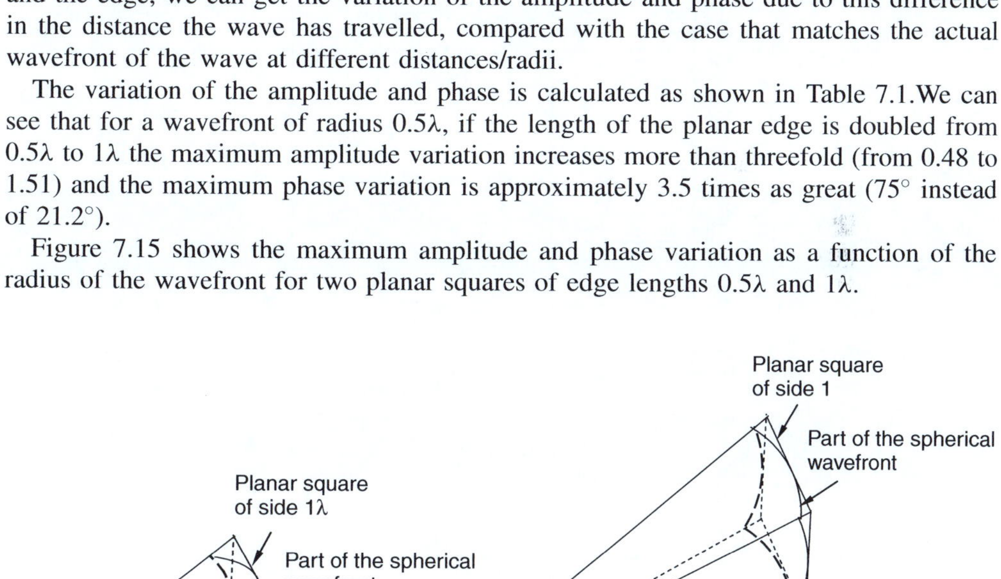

Figure 7.14 Curvature of the spherical wavefront at different distances from the source.

Table 7.1 Maximum amplitude and phase variation as a function of wavefront radius.

Radius of wavefront in wavelengths	Space loss in dB	Length of edge of plane 0.5 wavelength	Distance to outer edge of plane	Space loss in dB	Maximum amplitude variation	Difference between distance to centre and to edge of plane	Maximum phase variation
0.5	−18.97	0.5	0.56	−19.46	0.48	0.06	21.2
0.5	−18.97	1	0.71	−20.48	1.51	0.21	75
1	−21.98	0.5	1.03	−22.12	0.13	0.03	11
1	−21.98	1	1.12	−22.47	0.48	0.118	42
2	−24.99	0.5	2.02	−25.03	0.03	0.02	5.6
2	−24.99	1	2.06	−25.13	0.13	0.062	22
3	−26.76	0.5	3.01	−26.77	0.02	0.01	3.74
3	−26.76	1	3.04	−26.81	0.06	0.04	15

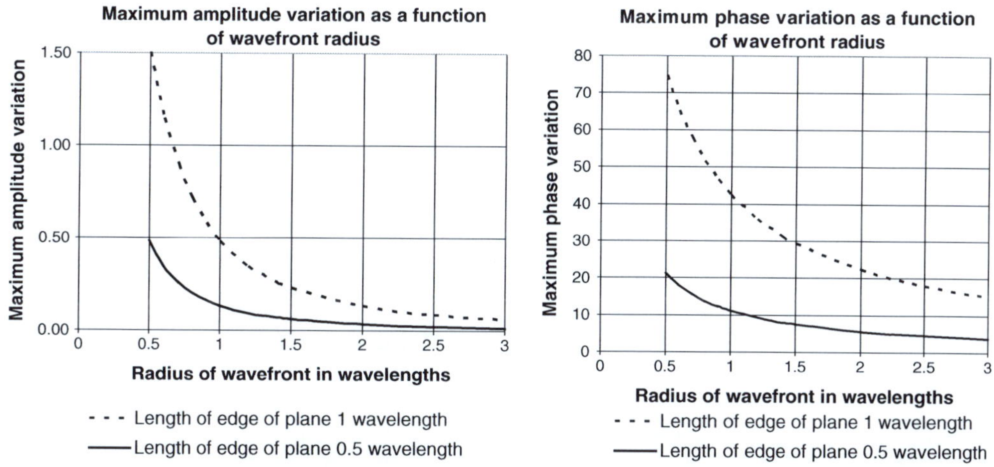

Figure 7.15 Variation of the amplitude and phase for different radii of wavefronts, with the length of the planar edge as a parameter.

7.6.5 Geometric Surface Models for GTD/UTD

For GTD/UTD the surface model of the airframe is made up of canonical models. Thus the fuselage could be a cylinder of circular or elliptical cross-section and the nose and tail could be a cone or part of an ellipsoid.

The canonical model of a large aircraft with a length of about 40 m, a wingspan of 38 m and a fuselage diameter of 3.12 m is shown in Figure 7.16 [8]. This airframe was considerably more difficult to replicate using canonical shapes since although the pressurized fuselage had a circular cross-section of about 3 m diameter, there was a bay below the pressurized cabin, enclosed by curved doors, resulting in a fuselage of almost elliptical cross-section, with a vertical major axis of about 4 m. The side elevation of the aircraft is shown in Figure 7.18.

In this case different canonical models were used for the same antenna, depending on whether an azimuth or elevation pattern was being produced. No predictions were obtained for the pitch plane patterns, since no measured pitch plane patterns were available for comparison.

A 3D canonical model of a Fokker 100 airframe was produced for use on the IPAS project and is described in IPAS Deliverable D16 [2] (see Figure 7.17a). This deliverable was used for the verification of modelling tools by demonstrating correlation between measurements on scaled models and computations. The CAD model is shown in Figure 7.17b to demonstrate the good representation that can be obtained with a canonical model for this particular airframe. A good representation of the airframe results in better correlation between the measured and predicted radiation patterns. The comparisons are shown in Section 7.6.7.

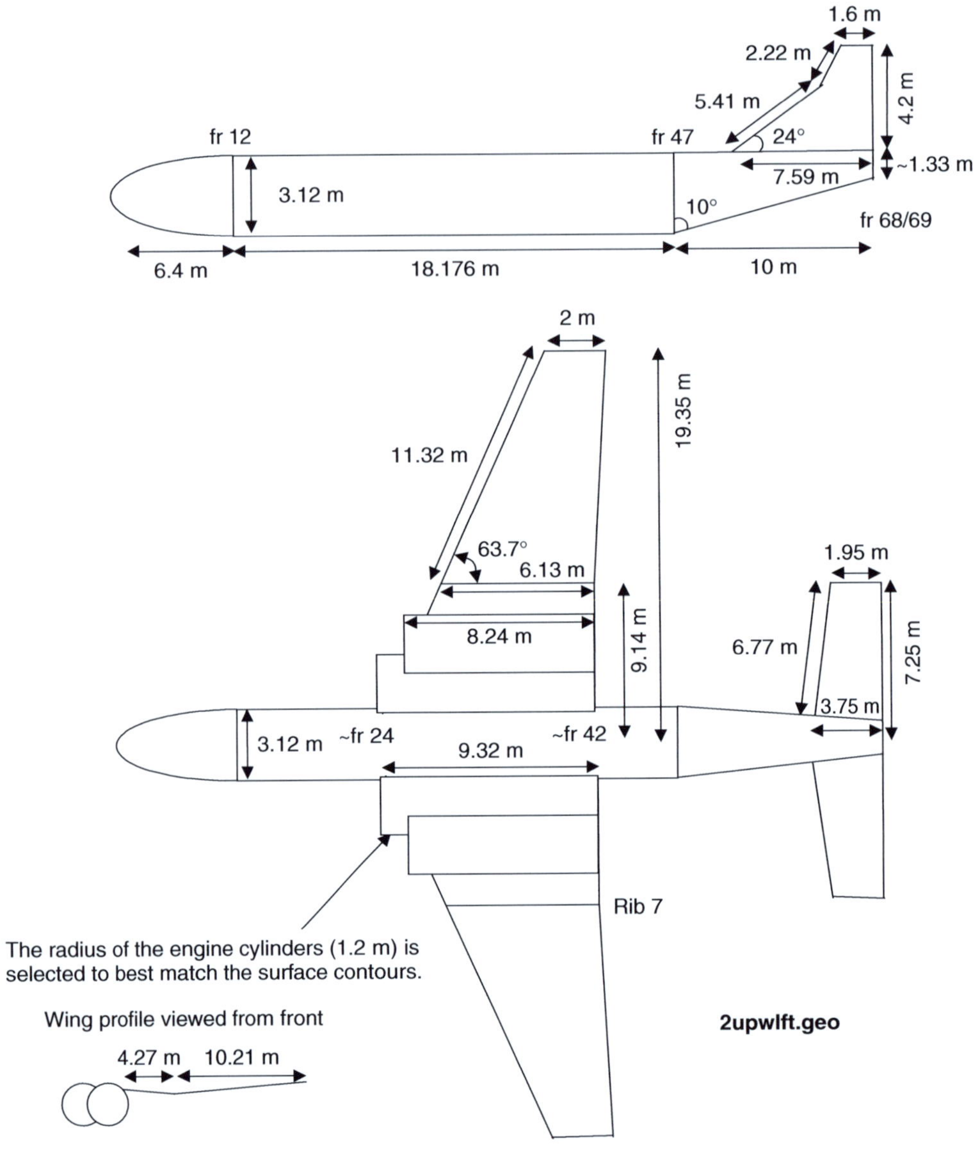

Figure 7.16 Canonical model of a large aircraft [8].

7.6.6 Higher-Order Interactions

A ray reflected off one surface can then be reflected or diffracted at another interaction with the structure. As the ray undergoes these multiple interactions its intensity is reduced and it contributes less to the final radiation pattern. Table 7.2 (from [3]) shows the levels of the ray compared with the incident ray as it undergoes interactions with a structure. This does not include the space attenuation to which a ray is subjected as it travels

Figure 7.17 3D view of the canonical (ALDAS) model of the Fokker 100 aircraft used in the IPAS project compared with the CAD model [2]. Reproduced by kind permission of NLR. Reproduced by kind permission of NLR

Table 7.2 The ratio of levels (in decibels) of a wave as it undergoes different interactions, compared with the initial incident ray [3].

Interaction	Level in dB compared with the initial incident ray
Reflection at flat plate	0.0
Reflection at curved surface	<−6.0
Diffraction at straight edge	−10.0
Diffraction at curved edge	<−16.0
Diffraction at vertex	<−16.0
Diffraction round cylinder	<−20.0
Diffraction round cone	<−15.0
Diffraction round curved surface	<−20.0
Multiple interactions	<−20.0

different distances. In the proprietary GTD/UTD software called ALDAS (Analysis of Low Directivity Antennas on Structures), an option exists to select the number of interactions used to obtain the far-field radiation pattern. The greater the number selected the longer is the processing time. It can be seen from the table that, apart from the direct wave, the reflection at a flat plate contributes the largest extent to the radiation pattern. The next largest contributor is the reflection at a curved surface.

7.6.7 Comparison of GTD/UTD with Scaled Model Measurements

The radiation patterns were obtained for a D-band antenna installed on the upper fuselage of the scaled model of a large aircraft and forward of the tail fin as shown in Figure 7.18 [8]. The comparison between the predicted and measured azimuth radiation patterns of the antenna is shown in Figure 7.19 for the straight and level and for the 2° nose-up attitudes of the aircraft. Note that the computed plot of Figure 7.19a shows a sharp null to the rear where the tail fin is, but in other angular sectors there is good correlation with the measured plot. The radiation patterns are not quite symmetrical because the antenna

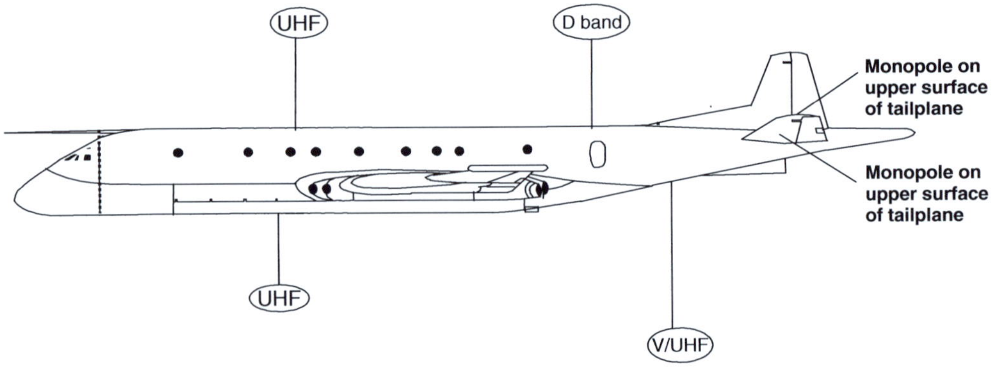

Figure 7.18 Positions of antennas installed on the scaled model of a large aircraft.

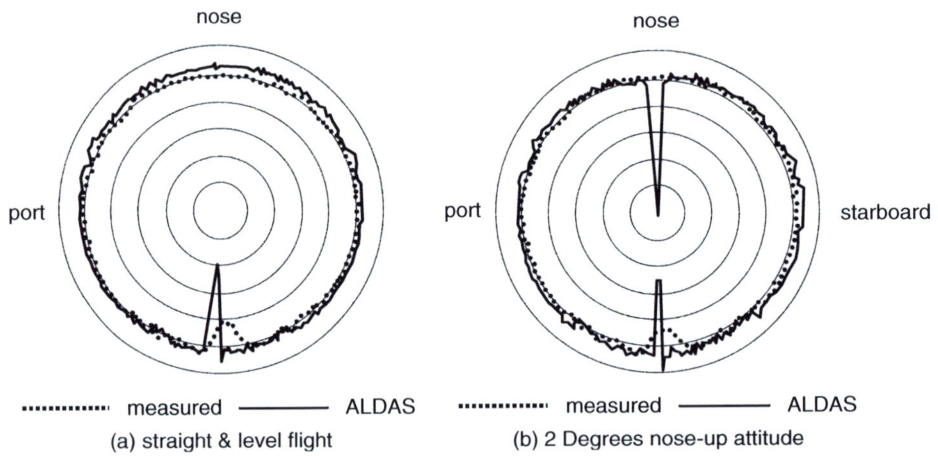

Figure 7.19 Azimuth plots for an antenna on the upper fuselage at 1215 MHz.

is mounted about 2 inches off-centre. Since the fuselage diameter was 3 m (118 inches) the circumference is about 371 inches, so that 2 inches represents just over 2°.

When the aircraft is in the 2° nose-up attitude, there is a sharp null forward. This is due to the geometric canonical model used in the ALDAS modelling (shown in Figure 7.16) which does not provide a good match to the real profile of the airframe. There is also a change in diffraction coefficient as the rays move from the circular fuselage to the semi-ellipsoid nose section.

Measurements on a 1/15th scaled model of a Fokker 100 scaled model aircraft were compared with the predicted results of the radiation patterns for positions 1 and 2 as depicted in Figure 7.4. The CAD model used for the GTD/UTD is shown in Figure 7.17. The azimuth radiation patterns depicted in Figure 7.20 show differences in the rear sector because the tail fin is not accurately modelled using canonical shapes. For position 1 it appears that specular reflections off the side of the tail fin combine with the direct rays

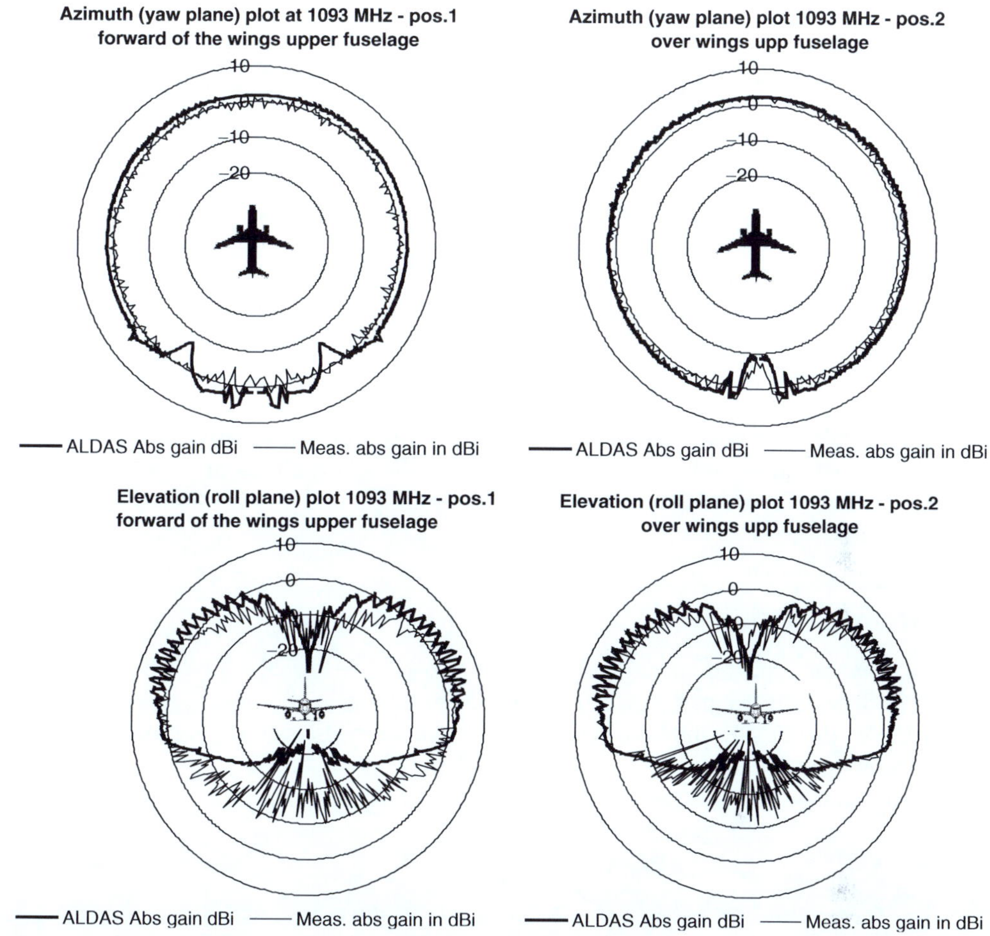

Figure 7.20 Fokker 100 Scaled model measured radiation patterns compared with computed patterns modelled using GTD/UTD at position 1 (forward of the wings) and position 2 (between/above the wings).

to give nulls either side of the tail. The same effect is not apparent for position 2, since the phase differences in this case are not the same as those for position 1.

In the roll plane patterns the lack of correlation in the lower hemisphere shows that the ALDAS software does not predict the creeping waves accurately. There always seems to be a discrepancy in the case of absolute gains between the measured and predicted values. In the cases shown the absolute data compares much better for position 2 than it does for position 1.

7.7 Physical Optics

In this case the waves are treated like rays of light with edge diffraction treated to a limited extent. The ray optics estimates the field on a surface and then integrates that field over the surface to calculate the transmitted or scattered field. It assumes that the currents induced on a metallic surface are the same as those induced on an infinite plane that is locally tangential to the surface. No allowance is made for multiple interaction, and edge diffraction is only partially treated. This software is used mainly for large reflector antennas where the interaction with the supporting surface is negligible and it is not accurate for low-gain antennas (such as monopoles) on the fuselage of an aircraft.

7.7.1 Comparison with MoM

The predicted patterns were compared on an antenna placed on the rear upper fuselage of the scaled model of the IPAS-1 aircraft (a single aisle aircraft) at 2 GHz (see Figure 7.21). Since the scale was 1/18th of the full-scale aircraft the equivalent aircraft frequency was 111 MHz. The correlation with MoM is not very good. The lack of higher-order interaction of the PO computation is manifest in pitch plane plots by the omission of beating especially in the lower hemisphere. This is due to the sharp cut-offs between the lit and semi-lit regions and can be seen in sharp drop-offs in the surface current densities in Figure 7.27b.

7.8 Hybrid Methods

There are several hybrid methods that combine two computational codes. These could be used to bridge the frequency gap or permit the modelling of different types of material. For instance, the FDTD method can be combined with the MoM to model antennas on structures that contain dielectrics. The dielectric volume and the surrounding air are modelled as cells.

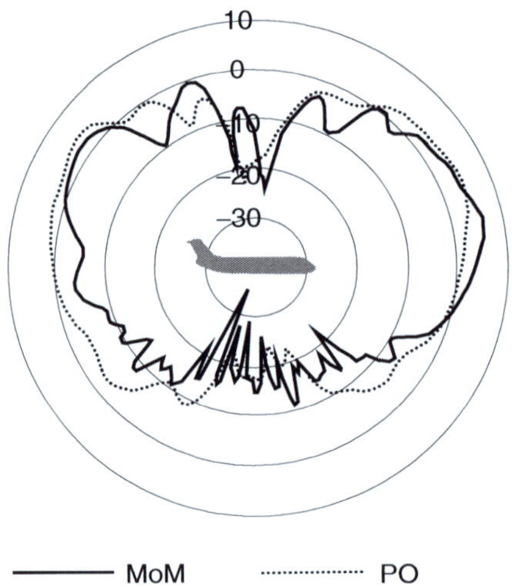

Figure 7.21 Comparison between the MoM and PO pitch plane radiation patterns on the scaled model of the IPAS-1 aircraft (a single aisle aircraft) at the equivalent aircraft frequency of 111 MHz. Reproduced by kind permission of EADS.

7.8.1 Multi-domain Method

The MD approach is based on the BEM, the finite element method and high frequency techniques applied to the analysis of antenna performance installed on aircraft and other structures. This method provides considerable reduction in the time taken compared with using MoM on its own. In IPAS Deliverable D3 on 'bridging the frequency gap' [9], the Office National d'Etudes et de Recherches Aérospatiales (ONERA) has described the FACTOPO subdomain method for electromagnetic computations based on BEM and high frequency techniques. One of the benefits of the subdomain method is the ability to divide the domain into several volume subdomains such as the antenna, airframe, engine, and so on, which can then be dealt with by several companies. The CAD files can also be decomposed into several portions using a tool such as GiD_CEM. The details of the antenna such as the radiating element and other physical characteristics are modelled and then the fields are computed on a fictitious enclosed surface. This surface is then used as the source and the combined field integral equation (CFIE) can then be used to compute the far-field radiation pattern as in the case of MoM. Asymptotic techniques (geometric optics, physical optics, physical theory of diffraction) can also be used in the same way.

There is also a reduction of the computation by one or more orders of magnitude especially in the context of parametric studies (i.e. the variation of one variable with another at fixed increments of a third variable). Global type basis functions are defined on surfaces surrounding the antennas. In Section 4.2.4 of IPAS Deliverable 16 [2], on scaled models, the accuracy and efficiency of this technique was assessed by computations of communication VHF and ATC monopole antennas mounted on Fokker 100 and ATR 42 scaled models. MD results obtained by combining two sets of subdomains were compared with direct BEMs and measurements provided by BAE Systems, NLR and ATR for several positions of the monopoles to show the benefits of the MD method.

One advantage of the MD method is that it allows antenna manufacturers to provide the radiating properties of their antennas without disclosing proprietary information about the actual physical structure of the radiating element.

7.8.1.1 Comparison with Measurements

The predicted radiation patterns obtained using FACTOPO at 3.6 GHz (aircraft frequency of 240 MHz) were compared with the measurements performed on an antenna placed on the Fokker 100 scaled model at position 2 on the upper fuselage above the wings (Figure 7.22). The correlation with the relative gain in the azimuth plane is very good, although there are variations in the absolute gain. The correlation with the roll plane measurements is very good in the upper hemisphere, showing almost identical levels in absolute gain. In the lower hemisphere the correlation is not very good in the number and positions as well as the depth of the nulls. These were also published at 120 MHz in a paper by Barka and Caudrillier [10].

7.8.2 Physical Optics and Hybrid MoM/PO

The PO method assumes that the currents induced on a metallic structure by an incoming field are the same as those induced on an infinite plane locally tangential to the surface.

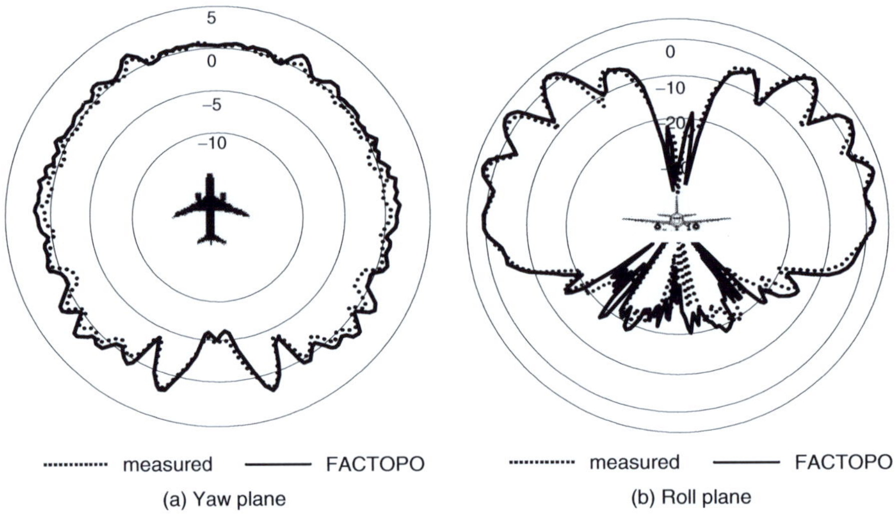

Figure 7.22 Comparison between FACTOPO and measurements on the Fokker 100 scaled model for an antenna on the upper fuselage above the wings at 240 MHz aircraft frequency. Reproduced by kind permission of ONERA.

In its simplest formulation multiple interactions (such as more than one reflection) are not considered and edge diffraction is only partially treated. To overcome these shortcomings the method has been hybridized with MoM and the equivalent edge current method. ASERIS/EMC2000 has been developed by EADS-CCR and NLR contributed to the development and valuation of the hybrid method in EMC2000. In the IPAS Deliverable D3 on 'bridging the frequency gap' [9], NLR has described a hybridization technique that involved combining the electric field integral equation with PO and the method of equivalent currents.

7.8.3 Hybrid of MoM and GTD/UTD

A combination of MoM and GTD/UTD was used in the case of a large aircraft where the possibility existed of installing antennas on the tailplane [8]. This tailplane had vertical stabilizers on its upper and lower surfaces and perpendicular to it. These were to be fabricated from fibreglass and were about 5 m outboard from the tail fin. The frequencies used by the V/UHF radios that were to be connected to these antennas were so low that the tailplane was very small in terms of wavelengths, and modelling the detail in the electrical vicinity of the antennas would necessitate the use of MoM code. However, because of the prohibitively large number of elements that would be required for the whole structure in MoM (in the late 1990s), a hybrid of MoM and GTD/UTD was used. The MoM software used was NEC2 and the GTD/UTD software was ALDAS. The side view of the aircraft is sketched in Figure 7.23.

Measurements were undertaken on a scaled model of the aircraft using scaled quarter-wave tuned monopoles at the positions of the vertical stabilizers on the upper and lower surfaces of the tailplane and perpendicular to the tailplane [8].

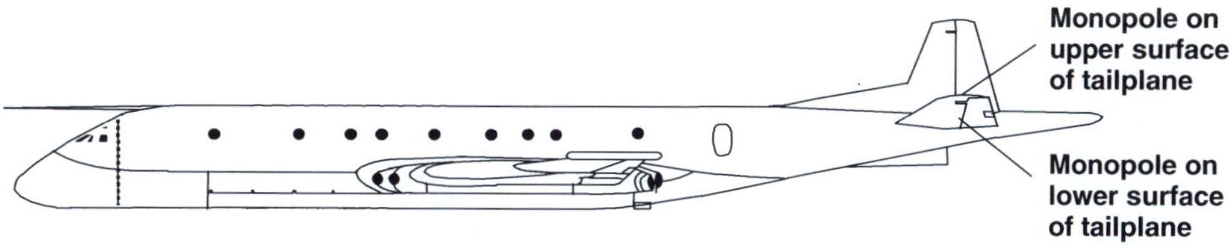

Figure 7.23 Tailplane of a large aircraft with monopoles. Reproduce by kind permission of EADS.

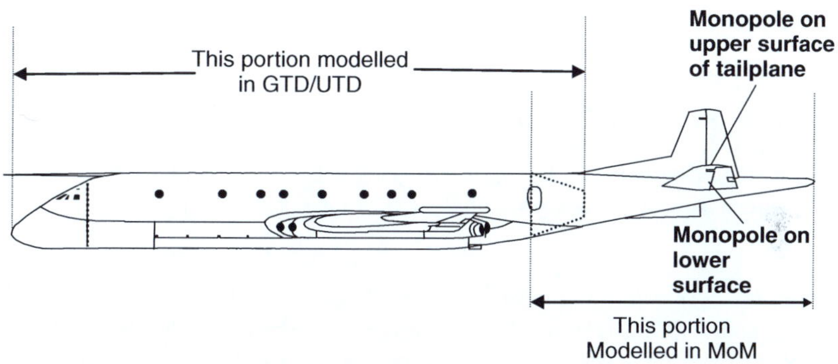

Figure 7.24 The portions of the airframe modelled in NEC and ALDAS.

One of the frequencies used by the V/UHF radios was 160 MHz and thus measurements were undertaken at this frequency. Since the wavelength is 1.88 m at this frequency, GTD/UTD could not be used because the fuselage was electrically too near and the size of obstacles were too small. Thus NEC2 code was used to model the detail of the tail fin, tailplane and rear fuselage in the electrical vicinity of the antennas. The resultant radiation pattern was then used as a feed pattern for the ALDAS package. It was expected that those parts of the aircraft modelled in MoM could be omitted from the ALDAS model (i.e. the ALDAS model of the aircraft should not include the tailplane, tail fin or the rear tail section of the aircraft). However, by a process of iteration it was found that a truncated cone, as shown in Figure 7.24, was required to give a closer correlation with the measured pattern. The ALDAS geometry used for these plots consisted of a four-part wing, a cylindrical fuselage of elliptical cross-section and a truncated tail cone approximately 6 m long.

7.8.3.1 Comparison with Measurements

The comparison between the measured and predicted radiation patterns at 160 MHz and 243 MHz for an antenna placed on the lower tailplane is shown in Figure 7.25. The nulls to starboard (between 0 and 30°) in the predicted plots at both frequencies for the monopole on the lower port tailplane are due to the discontinuity caused by the truncated cone section of the tail modelled in ALDAS.

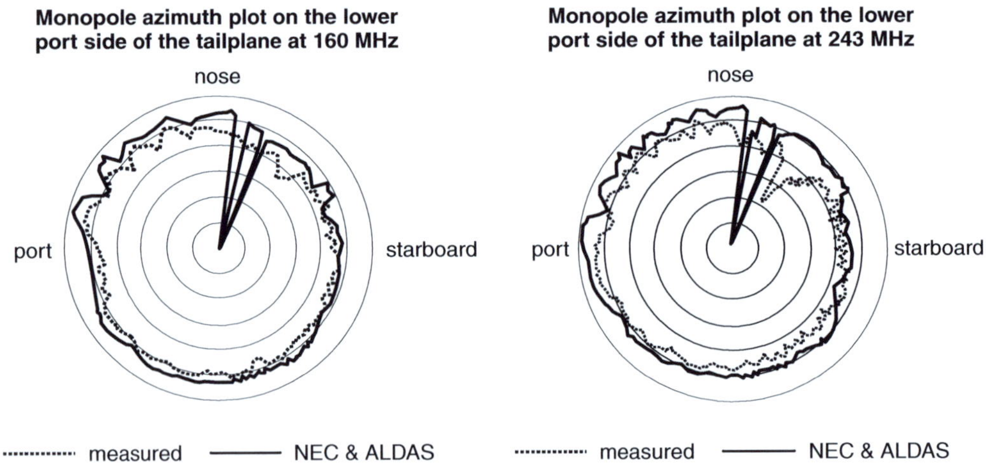

Figure 7.25 Comparison between the measured and predicted radiation patterns at 160 MHz and 243 MHz, for antennas on the lower port tailplane of a large aircraft.

The difference between the measured and predicted radiation patterns in the 0–45° sector, especially for 243 MHz, is due to the lack of detail in the ALDAS model. The scaled model of the real aircraft has a profile that is not accurately represented by the conical profile (of truncated tail cone) in the ALDAS canonical geometric model. The predicted and measured patterns show good correlation in the general shape over the rest of the angular sector, although the ALDAS patterns are smoother and do not show the same number of ripples as the measured ones, due to the lack of detail in the canonical model.

The comparison between the measured and predicted radiation patterns for the monopoles on the upper starboard tailplane are shown in Figure 7.26. Here the correlation is much better, although again the smaller numbers of ripples in the predicted patterns (especially for 243 MHz) are again due to the lack of detail.

7.9 Comparison of Predicted Surface Currents

It is normal to increase the density of the mesh near the antenna and have a looser mesh in areas far away from the antennas. However, the density of meshing required actually depends on the areas exposed to high EM fields. The higher the exposure the tighter the meshing (i.e. finer meshes) required. Areas that are exposed to lower EM fields can have looser meshing with greater grid spacing. These high fields induce high currents on the surfaces, so it is useful to obtain predictions of the current densities on the surface of the CAD model so that these areas can be identified. Graphic illustrations produced by the various computation packages are very useful in determining these areas that require special attention. These surface currents are used to calculate the near-field as well as the far-field radiation patterns. In the case of the near-field calculations, because the inductive effects have to be taken into account, the higher-order terms have to be retained. However, this is not necessary for the far-field radiation patterns.

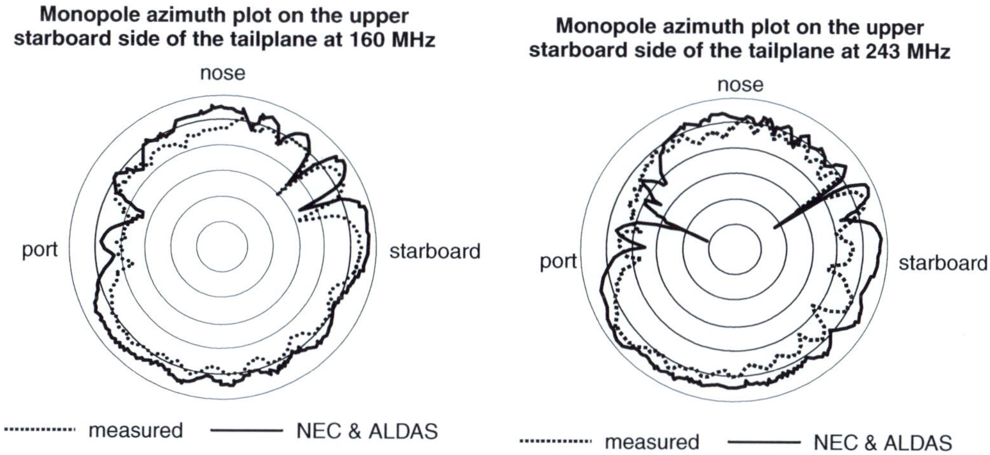

Figure 7.26 Comparison between the measured and predicted radiation patterns at 160 MHz and 243 MHz, for antennas on the upper starboard tailplane of a large aircraft.

7.9.1 Comparison between the Surface Currents on an Airframe using Different Codes

In order to qualitatively analyse the differences between codes, comparisons between the surface currents on a single aisle aircraft were obtained using different software. This task was undertaken as part of the IPAS project and reported in Deliverable D16 [2]. The antenna was placed on the rear upper fuselage as shown in Figure 7.27 and the white patch that shows the highest intensity appears near the antenna. The colours then vary to indicate decreasing order of magnitude, from red, orange, yellow, green and blue, down to the black which is the lowest (approaching zero) intensity. The scaled model aircraft was meshed for a frequency of 2 GHz and, since it was 1/18th scale, the equivalent aircraft frequency is 111 MHz.

(a) Surface currents using FMM (b) Surface currents using hybrid MoM/PO

Figure 7.27 Surface currents calculated with FMM and Hybrid MoM/PO [2]. Reproduced by kind permission of EADS. See Plate 10 for the colour figure.

The MoM currents vary smoothly over the surface of the airframe, but it can be seen from the PO prediction that the gradation in colour is a large step change to the black areas of low intensity. This is because PO is not good in the shadow boundaries such as the sides of the fuselage and other areas out of direct LOS of the antenna.

The surface currents shown in Figure 7.28 were obtained by DLR using PO, which only considers first-order reflections.

In Figure 7.29 the surface currents were predicted for an antenna placed in front of the tail fin of a Fokker 100. In these cases we can see the surface currents due to multiple reflections computed using FMM are not present in the case of the PO computations. Again we can see the step changes between the levels in the PO case, whereas there is a much smoother transition in the case of the FMM computations.

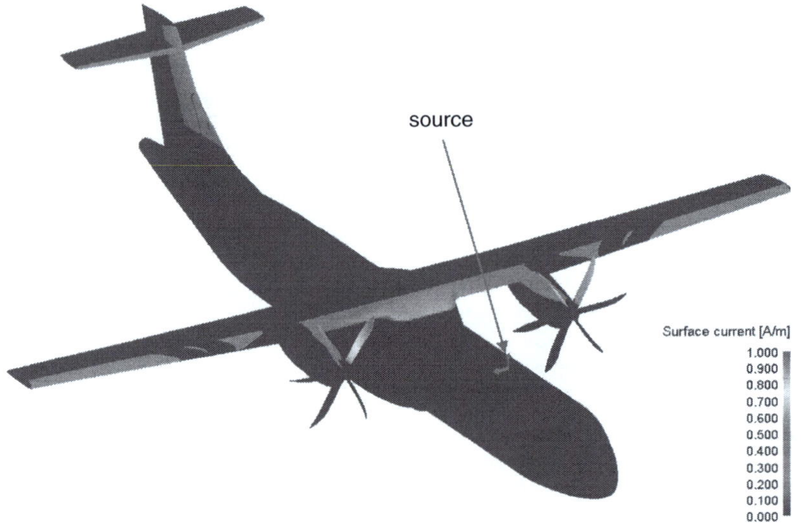

Figure 7.28 Surface currents on an ATR 72 airframe using PO [12]. ©2008 IEEE. Reproduced by kind permission of IEEE. See Plate 11 for the colour figure.

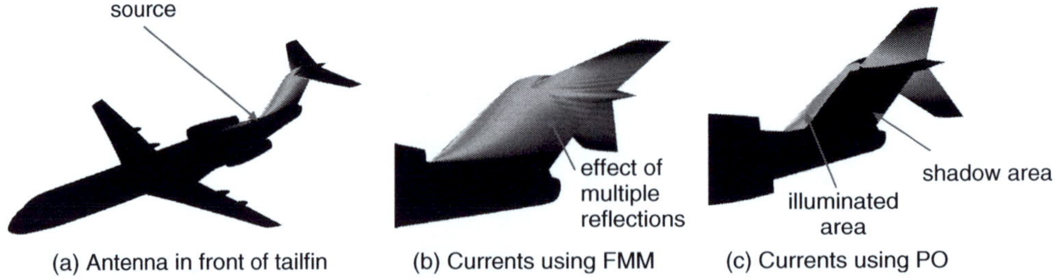

Figure 7.29 Surface currents on a Fokker 100 airframe using FMM and PO [13]. Reproduced by kind permission of IET. See Plate 12 for the colour figure.

Despite the differences in the surface currents calculated by MoM and the MoM/PO hybrid, there is surprisingly good correlation between the far-field radiation patterns produced using these different techniques.

7.9.2 Surface Currents on a Simplistic Airframe for Different Meshes

The greater the detail that is modelled in a structure, the more representative the structure is of the real outer profile and the better the radiation pattern is expected to correlate with the measured ones.

In order to model the detail in MoM the electrical lengths of the wire segments should be small. However, small segments length obviously increase the computation time. Under the IPAS project (Deliverable D17 [11]) an exercise was undertaken to alter the lengths and hence the number of segments used in different areas and compare the predicted surface currents induced on a simplistic 'aircraft'. One of the partners (EADS-CCR) modelled the canonical equivalent of a simplistic airframe that consisted of a 18 m long fuselage of 3 m diameter circular cross-section with hemispherical nose and tail sections. The wings were each 7.5 m long, giving a wingspan of 18 m, and the tail fin was 2.5 m high. The thicknesses of the wings and tail fin was 0.40 m. An antenna was placed (marked with a cross in Figure 7.30 on the nose 2 m aft of the nose tip.

The airframe was then meshed using three different configurations of wire grids, as follows:

1. Segments of around $\lambda/10$ for the entire airframe.
2. Segments of around $\lambda/10$ near the antenna, segments of around $\lambda/6$ segments for the rest of the fuselage further away from the antenna, the leading (forward) edges of the wings, and tail fin, and segments of around $\lambda/3$ for all the other areas.
3. Segments of around $\lambda/10$ near the antenna, and segments of around $\lambda/3$ for all the other areas.

In the case of configuration 2, although the leading (forward) edges of the wings and tail fin are quite far away from the antenna, they were modelled with $\lambda/6$ segments, rather

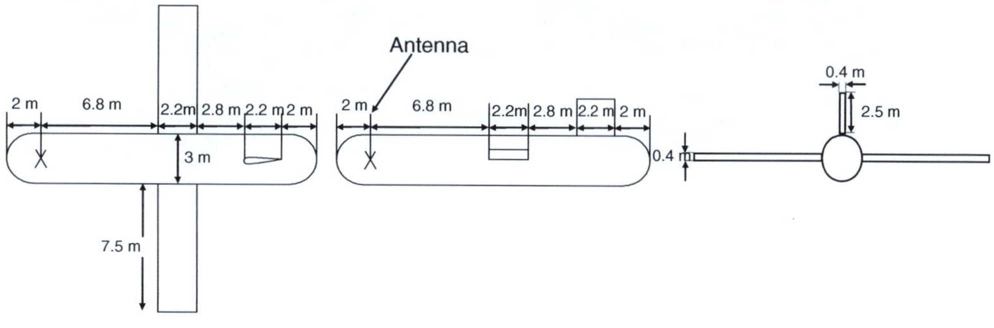

Figure 7.30 Simplistic airframe used to compare the current distributions obtained by varying the segment sizes of meshes. Reproduced by kind permission of EADS.

Table 7.3 The number of edges, assembly and processing times taken for each configuration of the wire gird model of the simplistic aircraft.

Configuration	Total number of edges	Assembly time in seconds	Total time in seconds using eight processors each with a clock speed of 2.4 GHz and 2 GB of RAM	Ratio of number of edges to the time taken
1	875 316	3 204	7 024	125
2	131 103	1 010	3 143	42
3	89 532	527	2 471	36

than λ/3 segments, because they were illuminated at low angles of incidence (i.e. almost normal incidence) by the radiation from the antenna.

The total number of edges, the time taken to assemble the wire grid model and the processing time are shown in Table 7.3, together with the ratio of the number edges to the processing time for each configuration of the simplistic aircraft. It can be seen that although the processing time increases with the number of edges, these times are not directly proportional to the number of edges.

The colour codes used to denote the areas of around λ/10, λ/6 and λ/3 segments were red, green and blue respectively. Figure 7.31a shows the aircraft in red all over for configuration 1, and the current distribution is shown in Figure 7.31b. Configuration 2 had a very small area in the vicinity of the antenna with λ/10 segments, a larger area on the front fuselage and the leading edges of the wings and tail fin with λ/6 segments, and the remainder of the aircraft with λ/3 segments. This is shown in Figure 7.32a and the corresponding current distribution is shown in Figure 7.32b. Configuration 3 had λ/10 segments in a very small area in the vicinity of the antenna, and λ/3 segments everywhere else, as shown in Figure 7.33a, while the current distribution is shown in Figure 7.33b.

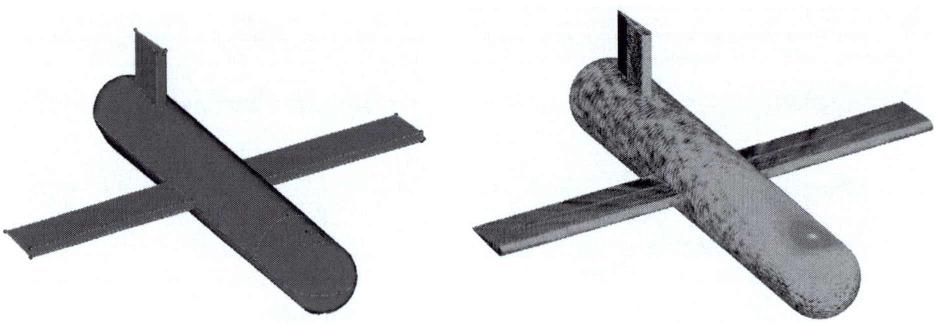

(a) Entire airframe with λ/10 segments

(b) Current distribution

Figure 7.31 Configuration 1 meshing and the corresponding current distribution. Reproduced by kind permission of EADS. See Plate 13 for the colour figure.

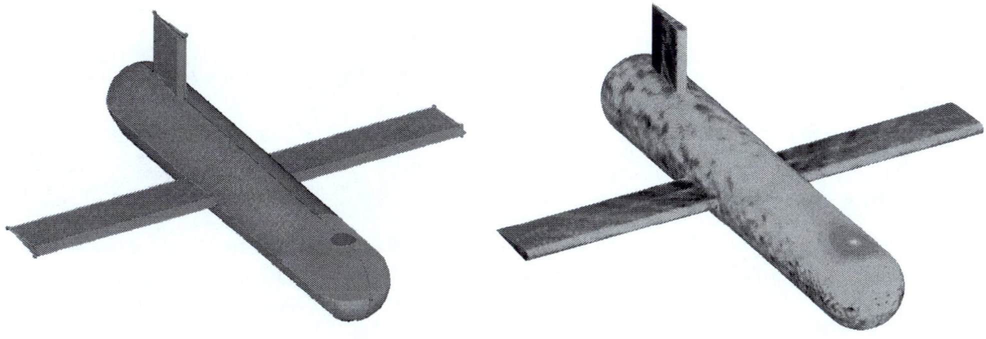

(a) Airframe with λ/10, λ/6 and λ/3 segments (b) Current distribution

Figure 7.32 Configuration 2 meshing (red = $\lambda/10$, green = $\lambda/6$, blue = $\lambda/3$) and the corresponding current distribution. Reproduced by kind permission of EADS. See Plate 14 for the colour figure.

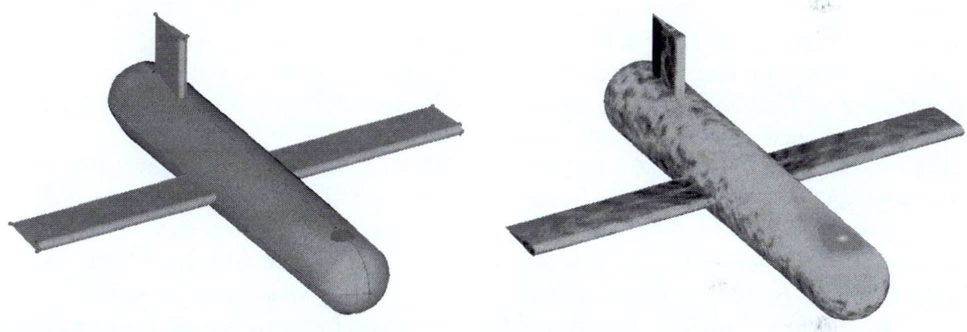

(a) Airframe with λ/10, and λ/3 segments (b) Current distribution

Figure 7.33 Configuration 3 meshing (red = $\lambda/10$, blue = $\lambda/3$) and the corresponding current distribution. Reproduced by kind permission of EADS. See Plate 15 for the colour figure.

Although the finer distribution of the current distribution is not captured in the case of configuration 3, the similarity can be seen between all the configurations. It was surprising to see that all three configurations gave almost identical radiation patterns.

7.10 Code-to-code Comparison of Radiation Patterns Predicted on the Simplistic Airframe

An antenna was placed on the nose, 2 m from the nose tip and the radiation patterns for the three principal planes obtained using FMM, MoM/PO hybrid and UTD. The FMM is considered to be the most accurate and therefore the other two codes were compared with it. The MoM/PO azimuth plane plot (Figure 7.34a) shows very good correlation with the

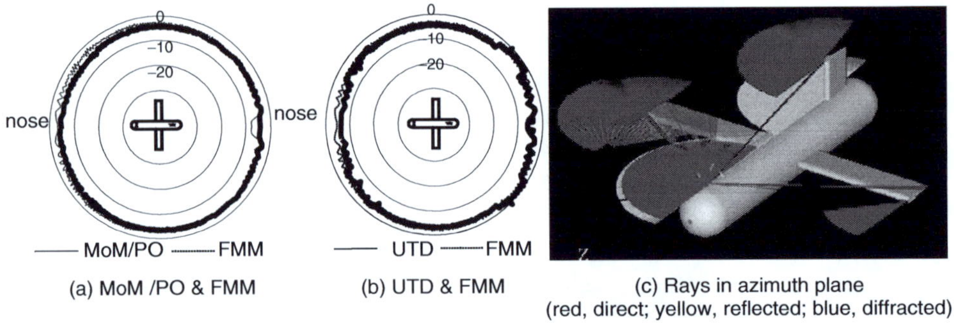

Figure 7.34 Comparison between the azimuth plane plots obtained using FMM, MoM/PO hybrid and UTD at 1 GHz for a simplistic aircraft. Reproduced by kind permission of EADS. See Plate 16 for the colour figure.

FMM plot, apart from the null due to the tail fin in the rear sector to the right. The UTD plot (Figure 7.34b) shows greater differences in the forward sector as well as two angular sectors to the left and right of the tail fin. The plot of rays generated by the ray tracer (Figure 7.34c) shows the reflected and diffracted rays from structure, as well as the direct rays in the azimuth plane. The creeping rays are not shown. For clarity only the rays to the starboard side of the aircraft are shown. The colour coding used is red for direct rays, yellow for reflected rays and blue for diffracted rays.

The roll plane plot for the MoM/PO (Figure 7.35a) shows very good correlation in the upper hemisphere, but the number and positions of nulls in the lower hemisphere do not correlate so well. In the case of the UTD plot (Figure 7.35b), the correlation is good to port and starboard but near zenith the levels are lower compared with the FMM plot and much smoother. In the lower hemisphere there are sharp peaks on either side as well a lack of any beating compared with the FMM plot. The plot of rays generated by the ray

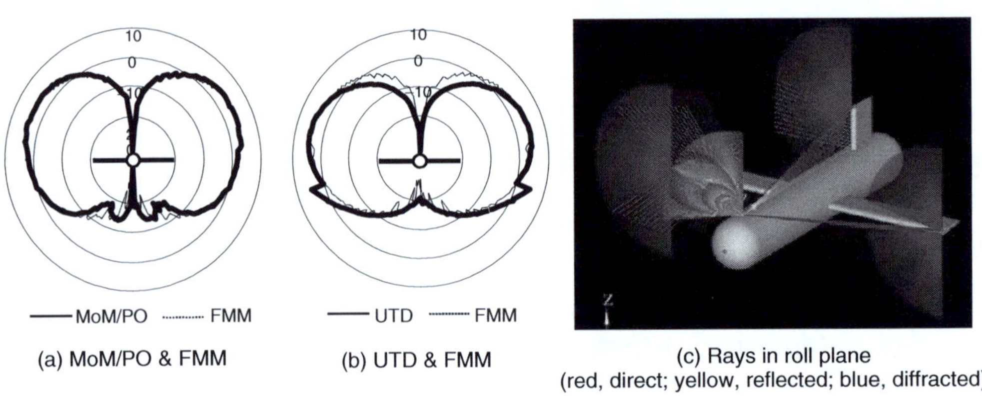

Figure 7.35 Comparison between the roll plane plots obtained using FMM, MoM/PO hybrid and UTD at 1 GHz for an antenna on the nose on a simplistic aircraft. Reproduced by kind permission of EADS. See Plate 17 for the colour figure.

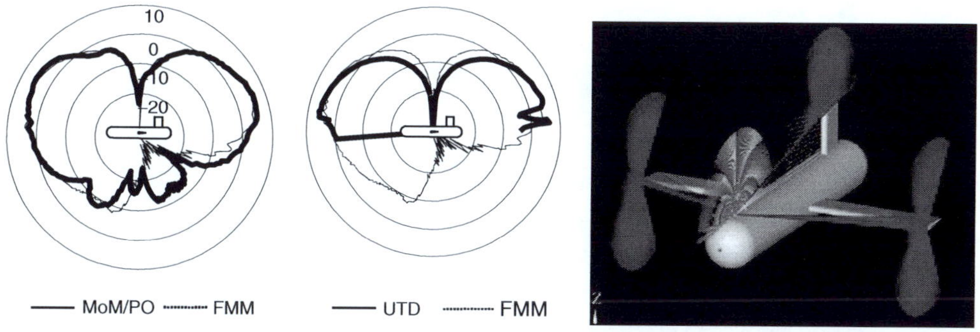

(a) MoM/PO & FMM (b) UTD & FMM (c) Rays in the pitch plane (red, direct; yellow, reflected; blue, diffracted)

Figure 7.36 Comparisons between FMM in the pitch plane and MoM/PO hybrid as well as UTD at 1 GHz for an antenna on the nose on a simplistic aircraft. Reproduced by kind permission of EADS. See Plate 18 for the colour figure.

tracer (Figure 7.35c) shows the reflected and diffracted rays from the structure, as well as the direct rays in the roll plane. The creeping rays are not shown. For clarity only the rays to the starboard side of the aircraft are shown. The colour coding used is the same as in Figure 7.34c.

The pitch plane plot for the MoM/PO (Figure 7.36a) shows very good correlation in the upper hemisphere, but the number and positions of nulls in the lower hemisphere do not correlate so well. There is an extra wide lobe in the case of the MoM/PO that does not appear in the FMM plot. In the case of the UTD (Figure 7.36b) the correlation in relative levels (relative gain) is quite good in the upper hemisphere although there are differences in the absolute gains. In the lower hemisphere there is a sharp cut-off for both the forward and rear sectors, so that there is no correlation in any area of the lower hemisphere. The plot of rays generated by the ray tracer (Figure 7.36c) shows the reflected and diffracted rays from structure, as well as the direct rays in the azimuth plane. The creeping rays are not shown. The colour coding used is the same as in Figure 7.34c. It can be seen that the lack of creeping rays results in no radiation in the lower hemisphere.

7.11 Relationship between Number of Unknowns and Surface Area

EADS-CCR have derived an empirical relationship between the number of unknowns and the surface area of different aircraft ([11], p. 148). The number of unknowns is proportional to the CPU runtime required to model the antenna performance on the aircraft as shown in Table 7.4. It can be seen that for a single aisle aircraft, the number of unknowns for modelling the antenna (if the segment length is $d = \lambda/10$) at 1 GHz (TCAS frequencies are 1.03 and 1.09 GHz) is 20 million. In Table 7.4, c is the velocity of EM radiation, approximately $3 \times 10^8\,\mathrm{m\,s^{-1}}$.

Table 7.4 An empirical relationship between the number of unknowns and the surface area for different aircraft.

Aircraft	Turbo prop	Single-aisle jet	Long-range twin-aisle	Super jumbo
Length (m)	27.05	31.4	75.3	73
Wingspan (m)	27.22	34.1	64.45	79.8
Height (m)	7.65	12.6	17.3	24.1
Surface area (m^2)	404	740	2537	4270
Number of surface unknowns	$1.90 \times 10^{-4}c^2/d^2$	$3.49 \times 10^{-14}c^2/d^2$	$1.20 \times 10^{-13}c^2/d^2$	$2.01 \times 10^{-13}c^2/d^2$
Number of unknowns at 100 MHz				
$d = \lambda/10$	19 000	35 000	120 000	200 000
$d = \lambda/6$	6800	13 000	43 000	72 000
$d = \lambda/3$	1700	3100	11 000	18 000
Number of unknowns at 300 MHz				
$d = \lambda/10$	170 000	310 000	1 100 000	1 800 000
$d = \lambda/6$	61 000	110 000	390 000	650 000
$d = \lambda/3$	15 000	28 000	97 000	160 000
Number of unknowns at 1 GHz				
$d = \lambda/10$	1 900 000	3 500 000	12 000 000	20 000 000
$d = \lambda/6$	680 000	1 300 000	4 300 000	7 200 000
$d = \lambda/3$	170 000	310 000	1 100 000	1 800 000

References

[1] Installed Performance of Antennas on AeroStructures (IPAS), an EU Specific Targeted Research Project, November 2003 to January 2007, Contract No AST3-CT-2003-503611

[2] Verpoorte, J. (2006) Verification of modelling tools by measurements and computation on scaled aircraft, Deliverable D16 Version 3, IPAS-TR_3/NLR/JV 061020-3, 20 October. Installed Performance of Antennas on AeroStructures (IPAS), an EU Specific Targeted Research Project, November 2003 to January 2007, Contract No AST3-CT-2003-503611.

[3] Antenna Software Ltd web site http://www.maasdesign.co.uk/WhyGTD/Tutorial.

[4] Macnamara, T.M. (1995) *Handbook of Antennas for EMC*, Artech House, Boston, ISBN 0-89006-549-7.

[5] Foster, P.R. (2000) CAD for antennas systems. *Electronics and Communications Engineering Journal*, 12, 3–14.

[6] Keller, J.B. (1962) Geometrical theory of diffraction. *Journal of the Optical Society of America*, 52, 116–130.

[7] Foster, P.R. (1996) "Region of application in GTD/UTD" – ALDAS manual, Microwave and Antenna Systems, UK. IEE Conference 360, CEM-96, Bath, pp. 382–395.

[8] Macnamara, T.M., Carnduff, C.M. and Foster, P.R. (2001) Comparison of measured and predicted aircraft patterns. ACES (Applied Computational Electromagnetics Society Monterey), March 2001, pp. 316–322.

[9] Barka, A. (2005) Bridging the computational frequency gap, Deliverable D3, IPAS/RP/ONERA/AB 050103, 3 October. Installed Performance of Antennas on AeroStructures (IPAS), an EU Specific Targeted Research Project, November 2003 to January 2007, Contract No AST3-CT-2003-503611.

[10] Barka, A. and Caudrillier, P. (2007) Domain decomposition method based on generalized scattering matrix for installed performance of antennas on aircraft. *IEEE Transactions on Antenna and Propagation*, 55, 1833–1842.

[11] Thain, A. (2006) Computation and analysis of existing and advanced tools, Deliverable D17, IPAS/RP/EADS/AT 061219, 19 December. Installed Performance of Antennas on AeroStructures (IPAS), an EU Specific Targeted Research Project, November 2003 to January 2007, Contract No AST3-CT-2003-503611.

[12] Heckler, M.V.T., Greda, L.A. and Dreher, A. (2008) Analysis of a navigation antenna installed on a civil airplane. IEEE Antennas and Propagation Society International Symposium Digest, San Diego, CA, July 2008, 4 pp. CD-ROM.

[13] Heckler, M.V.T. and Dreher, A. (2007) Analysis of planar 4×4 microstrip antenna arrays installed on airframes. 2nd European Conference on Antennas and Propagation (EuCAP 2007), Edinburgh, UK, Nov. 2007, 5 pp. CD-ROM.

8

Measurements

8.1 Introduction

This chapter describes the measurement facilities used for antenna diagnostics as well as radiated emissions for electromagnetic compatibility testing. As described in Chapter 3, measurements are performed on scaled models using scaled antennas in the initial design phases of the antenna layout and before any metal is cut, that is, before the antennas are physically installed on the aircraft. Radiation patterns are a measure of the power radiated in different directions. We need to measure the power at the same distance from the antenna in order to compare these levels. Ideally in the case of an azimuth cuts, for instance, we should keep the antenna under test (AUT) stationary and move the receive antenna around it in a circle at a fixed distance from it. The distance would be the far-field distance and would depend on the frequency of measurement. Thus several concentric circular paths of different radii would be required around the AUT, one for each band of frequencies. At the higher frequencies the dynamic range of the measurement set-up would be too small to permit the use of the longer ranges used at the lower frequencies, since the space loss at these larger electrical distances would be too high.

In practice, the best method of achieving the same result is:

1. to use the AUT in receive mode
2. to transmit from the range/facility antenna
3. to keep the transmit antenna stationary and rotate the AUT about its phase centre
4. to achieve the varying distances between the two antennas by keeping the receive antenna location fixed and having several transmit antennas at different distances.

The field over which the AUT moves during the measurement should ideally be uniform. This means that the amplitude as well as the phase of the incident wave from the transmit antenna has to be the same over the entire volume swept through by the rotating AUT. This is because the radiation pattern is the measurement of power at different angles, and if the field is not uniform at all positions of the AUT then an uneven field would give an incorrect radiation pattern. However, this ideal situation cannot be achieved in practice. The best facilities could have the amplitude of the electric field varying by less than 0.25 dB and phase variation less than 5°. In the case of the antennas on a scaled

Introduction to Antenna Placement & Installation Thereza M. Macnamara

model the volume swept out by the antenna (especially those further away from the centre of rotation of the scaled model) is very large and a field variation of 1 dB or even more could be expected. Probing of the received field is undertaken in order to ensure that the field is uniform over the entire volume that the receive antenna traverses.

Additionally the antenna would induce currents on the scaled model and part of the scaled model could be regarded as the AUT. At the lower frequencies where the wavelength is long, these currents would have higher values at the same physical distances as the currents at higher frequencies. This is because it is the electrical distance that determines the magnitude of the current, similarly to the way space loss is inversely proportional to the wavelength. Thus more of the scaled model will be illuminated at low frequencies than occurs at the higher frequencies, resulting in a larger effective AUT. To obtain the different cuts, the AUT has to be rotated in different orientations and about different axes. This is achieved by the use of different types of positioners on which the AUT is mounted.

Until the 1950s all antenna radiation patterns were undertaken in far-field outdoor ranges. In the 1950s indoor anechoic chambers were introduced and in the early 1960s compact ranges were used. In the 1970s near-field facilities began to become popular. The Airborne Near-field Test Facility (ANTF) being developed by EADS Astrium (under the EU IPAS research project 2003–2007 [1]) is a progression of a near-field facility to a portable one for outdoor use [2].

This chapter gives an overview of the most common measurement facilities as well as describing the other components of the facilities such as the positioners used for the AUT, radar absorbing material (RAM) used to line the indoor facilities, range and probe antennas and site attenuation factors used for radiated emission testing on open area test sites (OATSs). Scaled models which are used for measuring the installed antenna radiation patterns are also described in this chapter.

The use of RF rotary joints is avoided by rotating the positioner alternately in one direction and then in the opposite direction.

8.2 Positioners

The angle recorded on the positioner is the angle of the radiation pattern and is only the correct angle if the antenna is at the axis of rotation of the positioner. There are basically nine types of positioners used for the device under test (DUT):

1. azimuth
2. elevation
3. azimuth over elevation
4. elevation over azimuth
5. azimuth over elevation over azimuth
6. elevation over azimuth over elevation
7. roll over azimuth
8. roll over elevation over azimuth
9. roll over azimuth over elevation.

In addition there is often a polarization positioner used for the range antenna to change the polarization of the transmission from vertical to horizontal as well as to continuously

(a) Azimuth (b) Elevation (c) Roll over elevation over azimuth

Figure 8.1 Three types of positioners. Reproduced by kind permission of MI Technologies.

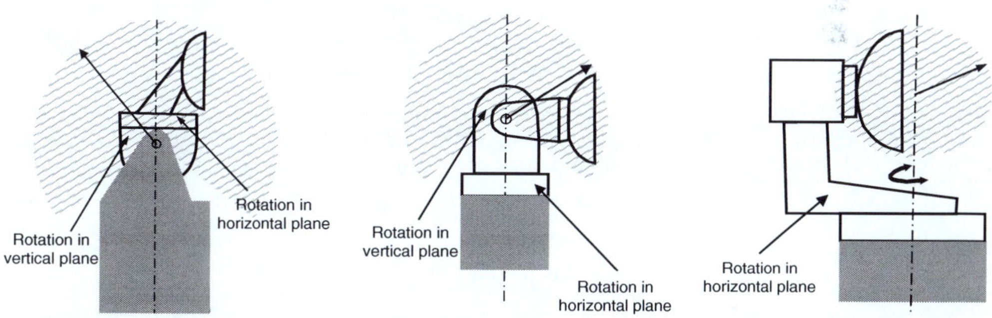

(a) Azimuth over elevation (b) Elevation over azimuth (c) Roll over azimuth

Figure 8.2 Volume swept out by the AUT for different types of positioners.

rotate the antenna for polarization purity measurements (spinning technique) as described in Section 1.4.4. Three of MI Technologies' commercially available positioners are shown in Figure 8.1.

The volume swept out by an AUT depends on the type of positioner used. In Figure 8.2 we can see from the shaded area that the azimuth over elevation positioners sweeps out a major part of sphere with the largest radius, the elevation over azimuth has the next largest radius of a sphere, and the roll over azimuth has the smallest spherical radius and hence the smallest swept volume. The field should be uniform over the swept volume.

8.2.1 Phase Centre Error

For uninstalled antennas the phase centre of the antenna can be positioned at or near the centre of rotation. However, for antennas on a scaled model, the axis of rotation would be near the centre of gravity of the scaled model and this could be some distance away from the antenna.

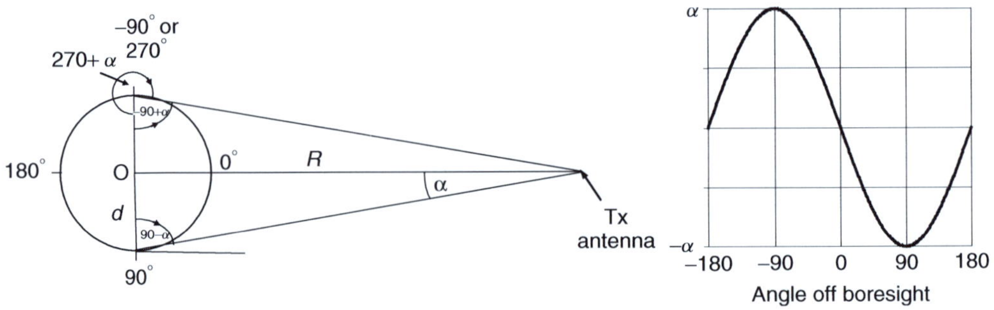

Figure 8.3 Phase centre error due to the offset d between the axis of rotation and position of the antenna under test.

Thus, referring to Figure 8.3a, instead of the AUT being at O (the centre of the circle), and rotating about its own axis, it would move around in a circle. If the AUT was at the centre, then $+90^\circ$ would be defined when the AUT had turned through 90° and it is facing the bottom in the figure so that the boresight of the AUT is making an angle of 90° with the line joining it to the transmit antenna. However, when the AUT is offset from the centre of rotation and the positioner/turntable has turned through 90°, the angle joining the AUT to the transmit antenna is $(90 - \alpha)^\circ$ instead of 90°. The error angle α varies sinusoidally over 360°, and we can see from Figure 8.3 that the angle α is given by

$$\tan\alpha = d/R,$$

where

d is the offset of the antenna from the axis of rotation of the positioner and

R is the range length, that is, the distance between the transmitting antenna and the centre of the positioner/turntable.

Thus it can be seen that the longer the range length, the smaller the angular error; and the larger the offset of the antenna from the centre of rotation, the larger the angular error. The maximum errors (of α degrees) are obtained when the antenna is at $+90^\circ$ and -90° (or 270°), and at 0 and 180° there is no error. The error varies sinusoidally over the 360° range as shown in Figure 8.3b. At 90° this angle (α) has to be subtracted from 90° whereas at 270° this angle has to be added to 270° to obtain the correct angle off boresight.

In external measurement ranges the distances between the transmit antenna and the AUT are larger and thus in general this angular error is small. However, in indoor anechoic ranges the distances are relatively small and hence this error could be significant. Indoor near-field facilities, however, can eliminate this error in the software, known as a parallax correction.

On a rectangular plot of a radiation pattern this error can be corrected fairly easily. However, on polar plots where the angles are divided equally over the 360° range this cannot be achieved.

8.3 Test Facility Antennas

In the case of anechoic chambers and compact ranges the facility antennas are used in transmit mode, as in the case of far-field ranges. However, in the case of near-field scanners the facility antennas are usually used in receive mode and are called probes. Occasionally the probes are used in transmit mode.

8.3.1 Range Antennas

The antennas used on outdoor ranges should have a narrow beamwidth and low sidelobes, so that the level of the radiation reflected off the ground and illuminating the DUT is very low. However, the variation of the field over the DUT should also be small. This is particularly important in the case of antennas installed on scaled models, since ideally the field should be uniform over the whole volume swept out by the scaled model. Thus we have conflicting requirements of narrow beams to avoid ground reflections and wide beamwidths for large DUTs, such as scaled model and ground test measurements on full-scale aircraft.

The range antenna should have the same polarization as the receive antenna. Additionally its cross-polar discrimination should be very good so as not to result in any errors. Most range antennas have a cross-polar discrimination of around 20 dB. This means that if a range antenna is orientated for vertical polarization, it can still detect horizontal polarization but at a level that is 20 dB less.

For instance, if we have a linearly polarized range antenna being used to measure the radiation pattern of a circularly polarized antenna, the gain measurement is usually performed by illuminating the AUT with horizontal and then vertical polarization. The gain for a left hand circular polarized and a right hand circular polarized antenna is then calculated from the individual gains measured. If the range antenna has an axial ratio of 20 dB the error in the gain measurement for a same-sense antenna is +0.828 dB, whereas for an opposite-sense antenna the error is 0.915 dB, as shown in Figure 8.4a. The range antenna used on a DLR range for measurements around 22 GHz (on the Fokker 100 scaled model for IPAS [3]) is shown in Figure 8.4b. It is a 30 cm parabolic reflector with

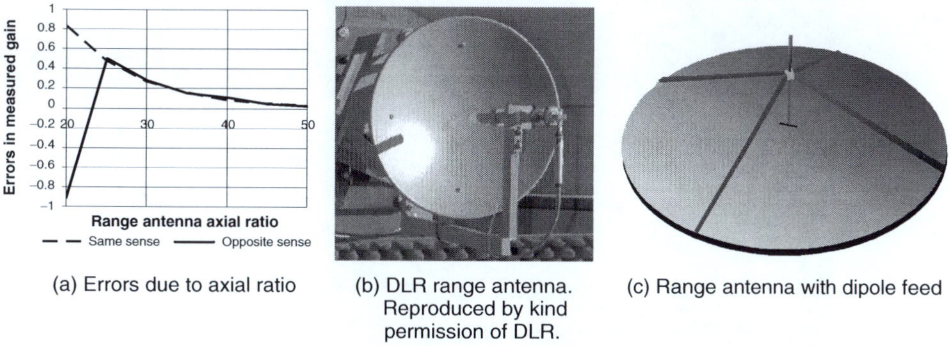

Figure 8.4 The errors due to the transmit antenna axial ratio and two far-field outdoor range antennas.

a circular waveguide feed. A reflector used at lower frequencies with a dipole feed is shown in Figure 8.4c.

8.3.2 Antennas Used in Near-Field Facilities

The antennas used in near-field facilities are usually called 'probes'. At frequencies below 1 GHz, the type of antenna used is usually a log-periodic, axial mode helix or a Yagi–Uda array, although the latter is quite narrowband. At higher frequencies they are usually open waveguides, with rectangular waveguides being the most common since they are cheaper to manufacture. Small open-ended circular waveguides and square waveguides are also used. Each probe antenna is calibrated at spot frequencies so that the absolute gain is known. A whole series of antennas are required to cover the frequency range of the facility. The circular waveguide could have pairs of orthogonal mounted pairs of probes, so that linear and vertical polarization can be measured simultaneously as shown in Figure 8.5. This is commonly called an orthomode transducer. The probe antenna is surrounded by RAM and has a choke flange around its periphery which reduces the back lobe of the radiation pattern of the probe. Pyramid RAM is used around the probe.

8.3.3 Antennas Used in Compact Ranges

The antennas used in compact ranges are effectively large reflector antennas based on optical systems with one or more reflectors, although in some cases dielectric lenses are used to produce the collimated beam. The feeds used tend to be open waveguides rather than horns, since a broad beamwidth is required to illuminate the reflector with uniform amplitude and phase in the case of single reflectors. The small aperture of open waveguides is also important, since ideally the antenna feeding the reflectors should be a point source. However, for dual reflector systems the feed has to have a narrow beamwidths to reduce spill over into the test zone – see Figure 8.12. Compact ranges are only used from about 1 GHz since the reflector physical dimensions (and those of the serrations) would be unacceptably large at lower frequencies.

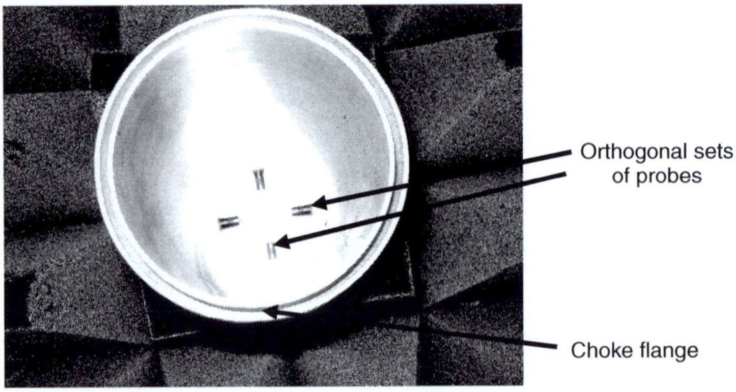

Figure 8.5 Open circular waveguide probe used in a near-field facility.

8.4 Scaled Models

The scaled models used for measuring the performance of antennas are subscale models. If the airframe is scaled to 1/10th of the full scale airframe, then the frequency has to be scaled up by a factor of 10. Thus, for instance, a VHF antenna operating at an aircraft frequency of 200 MHz would have to be measured at 2 GHz on a 1/10th scaled model and the antenna has to be tuned to this frequency.

In general, it is easier to scale down monopoles (blades) than printed circuit antennas such as spirals.

Another aspect to consider is the frequency range of the measurement facilities. Measurement facilities' frequency ranges vary between about 100 MHz and 26 GHz, although 40 GHz is now becoming more common. MI Technologies have indoor measurement facilities up to 94 GHz.

For a 1/10th scaled model, the highest aircraft frequency that can be used (in a 40 GHz facility) is 4 GHz and the lowest frequency is 100 MHz. Thus systems such as MLS (5.2 GHz) that would scale up to 52 GHz, and the lowest HF frequency (2 MHz) that would scale up to 20 MHz cannot be tested in most facilities. The higher the frequency the higher the cost, in general. RF cabling cannot be used for any long lengths at higher frequencies because of the losses caused by the dielectric between the centre and outer conductors. Waveguides are usually used and even they have to be silver plated. Similarly, the RF cables have to be low loss (usually implemented by using less dielectric) and the connectors are also gold plated. The higher costs are reflected in the higher charges for use of these facilities.

Another important feature to consider is the surface roughness of the scaled model. The airframe could have an uneven surface of 0.5 cm (5 mm), but on a 1/15th scaled model this would translate to a surface roughness of just 0.33 mm.

8.4.1 Use of Scaled Models

Ideally it is best to test the antennas with their systems on a real (full-scale) aircraft. However, the aircraft should be mounted away the ground and other structures. For a full-size aircraft this would entail the provision of massive positioners, lifting equipment, and so on, which, apart from the expense, is not feasible.

In addition, measuring the performance of antennas on the lower fuselage would involve inverting the aircraft, which would not be possible even for small fighter aircraft, although full-scale mock-ups or models could be used.

Scaled models reduce these problems to manageable levels. Furthermore, small to medium-sized scaled models can be measured in indoor test facilities, be they traditional anechoic facilities, compact ranges or near-field facilities, and are thus not subjected to the vagaries of the weather.

Scaled models can be tilted and rotated to obtain the required cut of the line radiation pattern. Most aircraft fly in a nose-up attitude. This varies from about 2° for a large airliner to about 15° for a fighter aircraft. In the case of fighter aircraft aerobatics are the norm and thus the antennas have to cope with pitch and roll angles that are continuously changing.

In the case of the communications systems there could be dropouts, but in the case of other systems that capture several samples before processing data this may not be so critical.

8.4.2 Characteristics of Scaled Models

Ideally from the logistics point of view, scaled models should be as small as possible. They are usually made from a lay-up of fibreglass materials made from a mould that is fabricated from a CAD model of the airframe, in a similar way to those used for wind tunnel testing. However, wind tunnel models do not require any internal access. In the case of scaled models used for antenna measurement, the model has to be hollow in the vicinity of the antenna sites. In addition, there has to be manual access at these sites in order to facilitate the connection of the RF cables to the scaled antennas. Some scaled models are constructed from wood, but the need for them to be hollow (for access to the scaled antennas) as well as the curved fuselage shapes make this choice not very popular.

The scaled model has a conductive coat applied to the surfaces that represent the metallic airframe. This is usually accomplished by use of a conductive spray paint. The areas such as the nose are usually left untreated to represent the radar radome.

The details that are included depend to a large extent on the positions of the antennas relative to the airframe. For instance, if the antenna is on the upper fuselage and the wings are almost at the same level as the antenna, then the wing flaps and the ailerons would be made as movable sections so that their effect on the antenna pattern can be evaluated at the most detrimental positions of these control surfaces. These configurations of the control surfaces depend on whether the aircraft is taking off, in straight and level flight, or landing. In the case of fighter aircraft the most critical manoeuvre has to be considered as well as the most critical systems (and antennas) for that manoeuvre.

Other appendages such as the undercarriages and pitot tubes also have to be considered. The undercarriages that are deployed during landing and take-off could have a very detrimental effect on lower fuselage antennas that are near them, especially if the antennas of these systems are used during landing.

The undercarriages are usually fabricated as metallized or metallic but the tyres that are made of solid rubber are usually omitted.

In the case of measurements in outdoor facilities, lifting gear would be required not only to raise the scaled model onto the positioner, but also to lift personnel to mount the scaled model on the positioner and connect all the RF cables.

8.4.2.1 Small Scaled Models

Small scaled models would be less expensive in terms of materials and can be made as an integral piece. They require smaller packing crates and are easier to handle and transport. In the case of measurement facilities, they require smaller quiet zones and have smaller phase errors for antennas away from the centre of rotation.

However, small scaled models would result in higher test frequencies. Since the wavelength is smaller this results in an increase in the precision requirements and tolerances. The body of the model has to be smoother, and some protuberances are reduced to such small dimensions that they are susceptible to breakage. The scaled antennas have to be very small. In the case of scaled monopoles, the antenna can be scaled down fairly easily and tuned, but for other types of antennas such as spirals, the scaling down is likely to be problematic from the point of view of fabrication as well as tuning and matching to the system. The smaller the degree of scaling down the greater the feasibility of obtaining a scaled antenna.

8.4.2.2 Large Scaled Models

Large scaled models have to be made in sections so that they can be disassembled for transportation. The cost of the materials is greater and they require larger packing crates, resulting in higher transportation and handling costs. The positioners used in test facilities need to be capable of handling the larger sizes and masses of these models. The quiet zones are larger and the phase error in measurements is also larger. If the scaled model has to be measured in outdoor facilities, consideration has to be given to the fact that these models cannot be left on the test tower overnight and there are additional requirements on the scaled model due to wind shear. The length of the boom used on the positioner, the positioner specifications and the height of the DUT tower are all peculiar to each measurement facility and have to be evaluated before a decision can be taken on overnight storage. In the case of a large scaled model (exceeding 3 m) a set-up time of 2 to 3 hours is to be expected. In most cases large models will be tested in outdoor facilities. With the risk of inclement weather in most countries, the time delays and the additional costs of down time could pose serious problems.

The benefits of a larger scaled model lie in the fact that the frequencies are lower, and hence the choice of facilities is likely to be greater. There are less stringent requirements on the tolerances and smoothness of surfaces. The phase centre errors are likely to be smaller (because of the longer ranges at the lower frequencies) and RF losses with the lower frequencies are also likely to be less. The scaling down of antennas that are monopoles is relatively straightforward. Other types of antennas may be difficult to scale down but less problematic than when using small scaled models.

8.4.2.3 Comparison between Small and Large Scaled Models

Table 8.1 shows qualitatively the main advantages and disadvantages of using the different sizes of scaled models. Obviously the priorities in any particular situation will dictate the final decision taken. For instance, if time is of the essence and a lower frequency range is the only one available in the time scale, then the decision would be taken to fabricate a large rather than a smaller scaled model.

8.5 Scaled Antennas

For scaled model measurements the antennas have to be scaled by the same factor. Whilst this is relatively easy to do for blades which can be represented by monopoles, it is quite difficult and expensive to do for other types of antennas, such as spirals, horns and other aperture antennas.

Spirals are usually implemented on printed circuit boards and, apart from the actual antenna being difficult to scale down, the balun and/or feed is almost impossible to scale down and match (to the antenna radiating element) at the scaled-up frequency because of

Table 8.1 The relative advantages and disadvantages of different sized scaled models.

	Small scaled models	Large scaled models
Cost of materials	Cheaper	More expensive
Can be fabricated	In a single piece	In several pieces
Smoothness	Must be smoother	Not so critical
Tolerances in absolute (physical not electrical) terms	Tighter	Not so critical
Protuberances	Subject to damage	Less fragile
Scaled antennas	Require more care in fabrication, especially for antennas other than monopoles	Less precise fabrication
Handling	Easier	More difficult
Phase error	Less for same length of range	Greater for same length of range
Mounting	Easier, assuming the mass is less	More difficult, assuming the mass is greater
Wind shear for outdoor ranges	Less	Greater
Quiet zone required	Smaller	Larger
Wing droop	Smaller	Larger
Vibration of wings	Less	More
Positioner specification for mass	Lower	Higher
Positioning accuracy of positioner	More stringent	Less stringent
Upper frequency limit	Higher	Lower
Availability of facilities	Large	Smaller

the minute dimensions involved. Patch antennas are easier to scale down and have been used in many cases.

Blades with top loading have also been represented as simplified scaled T or L antennas, but the actual details of some blades, especially broadband and active antennas, are not scaled down. Other aperture antennas like reflectors and phased arrays are impossible to scale down and full-scale antennas are used for measurements.

8.5.1 Monopole Antennas

When performing scaled model measurements the antennas are measured with only one antenna being installed at a time in each location. The other antennas are physically removed and the holes covered with conductive tape, so that there are no holes in the air-frame. There may be several frequency measurements performed for each antenna location and this will necessitate installing several tuned antennas at each location. Logistically this approach results is a large increase in the set-up time for each measurement run, since the RF cables have to inserted from inside the scaled model and connected to the antenna, which could be a modified sub-miniature Amphenol (SMA) flanged connector. It usually requires two test engineers to perform this task. This is particularly difficult on an elevated outdoor range and mechanical equipment such as 'cherry pickers' are required to raise the test engineers to change the antennas and/or their locations.

In order to save time BAE Systems at Great Baddow developed 'quick-fit' monopoles. The monopole antenna design was based on a modified female SMA connector with a threaded rod soldered to the centre conductor. At the highest frequency this threaded rod acted as the antenna (tuned to 16.4 GHz), but at the lower frequencies tapped rods were screwed onto the threaded rod and the lengths of the rod were cut to the correct length by installing the antenna on a square copper ground plane of side 300 mm as shown in Figure 8.6a and measuring the voltage standing wave ratio (VSWR) using a vector network analyser (VNA). Figure 8.6b shows the 'quick-fit' monopole and rod element. The square flange of the SMA connector was machined to a thickness of 0.5 mm to avoid large steps in the surface of the airframe, since the flange was on the surface of the scaled model. The quick-fit monopole was held in place on the outer surface of the scaled model, with a 0.1 mm thick circular copper disc of diameter 40 mm (cut from 50 mm wide Chomerics copper tape – part no. CH7043) that was pre-coated with conducting adhesive. The disc was applied over the flange as shown in Figure 8.6c, to provide a good physical and electrical bond. At 1, 1.8 and 3.6 GHz the monopoles were 71.3, 40 and 20 mm long with diameters of 2.36, 2.98 and 2.98 mm, respectively.

Using these quick-fit monopoles, the SMA connectors of scaled antennas could be left *in situ*, and different size rods screwed into place from the outside of the scaled model. Semi-rigid RF coaxial cables were used to connect all the antennas to a panel with back-to-back SMA female connectors. Flexible RF cables were used to connect each of the back-to-back connectors to the receiver. Unused antennas (for a particular measurement) were terminated in matched loads at the panel holding the back-to-back connectors rather than at the actual antennas (Figure 8.7b).

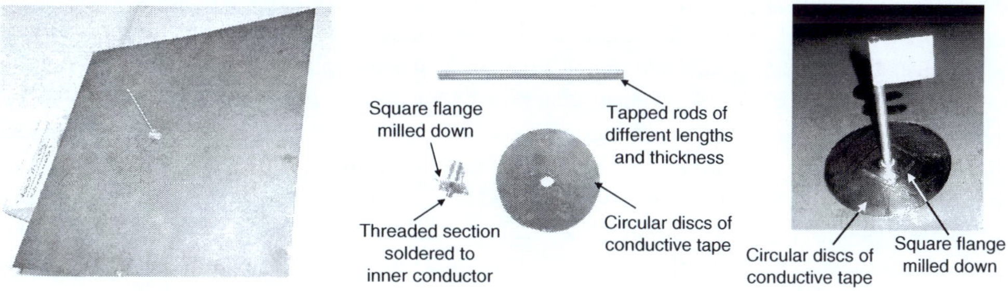

(a) Ground plane for tuning (b) Quick-fit SMA antennas (c) Quick-fit antenna in place

Figure 8.6 Ground plane and quick-fit monopole antenna with 40 mm rod element tuned to 1 GHz, SMA connector with threaded centre conductor and 40 mm diameter copper disc.

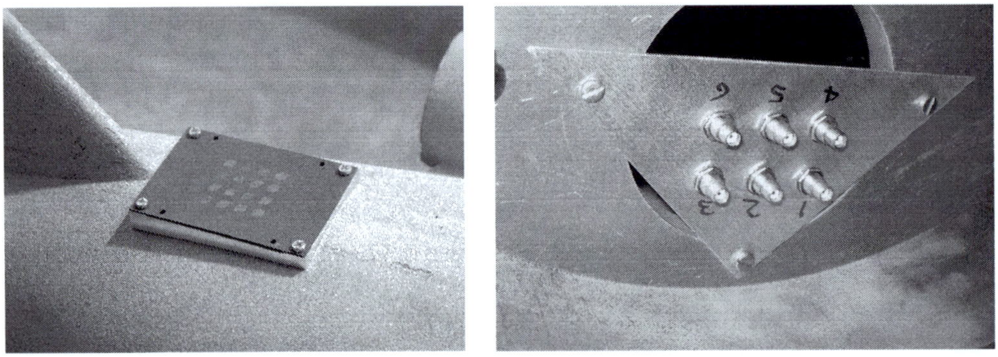

(a) DLR 22 GHz patch array. Reproduced by kind permission of DLR.

(b) Panel with SMA connectors

Figure 8.7 A 4 × 4 patch array installed on the Fokker 100 scaled model and the panel used for back-to-back connectors on the cylinder used for IPAS coupling measurements.

8.5.2 Patch Antennas

Patch antennas are usually implemented on printed circuit boards. They are easier to scale down than spiral antennas but not as simple as monopoles. They tend to be used at aircraft frequencies above 1.2 GHz. At this aircraft frequency if they were required for measurement on a 1/10th scaled model, the dimensions at 12 GHz could be implemented. In the IPAS project one of the partners, DLR, fabricated a 4 × 4 square array at around 22 GHz as shown in Figure 8.7a for measurements on a 1/15th Fokker 100 scaled model [3]. The array is shown installed on the upper rear fuselage of the scaled model.

8.6 Absorbers

Absorbers are commonly known as radar absorbing material (RAM). The most commonly used RAM consists of polyurethane foam which is loaded with a material such as carbon. The electric field is attenuated by dielectric losses in the loaded foam, and the reflected wave at the air–absorber interface is also low. The absorber is usually designed to be backed by a metal plate or foil, and the reflectivity of the absorber is compared to a perfect metal conductor. The reflectivity is quoted in dB (sometimes with a negative sign) and values between 20 and 45 dB can be expected at the specified frequencies. We should note that the negative sign here indicates that the reflected wave is 20 to 45 dB lower than the incident wave. The reflectivity in dB, R_{dB}, is related to the percentage of reflected power, R_{p}, by the formula

$$R_{\mathrm{p}} = \left[10^{(R_{\mathrm{dB}}/10)}\right] \times 100. \tag{8.1}$$

Thus a −45 dB reflectivity means that only 0.0032 % of the power is reflected.

The reflectivity depends on the permeability and permittivity of the material that define the intrinsic impedance, η, of the material:

$$\eta = \sqrt{\frac{\mu_r}{\varepsilon_r}}. \tag{8.2}$$

The reflectivity is defined as

$$R_{dB} = 20 \log \left[\frac{\eta_2 - \eta_1}{\eta_2 + \eta_1} \right] \tag{8.3}$$

where

η_1 is the impedance of the first medium, usually air
η_2 is the impedance of the material

These reflectivities are only obtainable at low angles of incidence, and incidence angles greater than about 45° should be avoided. The absorber may also change the polarization of the incident electric field but this is a second-order effect and will not be discussed here. The reflectivity of the absorber may also be a function of the incident wave polarization, depending on the composition of the absorbing material.

8.6.1 Carbon Loaded Foam RAM

This type of RAM is available in flat sheet form and in pyramid, convoluted (egg-box) or conical shapes. Its operation is based on loading the foam with conductive particles such as carbon to absorb the electric field.

8.6.1.1 Flat Sheet RAM

This absorber tends to have a broader frequency range than the shaped variety, but its reflectivity is not as good. It consists of a number of layers of polyurethane carbon loaded foam with different resistivities that are bonded together as shown in Figure 8.8a,

Figure 8.8 Carbon loaded radar absorbing material.

so that the resistance of successive layers changes in discrete steps. The absorber is not reversible – in other words, it can only be used facing one way. It has a front surface (which is often of a lighter colour) for the incident radiation and a back surface which should be backed by (or preferably bonded to) a metal surface for optimum performance. The lower the operating frequency of the absorber the greater is its thickness; at 600 MHz the thickness of the absorber is about 10 cm, but at 20 GHz the thickness is only about 0.5 cm. Absorbers can be used successfully at frequencies higher than those specified; the limit is only applicable to the low frequency end of the operating range. Reflectivities of the order of −25 dB are the best that can be expected, with much poorer performance (−10 dB) at the lower end of the operating frequency range. The front layer of the absorber, which receives the incident wave from free space, has less resistive loading than the other layers, and thus has the highest impedance. Ideally the impedance of this layer should be equal to the characteristic impedance of the wave that is incident on it – that is, in the case of a plane wave, it should have a real impedance or resistance of 377 Ω. The impedance of successive layers is gradually decreased, with the last layer having a very low impedance approaching that of a good metal conductor. The resistance is thus decreased in discrete steps, the number of these steps being equal to the number of layers.

8.6.1.2 Pyramid, Conical, Wedge and Convoluted (Egg-Box) Absorbers

Pyramid, conical or wedge RAM is characterized by continuous variation of the conductivity so there is a gradual change in the resistivity of the material, rather than the discrete step change obtained for the sheet RAM. For these absorbers, the variation in impedance is implemented by geometrically shaping the carbon loaded foam. This type of absorber has the best performance, with reflectivities of up to −50 dB when the thickness of the absorber is at least twice the wavelength of the incident radiation. The figure of −50 dB represents a reflection of only about 0.001 % of the incident power. The pyramid absorber shown in Figure 8.8b comprises four triangular inclined sides on a square base. The height of the pyramid (including the base) varies from around 10 cm to 4.6 m for frequencies down to 30 MHz ([4], p. 26). At 100 MHz, for instance, the length of the pyramid absorber is around 1.8 m. Several pyramids are joined together to form square or rectangular modular units. The pyramid taper provides a continuous impedance taper (to the incident wave), which varies from a high impedance at the tip of the taper to a very low impedance at the back where it is usually in contact with a good metal conductor. The incident wave is attenuated increasingly as it travels through the material. Note that in this case the variation of impedance is gradual, whereas in the case of the flat sheet RAM the impedance variation is implemented in discrete steps. The absorber also scatters or diffuses the incident energy in different directions (at the air–absorber interface) so that it is absorbed into the other surfaces of the absorber within the valleys of the material, as shown in Figure 8.8b. When used as floor material the valleys between the pyramids can be filled with rigid foam such as expanded polystyrene and then covered with a vinyl coating. Alternatively the RAM may also be covered by rigid glass fibre foam laminate to allow it to be used for the floor ([5], p. 356). The conical RAM available from Emerson and Cumming is hollow.

8.6.2 *Carbon Loaded Fibre Matting*

Carbon loaded fibre matting is also available which has similar performance to the foam loaded type but cannot be used at low frequencies; its low frequency limit is higher than that of the foam type, at around 1 GHz.

8.6.3 *Other Types of RAM*

Ceramic and ferrite sheets and Jaumann absorbers are resonant absorbers and are more narrowband with reflectivities of around −25 to −30 dB. They are sometimes used as backing for pyramid RAM. The other types of sheet absorbers are thin flexible sheets of lossy foams which used for special purposes and not as absorbers *per se*, and will therefore not be discussed here. Lossy resins are also available for moulding into custom built shapes or for spraying onto metal surfaces, but these types will not be discussed here since they are not used to line anechoic chambers.

8.7 Measurement Facilities

The measurement facilities available could be outside test ranges, covered or partly covered sites or completely enclosed. Some outdoor facilities are several kilometres long. Indoor facilities have to be lined (partly or wholly) with RAM. Those that are wholly lined effectively simulate free space facilities since they eliminate the reflected/diffracted waves from the walls of the chambers by absorbing them. Outdoor facilities ensure that plane wave conditions prevail by ensuring that the distance from the transmit antenna is large enough for this to occur. Compact ranges, on the other hand, attain plane wave conditions by using a collimated beam produced by lenses or reflectors.

The reflectivity levels and gain accuracies for the different types of facilities are compared in Table 8.2.

Table 8.2 Comparison of reflectivity levels and gain accuracies in measurement facilities.

	Reflectivity levels		Polarization purity of range antenna (dB)	Gain accuracy (dB)
	1 GHz (dB)	10 GHz (dB)		
Elevated far-field	−20 to −30	−45 to −50	−20 to −30	±1.5 to ±0.5
Ground reflection	−45 to −50	−35 to −45	−15 to −25	±1.5 to ±0.5
Rectangular anechoic	−20 to −35	−45 to −55	−20 to −30	±1.5 to ±0.5
Compact	−10 to −20	−45 to −50	−20 to −30	±2 to ±0.5
Spherical near-field	−40 to −50	−50 to −70	−40 to −60	±0.5 to ±0.25

After [6], p. 360.

8.8 Indoor Test Facilities

Chambers used for test facilities are either partly or fully lined with RAM. If the chamber is completely lined with RAM, then it is usually called an anechoic chamber. Partly lined chambers are used as a compromise solution for economic reasons since absorbers are relatively expensive. In addition, lining a room with absorbers reduces the effective volume of the room. Planar near-field test facilities are usually only partly lined.

8.9 Anechoic Chambers

When a screened room is lined with RAM it is known as an anechoic "no echo" chamber. Lining the room reduces the level of reflections from the walls, floor and ceiling. The level of these secondary or multipath reflections is not uniform throughout the volume of the room, and there is a relatively small volume within the room (called the quiet zone) where the level is at a minimum.

8.9.1 Quiet Zone

There is more than one definition of the quiet zone. Three possible definitions are:

1. the volume where rays enter after at least two reflections (Vol. 1 of [7], p. 670)
2. the volume within which the reflections from the internal surfaces of the chamber are a specified level below that of the direct ray from the transmitting antenna ([8], p. 381)
3. the volume within which the electric field (or power) varies by less than a specified amount.

The latter definition is the most common and most meaningful, but the second definition is also used. The level below the direct ray can be down to -60 dB for high-performance anechoic chambers, but is a function of frequency, the type of absorbers used and the position within the quiet zone.

Note that the quiet zone is not necessarily of a uniform shape such as a cuboid or sphere.

8.9.2 Rectangular Chambers

The early anechoic chambers were of uniform cross-section. The width of the chamber is chosen so that the angle of incidence of a ray (from the transmitting antenna) on the side walls is kept below a certain value. The maximum value of the angle of incidence depends on the RAM used to line the chamber, and values between 45° and 60° are quite common. The chamber should be long enough to ensure that far-field conditions prevail at the lowest frequency of operation. A length to width ratio of 3 is recommended to ensure that far-field (i.e. plane-wave) conditions prevail. However, if this ratio is reduced to 2 the side wall reflections are minimized. This is because fewer of the reflections from the side walls reach the quiet zone, and those that do enter the quiet zone are reflections as a result of small angles of incidence where the absorber has a better reflection performance. Chambers are often lined with the same type of absorbers on all the walls, but it is better

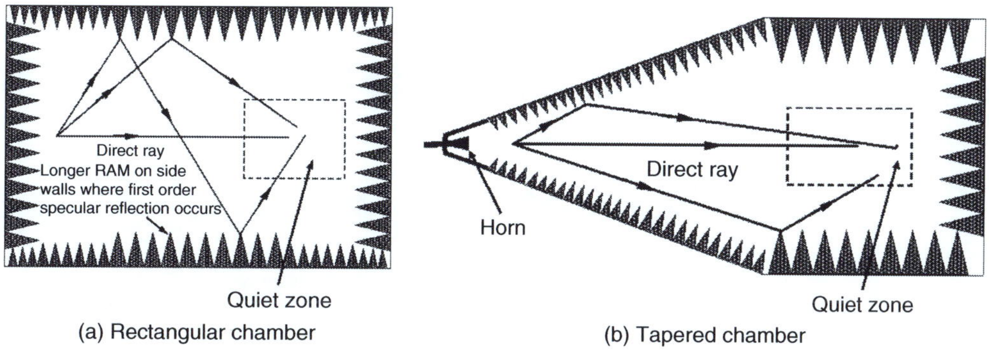

Figure 8.9 Anechoic chambers.

to line the end wall with higher performance RAM, since the main lobe of the transmitting antenna illuminates this wall. Higher performance RAM is also used for areas along the side walls where the first reflections from the transmitting antenna occur, as shown in Figure 8.9a. For reasons of cost the longer RAM is not used everywhere, since it is much more expensive than the shorter, higher frequency RAM.

8.9.3 Tapered Anechoic Chambers

The tapered chamber was first introduced in 1967. It is used for radiated emissions for systems/equipment testing rather than radiation patterns of antennas. It can be used at a lower frequency (below 1000 MHz) than a rectangular chamber of the same size. It also has a wider quiet zone than a rectangular chamber of the same width. It consists of a pyramidal or conical tapered section joined to a section of rectangular cross-section, as shown in Figure 8.9b. The tapered section can be considered as a horn which is radiating into a rectangular section. The tapered section is usually about twice as long as the rectangular section. In the tapered region the angle of incidence with the side walls is nearly 90° (grazing incidence) and thus RAM that has high absorption for large angles of incidence is required for these areas and wedge type RAM is usually used. The tapered chamber can be used down to 100 MHz if longer RAM is used on the back and side walls or if the back wall is completely removed and the radiation occurs into free space.

8.10 Compact Ranges

Compact ranges work on the principle of attaining a plane wave by the production of a collimated (parallel) beam, whereas in the case of a far-field range the planar wave is obtained by increasing the electrical distance so that the radius of curvature of the wave front is very large.

The collimated beam (as shown in Figure 8.10) is produced by using one or more reflectors based on optical systems in the same way as a spotlight or torch produces a collimated beam. Dielectric lenses are also used in some cases but these are less common.

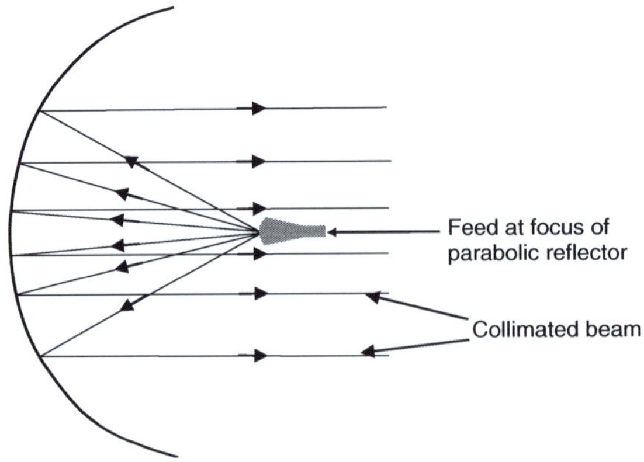

Figure 8.10 Single reflector system with on-axis feed.

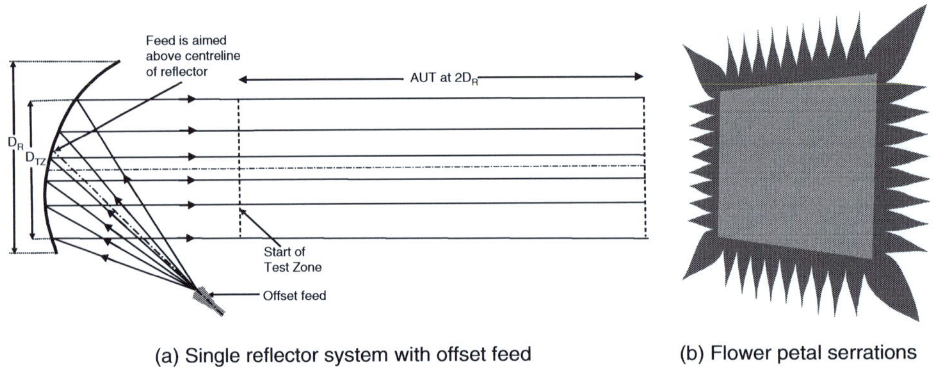

Figure 8.11 Single reflector system with offset feed and a reflector with flower petal serrations.

The on-axis feed causes aperture blockage and thus it is usually offset as shown in Figure 8.11a. The offset feed is aimed above the centreline of the reflector to compensate for the different path lengths of the waves to the reflector surface. If a single reflector is used it is usually parabolic and has a large radius of curvature as shown in Figure 8.11a, and the test zone starts beyond the position of the feed. The longer the focal length of the reflector the smaller is the variation in the amplitude taper across the aperture of the reflector. The edges of the reflectors have serrations (see Figure 8.11b) that reduce the scattering (known as edge diffraction) that would normally occur at the sharp edges.

The quiet zone is usually cylindrical in shape and large versions of this type of facility may have quiet zones of up to 5 m so that they can be used for most scaled model measurements. The diameter of the reflector, D_{R}, is normally 20λ larger than the diameter of the test zone, D_{TZ}, so that

$$D_{\mathrm{R}} = D_{\mathrm{TZ}} + 20\lambda.$$

The AUT is normally placed at a distance of $2D_{\mathrm{R}}$ away from the start of the test zone.

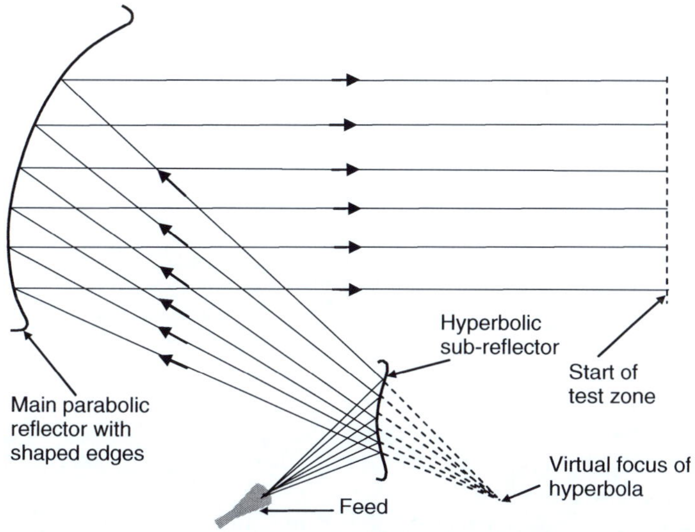

Figure 8.12 Cassegrain dual reflector system.

Dual reflector systems have a sub-reflector fed by a high-gain horn which has a narrow beamwidth thus preventing radiation spill over into the test zone, which would be in line of sight of the feed. A Cassegrain system has a flat or hyperbolic sub-reflector, whereas a Gregorian system has a concave sub-reflector. The alignment between the feed and the reflectors is performed by looking at the phase. In the case of the Cassegrain system the virtual focus of the hyperbolic sub-reflector and the real focus of the parabolic main reflector are coincident. The edges of the sub- and main reflectors are curved as shown Figure 8.12. The levels at the edges are around 30 dB less than that at the centre, thus obviating the need for any edge treatment such as serrations.

8.11 Near-Field Facilities

Near-field test facilities are usually incorporated into anechoic or semi-anechoic chambers, although some exist for outdoor use. It was shown in Chapter 5 that the Fourier transform of the aperture illumination produces the far-field pattern of an antenna. In near-field facilities the electric field is sampled at regular spatial intervals and the Fourier transformation is derived to obtain the far-field radiation pattern of the antenna. We can imagine the sampling plane as a series of Huygens sources separated by the sampling distance, usually $\lambda/2$. The near field is sampled over a surface that could be

1. a flat surface, in the case of a planar near-field facility
2. a cylinder in the case of a cylindrical near-field facility
3. a sphere in the case of a spherical near-field facility.

The chambers can be quite small compared to the dimensions required for far-field indoor measurement facilities. For instance, the MI Technologics small anechoic chamber that

Table 8.3 Comparison of indoor near-field facilities.

	Planar	Cylindrical	Spherical
Type of AUTs	High Gain > 15 dBi	Linear arrays	Low, medium and high gain
Positioner for AUT	Not required	Usually azimuth only	Two axes positioner
Angular extent of radiation pattern	Maximum of forward hemisphere	Azimuth 360° Elevation < ± 90°	Complete spherical
Computational time	Fastest (from 1 s)	Twice planar (from 2 s)	Longest (from 20 min)
Amount of RAM	Smallest	More RAM	Largest

After [6], p. 318.

can be used as a near-field facility is 4.6 m wide, 6.1 m long and 4.6 m high, whereas its large anechoic chamber that can be used as far-field facility or compact range is 7.3 m wide, 12.1 m long and 7.6 m high.

The advantages and disadvantages of planar, cylindrical and spherical near-field facilities are summarized in Table 8.3.

There are many errors in near-field facilities that have to be considered. These include mechanical errors, stability of the RF systems, probe interaction with the AUT being measured, background radiation, multipath and repeatability of RF connections. These errors do not necessarily apply in all cases or to all types of near-field facilities.

The positional and other mechanical errors are far less critical at the lower frequencies where the wavelength is quite long and therefore these physical errors are small in terms of the electrical distances involved. The probe is moved continuously but samples are only taken at half-wavelength intervals.

8.11.1 Planar Near-Field Facilities

Planar near-field facilities are mainly used for high gain antennas with most on the power in the main beam, and the levels of the sidelobes are low. The antenna is stationary and the back lobe is not measured. Usually only the wall behind the probe is lined with RAM. The power is sampled by the probe which is moved continuously over a planar surface, but samples are only taken at points separated by approximately half a wavelength in the horizontal and vertical directions. The front surface of the AUT is parallel to the scan area as shown in Figure 8.13a, with the measurement samples taken at the nodes where the lines cross. Typically the probe is at a distance of 3 to 5 wavelengths from the front surface of the antenna and a sampling area of between 5 and 10 wavelengths is required. The area over which the sampling is undertaken and its distance from the AUT determine the angular extent of the radiation pattern. Thus in order to obtain a full ±90° the scan area would have to be infinite. However, in practice the angle subtended by measurement surface at the edges of the antenna (and the line perpendicular to the surface) is the critical angle as shown in Figure 8.13b. This should generally be at least 70°. The greater the angle subtended by the antenna at the planar measurement surface, the greater the angle off boresight that can be derived for the radiation pattern. The larger planar near-field facilities can have scanning areas of 35 m by 16 m.

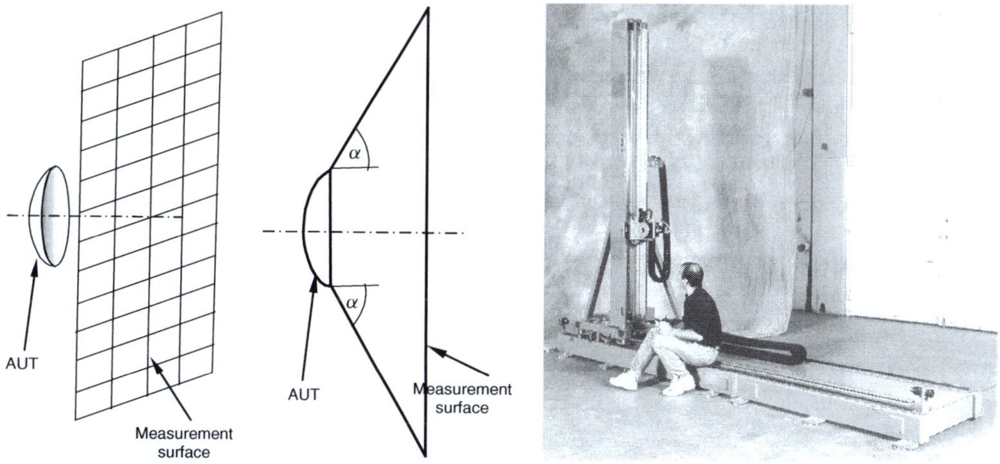

Figure 8.13 Planar near-field set-up for measuring aperture antennas. Reproduced by kind permission of MI Technologies–Ref [12].

8.11.2 Cylindrical Near-Field Facilities

In the case of the cylindrical facility an azimuth positioner is used for the AUT and the probe is moved vertically as shown in Figure 8.14a. The whole chamber has to be lined with RAM. The cross-section is not truly circular but a polyhedron. In order to obtain a spherical radiation pattern from a cylindrical field facility the measurements are performed with the DUT in the horizontal plane, and then the DUT is turned through 90° in the elevation plane and a second measurement undertaken. In the case of an aircraft scaled model, the first measurement would be undertaken with the aircraft horizontal, and this would give the azimuth plane cut. The aircraft would then be mounted with its longitudinal axis vertical, either nose or tail down. This would give the elevation roll plane cut. Interpolating between the two cuts would give the pitch plane cut, if the chamber was very long in the vertical plane. In the case of aircraft antennas the whole aircraft or scaled model is considered as the AUT and thus the subtended angles that have to be considered are from the nose and tail ends, when the aircraft is mounted vertically with its nose/tail down.

Ideally, in order to get a full spherical radiation pattern the cylindrical sampling areas should be infinitely long. In practice, this is of course not the case and thus there will be inaccuracies in the pitch plane plots depending on the height of the chamber. Typical plots obtained are shown in Figure 4.31 of Chapter 4, where some angular sections of the pitch plane plots were removed because they were inaccurate.

The cylindrical near-field test facility (CNTF) at BAE Systems in Great Baddow (UK) consists of an irregular dodecagon (polygon with 12 sides) anechoic chamber that is symmetrical about its central/vertical axis, and is 11 m high by 11 m wide by 11 m deep at its furthest points. One end of the chamber contains two electrically operated sliding doors that when open give an aperture 5 m wide by 7 m high. The walls, floor, ceiling and doors are covered with 25 000 (RAM) absorbing pyramids. The azimuth positioner used for the DUT is 1.1 m in diameter. The distance between the RF sampling probe and the positioner can be altered by moving the probe tower horizontally in 10 cm steps. Thus the vertical angle subtended by the vertical ends of the DUT can be varied depending on the swept volume of the DUT and the extent of the elevation angular coverage required. The maximum swept radius that can be accommodated is 4.5 m, giving a maximum AUT aperture size of 9 m. The RF sampling probe is driven up and down the probe tower by a chain driven winch. The RF sampling probe height is measured by a laser interferometer with a resolution of 80 nm. A detector is mounted under the RF sampling probe, which is in turn mounted on a servo controlled translation slide. This detector and its electronics lock onto a vertical laser beam and the servo loop ensures that the probe ascends the tower vertically. The result of these two laser systems means that the measurement system knows where the RF sampling probe is to better than 0.2 mm on any axis during a measurement run.

The number of measurement points depends on the frequency, since the relevant parameter is the electrical distance between the points. At low frequencies the points are physically further apart in the azimuth as well as the elevation planes. Thus, for instance, in the case of the BAE Systems facility at Great Baddow, below 3.72 GHz the number of 'rings' measured in the elevation plane is 64 and each ring has 128 measurement points in the azimuth plane. Each measurement set takes about 1 hour. When the frequency is above 13.08 GHz each ring has 512 measurement points in the azimuth plane and each measurement set takes about 2 hours. Figure 8.14b shows the Fokker 100 scaled model being measured in the cylindrical near-field chamber at BAE Systems, Great Baddow, UK.

Outdoor versions of the cylindrical near field are available like the one at BAE Systems on the Isle of Wight, UK (Figure 8.15). In this case the probe is mounted on a tower that is 30 m high. The azimuth positioner accommodates a DUT of up to 20 m diameter and therefore this can be used for large antennas, as well as small full-scale aircraft and

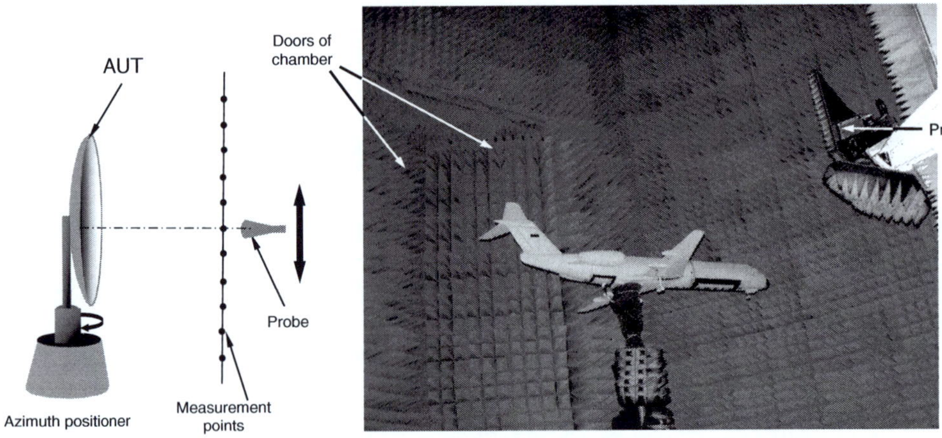

(a) Schematic (b) Fokker 100 scaled model under test

Figure 8.14 Schematic of a cylindrical facility and the Fokker 100 scaled model under test at BAE Systems' cylindrical near-field chamber at Great Baddow, UK.

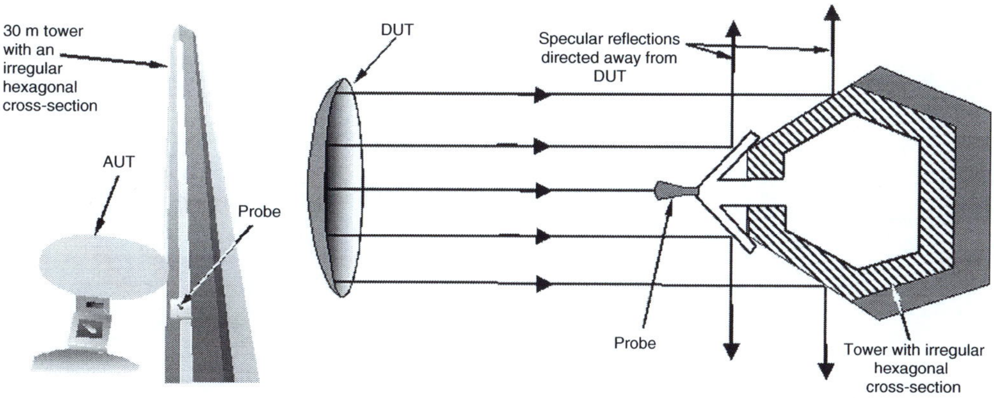

Figure 8.15 The BAE Systems cylindrical near-field outdoor facility on the Isle of Wight, UK.

scaled models – see [9]. The tower has a cross-ection of an irregular hexagon and the walls of the tower are angled so that any specular reflections from the walls are directed away from the DUT. This obviates the need for RAM on the outside of the walls which would not survive inclement weather.

8.11.3 Spherical Near-Field Facilities

The sampling surface is effectively a sphere, and this is implemented by rotating the AUT in azimuth and rotating the probe in the vertical plane by using a two-armed positioner (gantry) as shown in Figure 8.16 and then inverting the AUT to obtain the second hemisphere. Outdoor versions of the spherical near-field facility have been used for scaled model testing of vehicles and ships.

8.11.4 Advanced Antenna Near-Field Test System

EADS Astrium are currently (2009) developing an Advanced Antenna Near-field Test System (AANTS) with particular application to the measurement of the installed performance of antennas on spacecraft and aircraft (Figure 8.17). The system is based on a transformation algorithm that is capable of calculating the antenna far-field pattern from near-field data that is sampled at irregular spatial intervals along any non-uniform scanned surface. Furthermore, the applied algorithms permit the merging of different scan contours (e.g. a vertical curved cylindrical surface with the horizontal planar scan surface at the top of the cylinder), and also provide algorithmic echo suppression capabilities. These features enable the system to be used as a multi-contour near-field scanner in aircraft manufacturing and maintenance hangars which are already equipped with indoor cranes, thus providing a low-cost alternative to indoor near-field facilities that cannot accommodate large-scale DUTs such as full-scale aircraft. Current near-field facilities require sampling at highly precise regular spatial intervals over a uniform surface, are expensive, limited in size and require RAM on one or all of the walls. In thc casc of cylindrical and

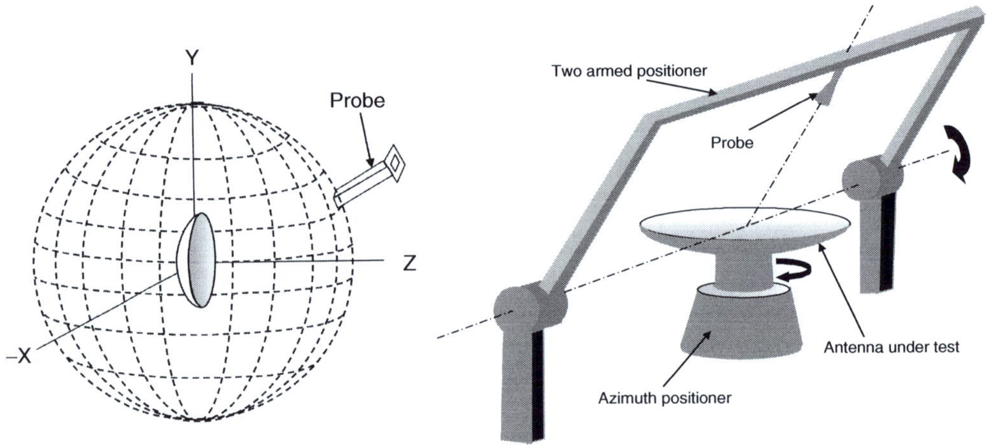

(a) Sampling of the near field. Reproduced by kind permission of MI Technologies.

(b) A typical implementation of the facility.

Figure 8.16 Spherical near-field facility.

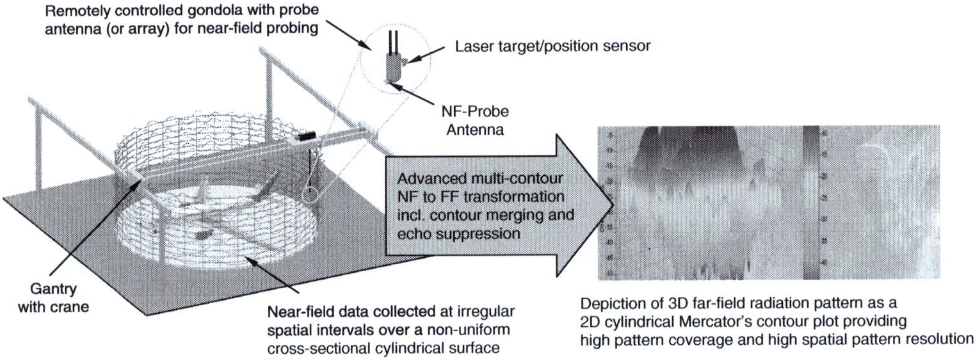

Figure 8.17 A schematic view of the AANTS used as a cylindrical near field facility. Reproduced by kind permission of Torsten Fritzel. See Plate 19 for the colour figure.

spherical facilities the DUT also has to be rotated, whereas in the case of the AANTS the DUT (aircraft) is stationary and the probe is moved in 3D space to sample the electric field over a surface that is not a smooth hemisphere in the case of antennas on the upper surface of an aircraft. Movement of the probe in 3D space obviates the requirement for a turntable to rotate the DUT which would pose an onerous task, especially in the case of full-scale aircraft.

Under the EU IPAS project [2] EADS Astrium conducted a feasibility exercise into using an airborne platform (such as a remotely piloted helicopter) to fly the probe over the sampling surface. However, the technology associated with the wireless transmission of true-in-phase RF measurement signals between the probe and the ground processing station still requires significant development. Thus the near-field facility is now being implemented as a crane-supported four-arm gantry from which a gondola is suspended. The gondola includes a near-field RF sensor, sensor axis stabilization, on-board RF instrumentation and laser targets that enable precise positional and orientation information of the RF sensor. The control signals can be hard-wired, thus obviating the requirement for any wireless transmission that would perturb and/or interfere with the RF field being sampled.

Reduction in the duration of the measurement time is being considered by the use of a multi-probe array especially at the high frequencies where the density of sampling is high.

It is estimated that using the crane-suspended advanced near-field measurement method, the measurement time for the radiation patterns of five TCAS antennas at seven spot frequencies in the 1030–1090 MHz band would take a team of two people about 5 days, using a probe antenna array of four elements. This time could include the collecting the data for other antennas in the same frequency band. This is a vast saving in time and cost compared with in-flight measurements, as well as a major improvement in the angular resolution and spatial coverage obtained.

During the production of this book, EADS Astrium was in the process of verification of the transformation of the measured irregular near-field data to the far field using an advanced algorithm. It is expected that this new antenna measurement method will reach sufficient technological maturity for wider application by 2013, possibly also paving the way for further applications such near-field radar cross-section measurement of large objects. The above method of performing large-scale near-field measurements with floating platforms has been internationally patented by EADS Astrium.

8.12 Outdoor Far-Field Ranges

The AUT is usually used in receive mode and the range antenna in transmit mode, although this is not always the case. The range antenna is held stationary and the AUT is rotated to obtain the azimuth cut. To get other great circle cuts the AUT is tilted to the required angle. The AUT is usually mounted on a non-metallic vertical post (called a mast or boom) so that it is held away from the metallic body of the positioner. This reduces the perturbation of the field in the measurement volume as well as facilitating rotation of the AUT in the vertical plane in the case of large AUTs such as antennas on scaled models. The three main types of outdoor test ranges for antenna measurements are

1. elevated
2. ground reflection
3. slant.

The elevated range is the most common. In all cases the field over the measurement volume should be uniform so that its amplitude as well as its phase are constant.

8.12.1 Measurement of the Variation of the Field over the Measurement Volume

The variation of the field over the measurement volume is checked by using an antenna mounted on a probe carriage similar to the one shown in Figure 8.18. The carriage used in this case is about 6 m long and the electric field is measured continuously as the antenna is moved. In Figure 8.18 the carriage is set up to probe the electric field in the horizontal direction and the antenna is a rectangular pyramid standard gain horn operating around 2 GHz. The horn has been set up so that its narrow dimension (and hence the electrical field) is horizontal and thus the horizontal component of the electrical field is being detected. Checks are also carried out for variations down range, that is, in the direction towards the transmit antenna.

The field variation for vertical polarization was measured in the horizontal direction for three different frequencies. The variation over the probe area is greater at the higher frequencies and, as in the case of radiation patterns, there are more ripples at the higher frequencies, as can be seen from Figure 8.19. The total amplitude variations at 2, 6 and 12 GHz are 0.933, 0.952 and 1.341 dB, respectively.

8.12.2 Elevated Ranges

These are the most common type of outdoor ranges (Figure 8.20a). The range length is around 200 m (but some are as long as 2 km) and both the transmit and receive antenna are at the same height of between 10 and 20 m.

The range transmit antenna should have a narrow beamwidth so that none of the power in the main beam reaches the ground. The sidelobes that could be reflected from the ground and into the AUT are prevented from so doing by the use of diffraction fences. These ground reflected waves from the transmitter would interact with the direct waves and result in an uneven field at the receive antenna.

Diffraction fences (Figure 8.20b) are used to deflect any waves that are reflected from the ground upwards towards the sky. This is the primary source of the electric field variation over the measurement volume. These diffraction fences are usually made of a

Figure 8.18 Antenna and carriage used to probe the electric field over the measurement volume.

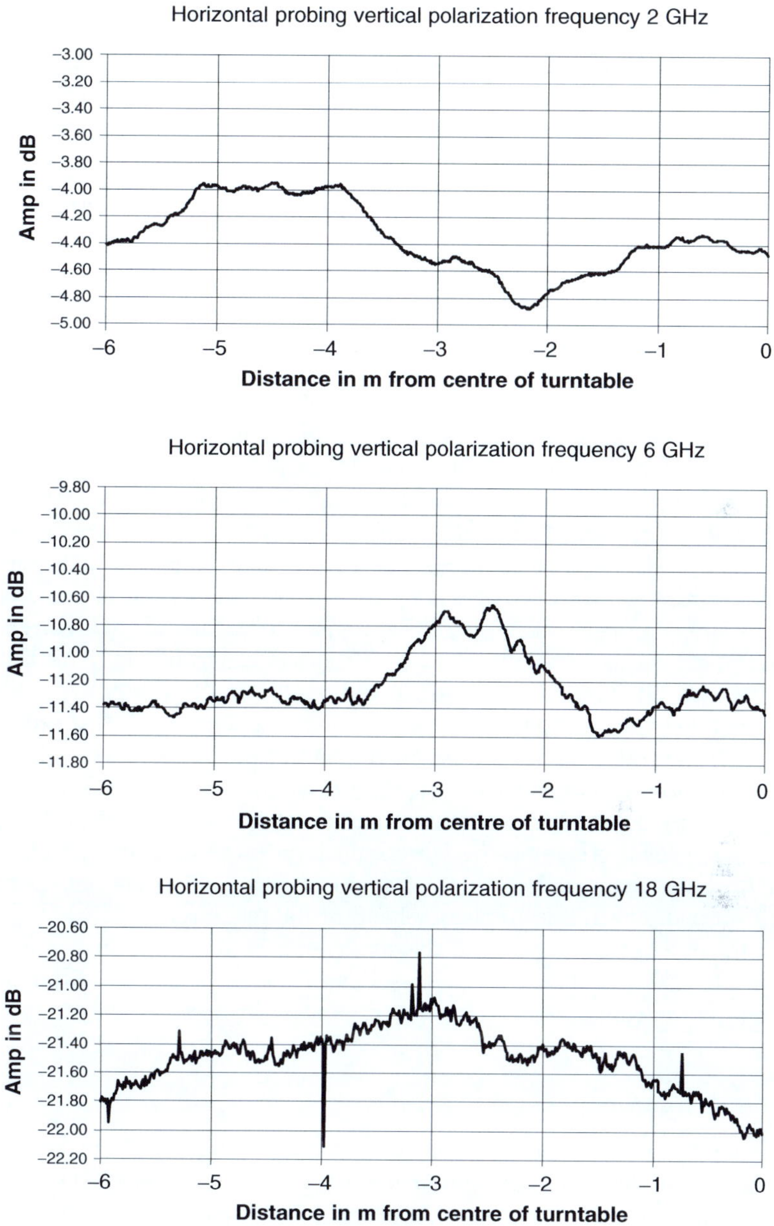

Figure 8.19 Field variation at the test zone of a far-field range, in the horizontal direction at three different frequencies.

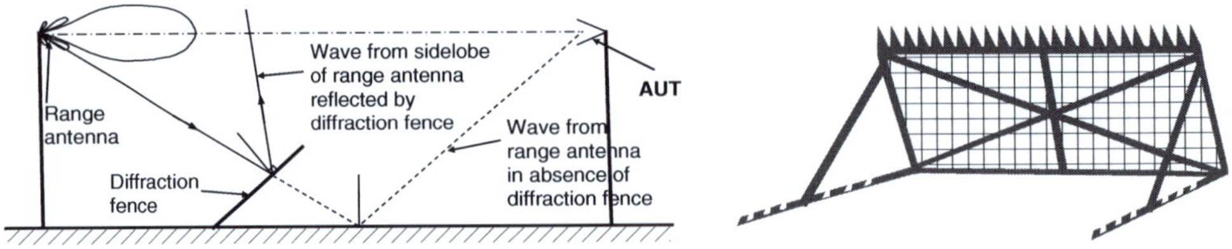

(a) Elevated range showing the position of diffraction fence (b) One type of diffraction fence

Figure 8.20 Elevated range geometry – not to scale.

wire mesh stretched onto a frame that can be tilted or adjusted so that the correct angle can be obtained for the characteristics of the particular test facility. The mesh is preferred to a solid surface because of the reduced weight and wind resistance it offers and mesh sizes of around $\lambda/20$ to $\lambda/50$ are usually used. These fences have triangular spikes called serrations, at the top to scatter the forward radiation. If the top of the diffraction fences did not have these spikes, 'knife edge' diffraction would occur, where the radiation would be forward scattered over the fences to the other side. More recently 'flower petal' serrations have been shown to work better on the edges as in the case of the outer periphery of compact range reflectors. A flat topped fence can reduce the reflectivity of the test zone field by up to 20 dB. Adding serrations can result in a further reduction of 10 dB ([6], p. 129). The initial positions and angles of the fences are established by sightlines from the transmit antenna side and judging where the first-order reflection would occur. If the transmit and receive towers are at the same heights and the ground between the two towers is level and flat, the diffraction fences would be placed about halfway between the two towers and inclined at about 45° to the vertical. In cases where the ground is not level obtaining the correct position is more difficult and requires the experience of skilled personnel. Several iterations may be required to obtain the optimum compromise over the operating frequency band of the measurement site.

If the diffraction fence is too high the main lobe of the transmit antenna could be blocked. In this case it is better to use a series of shorter fences at different distances from the transmit antenna tower.

8.12.3 Slant Ranges

In this case the distance between the transmit antenna and the AUT is nearly equal to the height of the tower. Thus the transmit antenna is almost below the AUT. These are not very common and not really suitable for large antennas or antennas installed on structures since the far-field criterion would necessitate the use of very tall towers, although they are used in some cases with the transmit antenna almost directly below the AUT tower.

8.12.4 Ground Reflection Ranges

Ground reflection ranges (Figure 8.21) are based on the direct wave and the ground reflected wave having a path difference of half a wavelength. The wave that is reflected

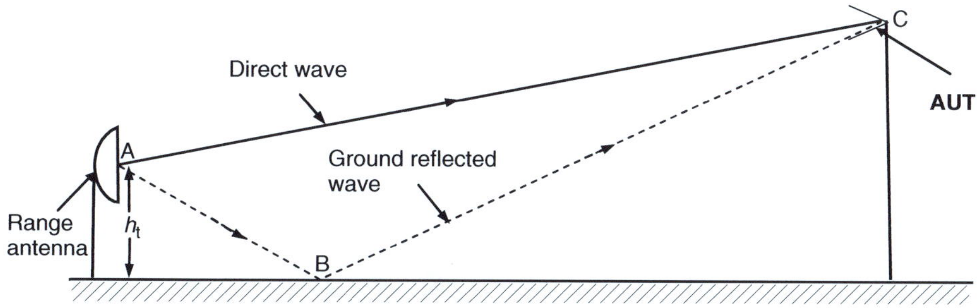

Figure 8.21 Ground reflection range.

off the ground has a phase difference of 180° so if the path length AB + BC is half a wavelength longer than the direct path AC, the reflected wave will arrive at the AUT with a phase difference of 360°, that is, it will be in phase and thus constructive interference will result. The path difference will only be half a wavelength for a particular frequency and the height h_t of the transmit tower has to be adjusted for each frequency. Thus this type of range is very narrowband. Additionally, in the case of scaled models the actual position of each antenna and distances from the transmit antenna will vary, making this type of range unsuitable for scaled model measurements.

8.12.5 Open Area Test Sites (OATS)

OATSs are mainly used for radiated emission testing of systems and equipment. They may be completely uncovered or enclosed by a material of low dielectric constant. The Federal Communications Commission (FCC) specify an optional ground plane, whereas CISPR (Center for International Systems Research – Department of State) specify a minimum size for a ground plane. A ground plane is considered a necessity especially for vertical polarization. Materials commonly used for ground planes are light gauge galvanized steel, light duty hardware cloth with 1/4 in. mesh, and heavy duty hardware cloth with 1/2 in. mesh [10]. The mesh is not as good as a solid ground plane but it provides better drainage. The ground plane should lie on the surface so that it can be inspected for signs of corrosion and non-continuity.

The size of the ground plane recommended by CISPR 22 is shown in Figure 8.22a, whereas the Fresnel ellipse shown in Figure 8.22b defines the approximate size of the ground plane recommended by the FCC Office of Science and Technology MP-4 procedure ([5], pp. 357 and 371). Note that the size of the ellipse is a function of the antenna separation D. The Fresnel ellipse is defined in the following manner. The distance D between the transmit and receive antennas must be chosen such that the path lengths between the reflected ray (AS + SB) and the direct ray (AB) are half a wavelength. Since we know that an ellipse is defined as the locus of a point that is a fixed distance from two other points (known as the foci, A and B in this case), all the reflected rays from the edges of the ellipse, such as (AS + SB) and (AT + TB) must be equal. If the transmit and receive antennas are each placed $D/2$ away from the nearest edge of the ellipse, then the reflected rays, such as AS + SB, will be equal to $2D$. It can be shown that if AB =

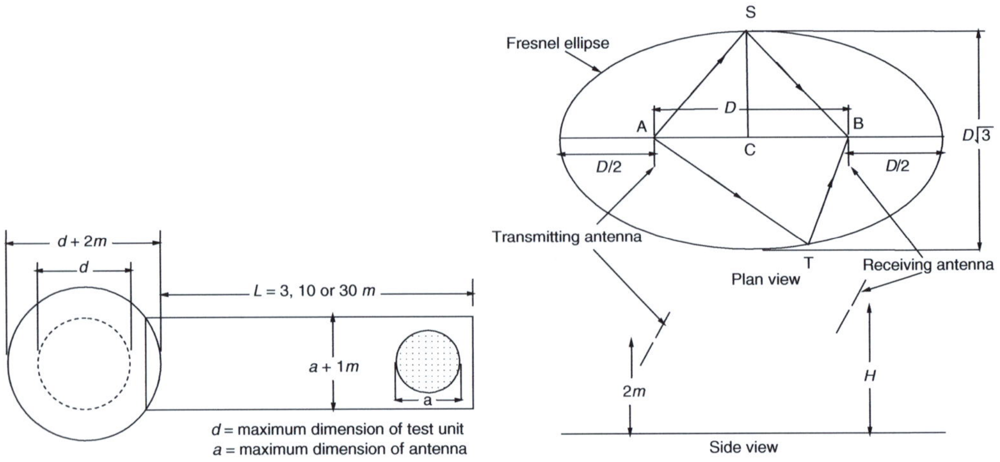

(a) Minimum size of ground plane CISPR 22 (b) FCC set-up for site attenuation

Figure 8.22 Open area test site.

SB (each will be equal to D), then by drawing a perpendicular line SC from S to AB and using Pythagoras' theorem, we get

$$D^2 = \mathrm{SC}^2 + D^2/4$$

$$\mathrm{SC}^2 = D^2 - D^2/4$$

$$\mathrm{SC} = \sqrt{D^2 - D^2/4} = D\sqrt{3/4}$$

This is the semi-minor axis of the ellipse, that is, the ellipse has minor and major axes of $D\sqrt{3}$ and $2D$ respectively.

The ellipse should be free of all obstructions, such as trees and bushes, and not have any EM reflecting objects above or below ground. It should be in an area of low RF ambients, and the terrain should be level to within about 2 in. ([5], p. 357). Bonding between the panels of the ground plane is of paramount importance and spacing intervals equal to 1/10th of the wavelength at the highest operating frequency are recommended. For a site that has an upper frequency limit of 1000 MHz, the wavelength is 30 cm, and thus the bonding interval should be 3 cm. However, bonding intervals of 12 cm (which represents 0.4 of the wavelength at 1000 MHz) have been used successfully [10]. The site itself produces attenuation and this attenuation factor must be measured. This factor is dependent on the transmit and receive antenna gains, their heights, the frequency of operation, and the ground reflected waves. There is also some reflection off other metal objects, but this is more difficult to quantify. The attenuation is measured using horizontal dipoles ([5], p. 358) at 10, 25 and 50 MHz intervals in the frequency ranges 25–100, 100–300 and 300–1000 MHz, respectively. The transmitting antenna is positioned at a fixed height of 2 m, whereas the receiving antenna height is adjusted for maximum reception. This height should be varied between 1 and 4 m for antenna separations of 3 and 10 m, and between 2 and 6 m for an antenna separation of 30 m. The peak reading, S_1, of the receiver (in dB) should be recorded at each frequency and then the antennas should be disconnected and the baluns connected together. The new readings, S_2, of the receiver at each frequency are compared with the original readings, and the difference $(S_2 - S_1)$ is the site attenuation factor. This value is added to the receiver reading when the transmit antenna is replaced by the DUT. In other words, allowance is made for the losses in the path length between the DUT and the receive antenna (which is a function of frequency, as well as the physical distance). Allowance is also made for the gains of the antennas and the constructive interference due to reflection from the ground plane.

8.12.5.1 Theoretical Site Attenuation Factors

The theoretical site attenuation may be calculated by taking into account the path losses, the gains of the antennas and the reflection from the ground plane. It is given by

$$A_s = 20\log_{10} D + 20\log_{10} F_m - G_t - G_r - 27.6 - R, \qquad (8.4)$$

where

D is distance between the transmit and receive antennas in m
F_m is the frequency in MHz
G_t is the gain of the transmitting antenna
G_r is the gain of the receive antenna
R is the contribution from the reflection in the ground plane, and is a function of the separation D between the antennas.

Average magnitudes for R are 4.5, 5.7 and 5.9 dB for 3, 10 and 30 m respectively, assuming a perfect ground plane. The theoretical equation (8.4) does not take into account the losses in the baluns of the antenna which are of the order of 0.5 dB at a frequency of 100 MHz ([5], p. 358).

For a half-wave dipole the magnitudes of G_t and G_r are each equal to 2.15 dBi. Note that the negative signs in Equation 8.4, indicate that the site attenuation factor is reduced because of these gains. The magnitudes of A_s are plotted in Figure 8.23 for values of D equal to 3, 10 and 30 m. Measured values may depart from these curves by up to 3 dB, but any marked deviations from a linear law usually indicate site problems ([5],p. 359).

8.13 Ground Test Measurements

Ground test measurements on the full-scale aircraft are usually performed for the following reasons:

1. to establish the level of coupling between different systems within the same fundamental frequency band
2. to perform radiation pattern measurements on certain antennas on the upper surface of the airframe.

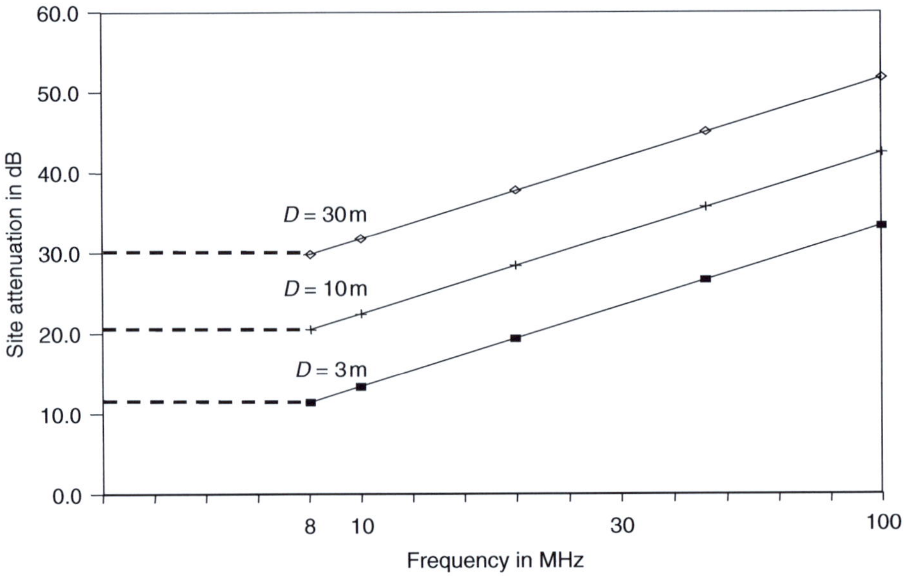

Figure 8.23 Theoretical site attenuation for distances of 3, 10 and 30 m.

8.13.1 Radiation Pattern Measurements

The radiation pattern measurements are performed by moving the transmit antenna in a circle at a fixed distance from the aircraft antenna, which is stationary. Alternatively, the aircraft antenna can be used in transmit mode and the portable antenna would be the receive antenna that is moved around in a circle. The measurements are usually taken at 5° or 10° intervals with the portable antenna at the same height as the aircraft antenna. This is thus equivalent to a mobile elevated range measurement. However, these measurements are usually undertaken on the apron of an airfield, so reflections from buildings and other aircraft affect the measurements. The coarse angular spacing results in a very rough measurement of the spatial coverage of the aircraft antenna. Avions de Transport Regional (ATR) undertook radiation pattern measurements on an air traffic control (ATC) antenna installed on ATR 72 aircraft (as shown in Figure 8.24b). The aircraft antenna was transmitting at 1 GHz, and since it was at a height of 4.4 m the receive tri-log antenna (Figure 8.24a) was on a movable platform at this height [11]. The overall length, wing span and tail height of the real aircraft are respectively 27.16, 27.06 and 7.65 m. The radiation patterns of the ATC antennas (at 1 GHz) and a VHF antenna at 126 MHz were measured using the same set-up.

The distance between the aircraft antenna and the movable receive antenna was kept at a constant distance of 37.5 m by using laser distance-measuring equipment. Samples were taken at 5° intervals in the azimuth plane only and compared with predictions using the method of moments as shown in Figure 8.24c,d. Since the angular resolution of the measurements was coarse (at 5° intervals) the correlation is not easy to discern. The sharp dip at around 315° in the measured plot could be attributed to a large aircraft on the apron of the airfield whose wing was located close to the receiving antenna. This could

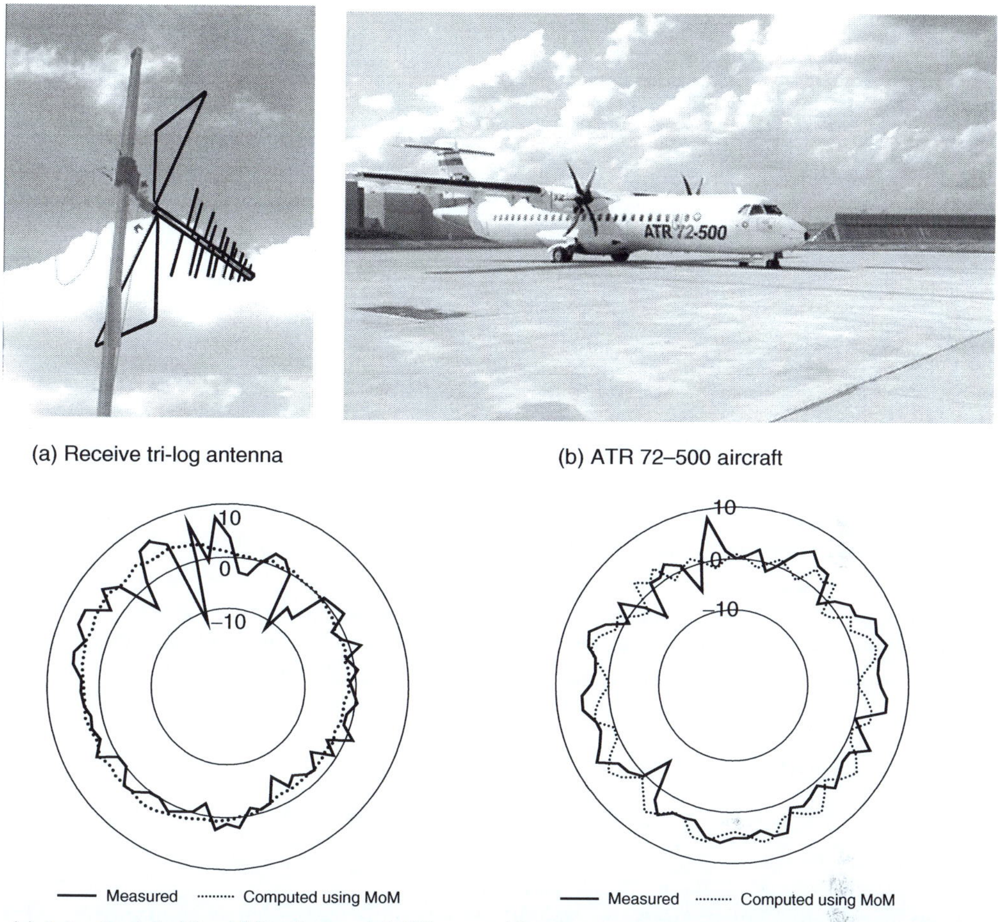

(a) Receive tri-log antenna

(b) ATR 72–500 aircraft

(c) Azimuth plot of the ATC antenna at 1 GHz

(d) Azimuth plot of the VHF antenna at 126 MHz

Figure 8.24 Comparison between the ground measurements and predicted radiation patterns of an ATC and VHF antenna on an ATR72 aircraft. Reproduced by kind permission of ATR.

not be avoided on the day the measurements were performed. This also demonstrates the inadequacies in the radiation pattern measurements carried out on the ground.

8.13.2 Coupling Measurements

The coupling between antennas on the top fuselage can be measured when the aircraft is on the ground, although the presence of other aircraft and buildings, for example, could have an influence on the measurements. Since the highest level (apart from the direct wave) occurs when the wave is subjected to specular reflection, this is the principal source of error. Physical inspection of likely specular reflection from the transmit antenna that could be incident on the receive antenna can prevent or account for any erroneous results.

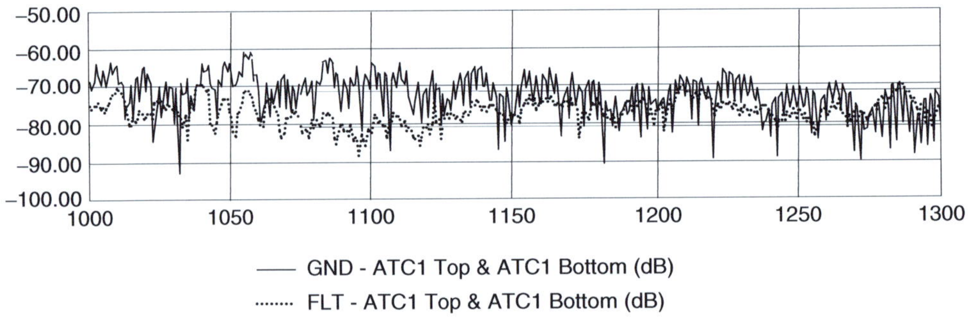

Figure 8.25 Comparison of the coupling between two ATC antennas on an ATR 72 measured on the ground and in-flight. Reproduced by kind permission of ATR.

The coupling between two ATC antennas on an ATR 72, one on top of the fuselage and one on the lower fuselage, was measured on the ground and also in flight – see [11]. The comparison between the two measurements is shown in Figure 8.25. It can be seen that the in-flight measurements show looser coupling (greater isolation) around the 1050–1150 MHz frequency range, indicating that there are some reflected waves from the lower antenna to the upper antenna when the aircraft is on the ground, and this reduces the isolation between the two antennas when the aircraft is on the ground.

8.14 In-Flight Measurements for System and Inter-System Testing

For flight test measurements of radiation patterns, the following usually occurs:

1. The ground station transmits a signal.
2. The aircraft flies away from the ground station until the signal is reduced to a very low level – this defines the range of the ground station.
3. The aircraft flies back about two-thirds of this distance.
4. A dipole is deployed from the aircraft into the air space.
5. The aircraft flies around this area within the volume over which the measurement is to take place.
6. The received signal strength or power is measured to ensure that the variation is within specified limits.
7. The aircraft is flown in several figures of eight forming a ‘clover leaf’ path, taking about 1 hour to obtain one radiation pattern cut at one frequency.
8. Thirty-six readings are taken at different headings to get 36 points on the 360° plane. This would give a pattern like the ones shown in Figure 8.24c,d, but with a coarser angular resolution.

Because the aircraft and ground transmitter are not at the same height the cut obtained is a conical cut rather than a true azimuth cut.

The number of personnel required for the flight test would most probably be one pilot, one navigator, one flight test engineer and two measurement systems engineers on the aircraft, and two test engineers and one measurement systems engineer on the ground. Thus approximately eight personnel would be required per hour for each single cut (line radiation pattern) at each frequency. Additionally there would be a measurement set-up time.

Certain frequencies that would interfere with navigational or landing systems such as ILS and ATC frequencies cannot be broadcast, so other frequencies near the frequency bands are used instead. Before broadcasting any frequencies the regulatory bodies have to be contacted and permission obtained.

Coupling measurements could also be performed in flight. These could be done in a few minutes over an entire frequency band, but the physical connections between antenna pairs might be quite time-consuming.

References

[1] Installed Performance of Antennas on AeroStructures (IPAS), an EU Specific Targeted Research Project, November 2003 to January 2007, Contract No AST3-CT-2003-503611.

[2] Fritzel, T. (2006) ANTF feasibility report, Deliverable 15 Version 1, IPAS/Tr_4_1/AST/TF 061222-1, 22 December. Installed Performance of Antennas on AeroStructures (IPAS), an EU Specific Targeted Research Project, November 2003 to January 2007, Contract No AST3-CT-2003-503611.

[3] Verpoorte, J. (2006) Verification of modelling tools by measurements and computation on scaled aircraft, Deliverable D16 Version 3, IPAS-TR_3/NLR/JV 061020-3, 20 October. Installed Performance of Antennas on AeroStructures (IPAS), an EU Specific Targeted Research Project, November 2003 to January 2007, Contract No AST3-CT-2003-503611.

[4] Belling & Lee Intac Ltd (1986) Electromagnetic Shielded Enclosures, p. 26.

[5] Keiser, B. (1987) *Principles of Electromagnetic Compatibility*, 3rd edn, Artech House, Norwood, MA.

[6] Joy, E. (1999) Far-field, Compact and Near-Field Antenna Measurement Techniques course, School of Electrical and Computer Engineering, Georgia Institute of Technology, Atlanta, Georgia 30332-0250, February 23–26 1999, Nice, France.

[7] Rudge, A.W. (1982) *The Handbook of Antenna Design*, vol. I, Peter Peregrinus, Stevenage.

[8] Bailey, A.E. (1985) *Microwave Measurement*, Peter Peregrinus, London.

[9] Steiner, H.-J. and Fritzel, T. (1993) Concept and design of a cylindrical outdoor near field range for high precision RF-measurements. Proceedings of the 15th AMTA 1993, Dallas.

[10] Tarico, F. (1989) Experiences in building an open area test site. IEEE National Symposium on Electromagnetic Compatibility 1989, Denver, pp. 157–162.

[11] Clerc, D. (2006) Comparison between measured ATR72 flight test aircraft and predicted data, Deliverable 19, IPAS/RP_4.4/ATR/DC. Installed Performance of Antennas on AeroStructures (IPAS), an EU Specific Targeted Research Project, November 2003 to January 2007, Contract No AST3-CT-2003-503611.

9

Reference

9.1 Centigrade to Fahrenheit Temperature Conversion

The Centigrade or Celsius scale of temperature has 100° between the freezing point of water (taken as 0°) and boiling point of water (taken as 100°), whereas the Fahrenheit scale has 180° between the freezing point of water, which is 32°, and boiling point of water, which is 212°.

To convert degrees Centigrade to degrees Fahrenheit the following formula should be used:

$$F = 32 + 1.8 \times C,$$

where

C is the temperature in degrees Centigrade and
F is the temperature in degrees Fahrenheit.

To convert degrees Fahrenheit to degrees Centigrade the following formula should be used:

$$C = (F - 32)/1.8,$$

where

F is the temperature in degrees Fahrenheit and
C is the temperature in degrees Centigrade.

Table 9.1 shows the conversion between the two temperature scales. Note that at −40° both the scales are at the same temperature. Normal body temperature is 37 °C or 98.4 °F.

9.2 Conductivity of Common Metals

Copper (Cu) is taken as the standard and the conductivities of all other metals are related to it as shown in Table 9.2. Thus Al, for instance, has a conductivity of 0.61 relative to copper, or 61% times the conductivity of copper. Note that if the conductivity relative to

Table 9.1 Conversion table between Centigrade and Fahrenheit scales of temperature.

Centigrade	−55	−50	−40	−30	−20	−10	0	10	20	30	40	50	55
Fahrenheit	−67	−58	−40	−22	−4	14	32	50	68	86	104	122	131

Table 9.2 Absolute and relative conductivity of common metals.

Metal	Conductivity in S-m	Conductivity relative to Cu
Aluminium	3.538×10^7	0.61
Beryllium	5.8×10^6	0.10
Brass (70% Cu)	1.450×10^7	0.25
Brass (90% Cu)	2.41×10^7	0.42
Cadmium	1.334×10^7	0.23
Copper	5.8×10^7	1.00
Gold	4.06×10^7	0.70
Iron	9.86×10^7	0.17
Lead	4.64×10^6	0.08
Magnesium	2.204×10^7	0.38
Mu-metal	1.74×10^6	0.03
Nickel	1.16×10^6	0.20
Permalloy	1.74×10^6	0.03
Phosphor-bronze	5.8×10^6	0.18
Silver	6.09×10^7	1.05
Stainless steel	1.16×10^6	0.02
Tin	8.7×10^6	0.15
Zinc	1.68×10^7	0.29

copper is greater than 1, the metal has better conductivity than copper. For instance, the conductivity of silver is 1.05 relative to copper.

9.3 Degrees to Radians and Radians to Degrees

A circle has 360° or 2π radians, thus 1° is $2\pi/360$ or $\pi/180$ radians. Thus the conversion factor is $1° = 0.01745$ radians, since $\pi = 3.142$.

To convert radians to degrees we must multiply by $180/\pi$ or 57.3, that is, 1 radian = 57.3°. Tables 9.3 and 9.4 show the conversion to and from radians and degrees.

9.4 Dielectric Constants and Loss Tangents of Common Materials

The dielectric constant is the same as the relative permittivity ε_r of a material, that is, the permittivity of a material ε, relative to the permittivity of a vacuum ε_0. Thus

$$\varepsilon_r = \varepsilon/\varepsilon_0.$$

Table 9.3 Degrees to radians.

Degrees	10	20	30	40	50	60	70	80	90	100	110	120	130
Radians	0.17	0.35	0.524	0.7	0.87	1.05	1.22	1.4	1.57	1.75	1.92	2.09	2.27
Degrees	140	150	160	170	180	190	200	210	220	230	240	250	260
Radians	2.44	2.62	2.79	2.97	3.14	3.32	3.49	3.665	3.84	4.01	4.19	4.36	4.54
Degrees	270	280	290	300	310	320	330	340	350	360			
Radians	4.71	4.89	5.06	5.24	5.41	5.59	5.76	5.93	6.11	6.28			

Table 9.4 Radians to degrees.

Radians	0.2	0.4	0.6	0.8	1.0	1.2	1.4	1.6	1.8	2.0	2.2	2.4	2.6	2.8	3.0	3.14
Degrees	11.5	22.9	34.4	45.8	57.3	68.8	80.2	91.7	103	115	126	138	149	160	172	180
Radians	3.2	3.4	3.6	3.8	4.0	4.2	4.4	4.6	4.8	5.0	5.2	5.4	5.6	5.8	6.0	6.28
Degrees	183	195	206	218	229	241	252	264	275	286	298	309	321	332	344	360

The dielectric constant of a material changes with frequency, so the frequency must quoted when stating the magnitude of the dielectric constant.

There are two parts to the dielectric constant, a real part ε_r' and an imaginary part ε_r''. The real part is associated with the change in wavelength, whereas the imaginary part is associated with the loss in the material and the phase delay experienced by the wave when passing through the dielectric. The ratio of the imaginary to real parts of the relative permittivity is known as the loss tangent tan δ. Thus

$$\tan \delta = \varepsilon_r''/\varepsilon_r'.$$

Table 9.5 shows the real parts of the dielectric constants ε_r' and the values of tan δ of some materials at 3 and 10 GHz. Most of these materials are measured at approximately 25 °C and zero humidity.

9.5 Electrochemical Series

When dissimilar materials are in contact with each other in the presence of moisture, electrolytic corrosion occurs. It is important for aircraft engineers to be aware of the electrochemical properties of materials so that they do not use materials that will be subjected to corrosion. The electrochemical series is the classification of elements depending on the electrode potential that is developed when the element is immersed in a normal solution of ionic concentration.

The greater the difference in the electromotive force (EMF) levels in the electrochemical series between the two elements, the greater is the level of corrosion. For instance, if the

Table 9.5 The real part of the dielectric constant and loss tangent for different materials at two frequencies.

Material	Dielectric constant ε_r' at 3 GHz	tan δ at 3 GHz	Dielectric constant ε_r' at 10 GHz	tan δ at 10 GHz
Alumina (99%) ceramic	8.7	–	9.7–10.3	0.0004
Corning glass 1990	7.9	0.00199	7.94	0.0032
Dow Corning no. 500	2.20	0.00145	–	–
Teflon Poly F-1114	2.1	0.00015	2.08	0.00037
Fibreglass BK-164	3.88	0.01	3.99	0.0131
Ice (−12 °C)	3.20	0.0009	3.17	0.0007
Ignition Sealing Compound No. 4	2.77	0.010	–	–
Fibreglass Laminate ECC-11-148	3.78	0.0140	–	–
Fibreglass Laminate BK-164	3.88	0.0120	3.99	0.0131
Mica, glass-bonded	–	–	7.5	0.0020
Neoprene GN	2.84	0.0480	–	–
Nylon no. 610	2.84	0.0117	–	–
Plexiglass	2.60	0.0057	2.59	0.0067
Polyethylene	2.32	0.0050	2.31	0.0044
Polystyrene foam	1.05	0.00003	–	–
Polystyrene hydrogenated	2.25	0.00016	2.25	0.00041
Polyglass D+ Monsanto	3.22	0.00120	3.22	0.00130
Rubber GR-S BXG-117G compound	2.75	0.027	6.3	0.019
Sodium chloride solution	70.8	0.29	–	–
Teflon (PTFE), unreinforced	2.1	0.00015	2.08	0.00037
Quartz, fused	3.80	0.0001	3.78	0.0001
Water (at 25 °C)	77.11	0.15	55	0.54

materials are beryllium and silver, the difference is −1.70–0.8, that is, 2.5 V, and this is a corrosive combination. An EMF greater than 0.25 V should be avoided on outside installations. Hydrogen is taken as the zero reference level. Table 9.6 lists the metals by the magnitude of their EMF levels.

Since most airframes are made from aluminium, the difference between the EMF level of the element and the EMF level of Al is listed alphabetically in Table 9.7.

9.6 Electromagnetic Spectrum and Frequency Bands for Different Nomenclatures

The electromagnetic spectrum shown in Table 9.8 covers all the EM waves, but the frequency bands shown in Table 9.9 only apply to those used at RF and microwaves in the aircraft industry.

Table 9.6 Electrochemical series by EMF levels.

Element	Symbol	EMF level
Lithium	Li	−2.959
Sodium	Na	−2.712
Magnesium	Mg	−2.34
Beryllium	Be	−1.70
Aluminium	Al	−1.67
Manganese	Mn	−1.05
Zinc	Zn	−0.762
Chromium	Cr	−0.71
Iron	Fe	−0.44
Cadmium	Cd	−0.402
Cobalt	Co	−0.277
Nickel	Ni	−0.250
Tin	Sn	−0.136
Lead	Pb	−0.126
Hydrogen	H	0.0
Copper doubly charged	Cu^{++}	+0.345
Copper singly charged	Cu^{+}	+0.522
Mercury	Hg	+0.799
Silver	Ag	+0.800
Platinum	Pt	+1.2
Gold	Au	+1.62

9.6.1 EM Spectrum

EM waves are transverse waves, unlike sound and ultrasonic waves, which are longitudinal waves that require a medium. These sonic waves cannot be transmitted in a vacuum, whereas EM waves do not require a medium and can be transmitted in a vacuum, such as deep space.

The RF and microwave region only covers a very small part of the EM spectrum as shown in Table 9.8.

9.6.2 Frequency Bands and Wavelengths for Different Nomenclatures

Most frequency bands are denoted by acronyms or single capital letters, with subscripts in some cases. However, in many cases the same letters are used by different organizations to denote different frequency bands. For instance, the letter 'L' can refer to the frequency band 20–40 GHz using the NATO designation, 1–2 GHz using the IEE designation, and 0.39–1.55 GHzusing the US designation. Thus it is important for the name of the organization to be quoted in each case. The RF and microwave frequencies are shown in Table 9.9.

Table 9.7 Electrochemical series listed by the elements alphabetically and compared with aluminium.

Element	Symbol	EMF level	Difference in EMF level compared with Al
Aluminium	Al	−1.67	0
Beryllium	Be	−1.7	0.03
Cadmium	Cd	−0.402	1.268
Chromium	Cr	−0.71	0.96
Cobalt	Co	−0.277	1.393
Copper doubly charged	Cu^{++}	0.345	2.015
Copper singly charged	Cu^{+}	0.522	2.192
Gold	Au	1.62	3.29
Hydrogen	H	0	1.67
Iron	Fe	−0.44	1.23
Lead	Pb	−0.126	1.544
Lithium	Li	−2.959	1.289
Magnesium	Mg	−2.34	0.67
Manganese	Mn	−1.05	0.62
Mercury	Hg	0.799	2.469
Nickel	Ni	−0.25	1.42
Platinum	Pt	1.2	2.87
Silver	Ag	0.8	2.47
Sodium	Na	−2.712	1.042
Tin	Sn	−0.136	1.534
Zinc	Zn	−0.762	0.908

9.7 Formulas

Useful formulas are included in this section. Some are described in detail in the main chapters of the book.

Angular frequency

Angular frequency in radians per second is given by

$$\omega = 2\pi f$$

where $\pi = 3.142$ and f is the frequency in Hz.

Capacitance of two or more capacitors in series

The capacitance C_t of two or more capacitors in series is given by

$$C_t = \frac{1}{\dfrac{1}{C_1} + \dfrac{1}{C_2} + \dfrac{1}{C_3} + \cdots + \dfrac{1}{C_n}}$$

where $C_1, C_2, \ldots, C_n$ are the individual capacitances in farads.

Table 9.8 The electromagnetic spectrum.

Frequency	Band	Wavelength
DC 0		
AC 50/60 Hz		5000/6000 km
	RF	
500 MHz 5×10^{8} Hz		60 cm
	Microwaves	
30 GHz 3×10^{10} Hz		1 cm
	mm waves	
300 GHz 3×10^{11} Hz		1 mm
	Infrared	
430 THz 43×10^{13} Hz		0.7 µm 7×10^{-7} m
	Visible light	
750 THz 75×10^{13} Hz		0.4 µm 4×10^{-7} m
	Ultra violet	
60 PHz 6×10^{16} Hz		5 nm 5×10^{-9} m
	X-rays	
50 EHz 5×10^{19} Hz		6 pm 6×10^{12} m
	Gamma rays	
1000 EHz 10^{21} Hz		0.3 pm 3×10^{-13} m
	Cosmic rays	
10^{7} EHz 10^{25} Hz		30 am 3×10^{-17} m

Table 9.9 RF and microwave Frequency bands and wavelengths.

Aircraft bands

Band	HF	VHF	UHF		L		S		C		
Frequency	2 MHz	30 MHz	225 MHz	400 MHz	1 GHz		3 GHz	3.5 GHz	7.5 GHz		
Wavelength	150 m	10 m	1.55 m	75 cm	30 cm		10 cm	8.6 cm	4 cm		

Band	S		X	K	
Wavelength	12 cm	7.5 cm	5 cm	2.4 cm	7.5 mm
Frequency	2.5 GHz	4 GHz	6 GHz	12.5 GHz	40 GHz

Band	Q	
Frequency	33 GHz	50 GHz
Wavelength	9.1 mm	6 mm

Common Usage

Band	RF	Microwaves	mm waves	sub mm waves	
Frequency	10 kHz	300 MHz	30 GHz	300 GHz	3000 GHz
Wavelength	30 km	1 m	1 cm	10 mm	100 μm

IRR

Band	ELF	VLF	LF	MF	HF	VHF	UHF	SHF	EHF	
Frequency	3 Hz	3 kHz	30 kHz	300 kHz	3 MHz	30 MHz	300 MHz	3 GHz	30 GHz	300 GHz
Wavelength		100 km	10 km	1 km	100 m	10 m	1 m	10 cm	1 cm	1 mm

ITU

Band	ELF	SLF	ULF	VLF	LF	MF	HF	VHF	UHF	L	S	C	X	Ku	`K	Ka	mm	
Frequency	3 Hz	30 Hz	300 Hz	3 kHz	30 kHz	300 kHz	3 MHz	30 MHz	300 MHz	1 GHz	2 GHz	4 GHz	8 GHz	12 GHz	18 GHz	27 GHz	40 GHz	300 GHz
Wavelength	100 Mm	10 Mm	1 Mm	100 km	10 km	1 km	100 m	10 m	1 m	30 cm	15 cm	7.5 cm	37.5 mm	25 mm	16.7 mm	11 mm	7.5 mm	1 mm

NATO

Band	A	B	C	D	E	F	G	H	I	J	K	L	M	N	
Frequency GHz	0	0.25	0.5	1	2	3	4	6	8	10	20	40	60	100	200
Wavelength		1.2 m	60 cm	30 cm	15 cm	10 m	7.5 m	5 cm	37.5 mm	3 cm	15 mm	7.5 mm	5. mm	3 mm	1.5 mm

UK IEE/IET

Band	L	S	C	X	J	K	Q	V	O	
Frequency GHz	1	2	4	8	12	18	27	40	60	90
Wavelength	30 cm	15 cm	7.5 m	37.5 mm	25 mm	16.7 mm	11 mm	7.5 mm	5. mm	3.3 mm

USA

Band	I	G	P	L	S	X	K	Q	V	W	
Frequency GHz	0.1	0.15	0.225	0.39	1.55	5.2	10.9	36	46	56	100
Wavelength	3 m	2 m	1.55 m	77 cm	19.4 cm	5.77 cm	2.75 cm	8.3 mm	6.5 mm	5.4 mm	3 mm

Capacitance of two or more capacitors in parallel

The capacitance C_t of two or more capacitors in parallel is given by

$$C_t = C_1 + C_2 + \cdots + C_n$$

where $C_1, C_2, \ldots, C_n$ are the individual capacitances in farads.

Characteristic or intrinsic impedance of free space

The characteristic or intrinsic impedance of free space Z_0 is given by

$$Z_0 = \sqrt{\frac{\mu_0}{\varepsilon_0}}$$

where $\mu_0 = 4\pi \times 10^{-7}\,\text{H m}^{-1}$ is the permeability of free space and $\varepsilon_0 = 8.85 \times 10^{-12}$ F m^{-1} is the permittivity of free space.

Characteristic impedance of a TEM wave in a dielectric

The characteristic impedance of a TEM wave in a dielectric is related to the characteristic impedance of a TEM wave in free space by the formula

$$Z = \frac{Z_0}{\sqrt{\varepsilon_r}}$$

where

Z is the impedance in the dielectric in ohms

$Z_0 = 377\,\Omega$ is the impedance in free space and

ε_r is the dielectric constant or relative permittivity of the dielectric.

Charge in terms of current and time

The charge in terms of current and time

$$Q = It$$

where

Q is the charge in coulombs

I is the current in amperes and

t is the time in seconds.

Conductance in terms of resistance

The conductance is related to resistance by

$$G = \frac{1}{R}$$

where

G is the conductance in siemens and

R is the resistance in ohms.

Conductivity of a resistor

The conductivity of a resistor is given by

$$\sigma = \frac{l}{RA}$$

where

σ is the conductivity in S-m

l is the length in m

R is the resistance in ohms and

A is the area of the cross section of the material in m^2.

Electric field strength or intensity due to an antenna

The electric field strength or intensity in the far field of an antenna is given by

$$E = \sqrt{\frac{30PG}{d}}$$

where

E is the electric field in V/m

P is the power of the transmitter in watts

G is the gain of the antenna in linear terms and

d is the distance from the antenna in metres.

Note that this formula does not allow for any attenuation between the transmitter and the input to the antenna.

Electric field in terms of the potential difference

The electric field is expressed in terms of the potential difference by

$$E = \frac{V}{d}$$

where

E is the electric field in V/m

V is the potential difference in volts and

d is the distance in metres.

Energy to power conversion

Energy E_n in joules is related to power in watts by

$$E_n = Pt = VIt$$

where

P is the power in watts

t is the time in seconds

V is the voltage in volts and

I is the current in amperes.

Far field or Fraunhofer region

The far-field or Fraunhofer region for an antenna depends on whether it is a wire or aperture antenna. For aperture antennas, such as horns and reflectors, this distance R_f is given by

$$R_f = \frac{2D^2}{\lambda}$$

where
D is the largest dimension of the antenna and
λ is the operating wavelength.

In order to ensure that a plane wave is present, the value of R_f is sometimes taken as

$$R_f = \frac{4D^2}{\lambda}$$

For wire antennas such as monopoles and dipoles, this distance is given by

$$R_f = \frac{\lambda}{2\pi}$$

where $\pi = 3.142$. Sometimes, to be absolutely certain that a plane wave is present, this distance is taken as

$$R_f = \frac{\lambda}{\pi}$$

Lenz's law

Lenz's law states that the induced EMF V in volts is given by

$$V = -\frac{d\Phi}{dt}$$

where $d\Phi/dt$ is the rate of change of the magnetic flux linkage in Wb per second.

Note that the negative sign indicates that the induced EMF acts in opposition to the EMF causing it.

For an inductor or coil,

$$V = L\frac{dI}{dt}$$

where
V is the induced EMF in volts
L is the self inductance in henries and
dI/dt is the rate of change of current.

For a transformer or two coils a and b in close proximity the EMF induced in the second coil is given by

$$V_a = -M\frac{dI_b}{dt}$$

where
V_a is the induced EMF in volts in coil A
M is the mutual inductance in henries and
dI_b/dt is the rate of change of current in coil B.

Magnetic field strength or intensity

The magnetic field strength or intensity at a distance d from an infinitely long straight wire is given by

$$H = \frac{2I}{d}$$

where
H is in A/m
I is the current through the wire in amperes and
d is the distance from the wire in metres.

The magnetic field strength or intensity along the axis of a long solenoid of n turns is given by

$$H = nI$$

where
H is in A/m
I is the current through the wire in amperes and
n is the number of turns per metre.

The magnetic field strength or intensity at the centre of a short circular coil of n turns is given by

$$H = \frac{nI}{2r}$$

where
H is in A/m
I is the current through the wire in amperes and
r is the radius of the coil in m

Magnetic field intensity to electric field intensity

The electric field intensity is related to the magnetic field intensity (in the far field of an antenna only) by

$$\frac{E}{H} = Z_0 = 120\pi = 377\,\Omega,$$

that is,

$$E = Z_0 H,$$

where
E is the electric field intensity in V/m
H is the magnetic field intensity in A/m and
π is 3.142.

Ohm's law

Ohm's law states that the voltage is proportional to the current. The constant of proportionality is called the resistance. Thus Ohm's law can be written as

$$V = IR$$

where
V is the voltage in volts
I is the current in amperes and
R is the resistance in ohms.

Power in terms of current and resistance

Power is expressed in terms of current and resistance by

$$P = I^2 R$$

where
P is the power in watts
I is the current in amperes
and R is the resistance in ohms.

Power density for a plane wave

The power density for a plane wave is given by

$$P_d = \frac{E^2}{Z_0}$$

where
P_d is the power density in W/m^2
E is the electric field intensity in V/m and
Z_0 is 120π or 377 Ω.

Power factor

The power factor P_f takes into account the phase between the voltage and the current and is defined as

$$P_f = VI \cos\phi$$

where
P_f is the factor power in watts
V is the voltage in volts
I is the current in amperes and
ϕ is the phase angle between the current and the voltage.

Reactance of a capacitor

The reactance of a capacitor X is given by

$$X = \frac{1}{\omega C}$$

where
X is the reactance in ohms
ω is the angular frequency in radians per second and
C is the capacitance in farads.

Reactance of an inductor

The reactance of an inductor X is given by

$$X = \omega L$$

where
X is the reactance in ohms

ω is the angular frequency in radians per second and

L is the inductance in henries.

Resistance of two or more resistors in series

The total resistance R_t of two or more resistors in series is given by

$$R_t = R_1 + R_2 + \cdots + R_n$$

where

R_t is the total resistance in ohms and

R_1 to R_n are the resistances of the individual resistors in ohms.

Resistance of two or more resistors in parallel

The total resistance R_t of two or more resistors in parallel is given by

$$R_t = \frac{1}{\dfrac{1}{R_1} + \dfrac{1}{R_2} + \dfrac{1}{R_3} + \cdots + \dfrac{1}{R_n}}$$

where

R_t is the total resistance in ohms and

R_1 to R_n are the resistances of the individual resistors in ohms.

Resistivity of a conductor

The resistivity ρ of a conductor is given by

$$\rho = \frac{RA}{l}$$

where

ρ is the resistivity in Ω m

R is the resistance in Ω

A is the cross-sectional area of the conductor in m^2 and

l is the length of the conductor in m.

Resonant frequency of a parallel or series LCR circuit

The resonant frequency of a parallel or series LCR circuit is given by

$$f = \frac{1}{2\pi\sqrt{LC}}$$

where

f is the frequency in Hz

L is the inductance in henries and

C is the capacitance in farads.

Skin depth

The skin depth δ is given by

$$\delta = \sqrt{\frac{2}{\omega\mu\sigma}}$$

where
δ is the skin depth in m
ω is the frequency in radians per second
μ is the magnetic permeability in H/m and
σ is the conductivity in S m.

Speed of electromagnetic waves in a coaxial line

The speed of a TEM wave in a coaxial line having a dielectric of relative permittivity is given by

$$v = \frac{c}{\sqrt{\varepsilon_r}}$$

where
v is the velocity of the EM wave in the dielectric in m/s
c is the velocity of the EM wave in free space in m/s and
ε_r is the relative permittivity or dielectric constant in the dielectric.

Speed of EM radiation

The velocity of EM radiation c is given by

$$c = f\lambda$$

where
c is the velocity of EM radiation in m/s
f is the frequency in Hz and
λ is the wavelength in m.

Voltage–watts conversion

To convert wattage to voltage, the impedance has to be known. In RF coaxial cable the characteristic impedance R is 50 Ω.

The relationship between voltage V and power P is

$$V = \sqrt{PR}$$

where
V is the voltage in volts
P is the power in watts and
R is the resistance in ohms.

The relationship in decibels, for voltage in dBV and power in dBW, is

$$20 \log V = 10 \log P + 10 \log R$$

Thus 1 W equates to 17 dBV or 7.07 V, and 1 mW equates to −13 dBV (107 dBμV) or 223.9 mV.

Wave impedance

For a plane wave propagating through free space, the wave impedance Z_0 is given by

$$Z_0 = \sqrt{\frac{\mu_0}{\varepsilon_0}}$$

where
Z_0 is the wave impedance in ohms
μ_0 is the permeability of free space and
ε_0 is the permittivity of free space.

9.8 Frequency to Wavelength

The frequency and wavelength are related by the formula

$$\lambda = c/f$$

where

λ is the wavelength in m
c is the velocity of light/EM radiation in m/s and
f is the frequency in Hz.

The velocity of light is nearly 300 000 000 m/s.

If the frequency is in MHz, then the above formula reduces to

$$\lambda = 300/F$$

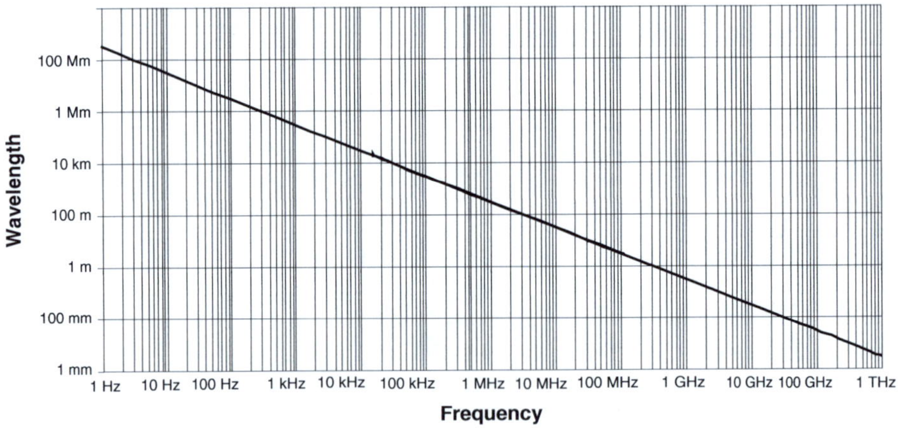

Figure 9.1 Wavelength against frequency up to 1000 GHz.

Table 9.10 Wavelength for frequencies up to 1000 GHz.

Frequency	3 Hz	10 Hz	30 Hz	100 Hz	300 Hz	1 kHz	3 kHz	10 kHz	30 kHz	100 kHz	300 kHz
Wavelength	100 Mm	30 Mm	10 Mm	3 Mm	1 Mm	300 km	100 km	30 km	10 km	3 km	1 km
Frequency	1 MHz	3 MHz	10 MHz	30 MHz	100 MHz	300 MHz	1 GHz	3 GHz	10 GHz	100 GHz	1 THz
Wavelength	300 m	100 m	30 m	10 m	3 m	1 m	30 cm	10 cm	3 cm	1 cm	3 mm

where F is the frequency in MHz. This is very easy conversion to perform, since it can be seen that 300 MHz gives us a wavelength of 1 m.

If the frequency is in GHz, then the formula can be written as

$$\lambda_{\text{cm}} = 30/F_{\text{GHz}}$$

where

λ_{cm} is the wavelength in cm and
F_{GHz} is the frequency in GHz.

It can be seen that a frequency of 30 GHz gives us a wavelength of 1 cm. The conversions are shown in Figure 9.1 and Table 9.10.

9.9 Gain in dB and Gain as a Linear Ratio

The values of gain are sometimes very high and therefore they are usually quoted in decibels. Thus a linear gain of 1000 can be written as 30 dB. Gains in decibels are also more convenient because they can be added together, whereas linear gains have to be multiplied. When the gain is quoted in dBi it is related to that of an isotropic antenna.

The gain in dBi, G_{dBi}, is related to the linear or numeric gain G by the formula

$$G_{\text{dBi}} = 10 \log G$$

where log denotes logarithm to the base 10.

To convert the linear or numeric gain G to the gain in dBi, the formula is

$$G = 10^{G_{\text{dBi}}/10}.$$

To obtain intermediate values of linear gain from the gain in dBi multiply linear gains of the whole number by the linear gain of the decimal fraction. For example, to convert 5.6 dBi to linear gain, look up 5 dBi (=3.162; see Table 9.11) and multiply this by the 0.6 dBi value (=1.148) to give $3.162 \times 1.148 = 3.26998$.

Table 9.11 Conversion between gain in dB and gain as a linear ratio.

Gain in dBi to linear gain										
Gain in dBi	−10	−9	−8	−7	−6	−5	−4	−3	−2	−1
Linear gain	0.1	0.126	0.158	0.2	0.251	0.316	0.398	0.501	0.631	0.794
Gain in dBi	−0.9	−0.8	−0.7	−0.6	−0.5	−0.4	−0.3	−0.2	−0.1	0
Linear gain	0.813	0.832	0.851	0.871	0.891	0.912	0.933	0.955	0.977	1
Gain in dBi	0.1	0.2	0.3	0.4	0.5	0.6	0.7	0.8	0.9	1
Linear gain	1.023	1.047	1.072	1.096	1.122	1.148	1.175	1.202	1.230	1.259
Gain in dBi	2	3	4	5	6	7	8	9	10	
Linear gain	1.585	1.995	2.512	3.162	3.981	5.012	6.310	7.943	10	
Linear gain to gain in dBi										
Linear gain	0.1	0.2	0.3	0.4	0.5	0.6	0.7	0.8	0.9	1
Gain in dBi	−10	−7	−5.2	−4	−3	−2.2	−1.5	−1	−0.5	0
Linear gain	2	3	4	5	6	7	8	9	10	
Gain in dBi	3.01	4.77	6.02	6.99	7.78	8.45	9.03	9.54	10	
Linear gain	20	30	40	50	60	70	80	90	100	
Gain in dBi	13.01	14.77	16.02	16.99	17.78	16.45	19.03	19.54	20	

9.10 Greek Alphabet

The Greek alphabet is used in many scientific equations, and as symbols to denote physical constants and variables. Both upper and lower case are used and their pronunciations are listed in Table 9.12.

9.11 Imperial to Metric Conversions – Distance, Area, Volume, Speed

Although the Système International (SI) system of units is universally accepted nowadays, the aircraft industry has used a mixture of units. For instance, altitude (distance above ground) is usually measured or quoted in feet, while speed is measured or quoted in knots (nautical miles per hour), mph (miles per hour) or kilometres per hour. Knots are usually in international units (1 international nautical mile = 6080 ft) but note that the UK nmi and US nmi differ fromthe international nautical mile. Nautical miles are sometimes abbreviated to NM.

Aircraft equipment sizes are usually quoted in millimetres or inches.

9.11.1 Conversion Factors for Lengths/Distances

The factors used to convert distances to and from metric units are shown in Table 9.13.

9.11.2 Conversion Factors for Areas

The factors used to convert areas to and from metric units are shown in Table 9.14.

Table 9.12 The Greek alphabet and pronunciation.

Lower case	Upper case	Name
α	A	Alpha
β	B	Beta
γ	Γ	Gamma
δ	Δ	Delta
ε	E	Epsilon
ζ	Z	Zeta
η	H	Eta
θ	Θ	Theta
ι	I	Iota
κ	K	Kappa
λ	Λ	Lambda
μ	M	Mu
ν	N	Nu
ξ	Ξ	Xi
o	O	Omicron
π	Π	Pi
ρ	P	Rho
σ	Σ	Sigma
τ	T	Tau
υ	Y	Upsilon
ϕ	Φ	Phi
χ	X	Chi
ψ	Ψ	Psi
ω	Ω	Omega

9.11.3 Conversion Factors for Volume and Capacity

The factors used to convert volume and capacity to and from metric units are shown in Tables 9.15 and 9.16.

9.11.4 Conversion Factors for Speed

The speed used for aircraft is usually in kilometres per hour or miles per hour, but other conversions such as feet per second are also shown in Table 9.17.

The Imperial system measures energy in British thermal units (BTUs) and calories, but the SI unit is the joule. The conversion factors to be used are shown in Table 9.18.

9.12 Periodic Table Alphabetically

The periodic table is usually depicted as a list of elements in order of the group to which they belong. However this makes it difficult to identify elements when their symbols are given. Thus the Table 9.19 below shows the periodic table listed alphabetically according to the chemical symbol used for each element.

Table 9.13 Imperial to metric conversion factors for distances.

Conversion factor	To convert	To	Multiply by	Conversion factor	To convert	To	Multiply by
1 inch = 25.4 mm	inches	mm	25.4	1 mm = 0.0394 inches	mm	inches	0.0394
1 inch = 2.54 cm	inches	cm	2.54	1 cm = 0.3937 inches	cm	inches	0.3937
1 inch = 0.0833 cm	inches	ft	0.0833	1 ft = 12 inches	ft	inches	12
1 inch = 0.0254 cm	inches	m	0.0254	1 m = 39.37 inches	m	inches	39.37
1 ft = 30.48 m	ft	cm	30.48	1 cm = 0.0328 ft	cm	ft	0.0328
1 ft = 0.3048 m	ft	m	0.3048	1 m = 3.28 ft	m	ft	3.2808
1 ft = 0.000189394 miles	ft	miles	0.000189394	1 mile = 5280 ft	miles	ft	5280
1 ft = 0.000166667 UK nmi	ft	UK nmi	0.000166667	1 UK nmi = 6000 ft	UK nmi	ft	6000
1 ft = 0.000164474 Intl. nmi	ft	Intl. nmi	0.000164474	1 intl. nmi = 6080 ft	intl. nmi	ft	6080
1 ft = 0.000164466 Intl. nmi	ft	US nmi	0.000164466	1 US nmi = 6080.28 ft	US nmi	ft	6080.28
1 ft = 0.0003048 km	ft	km	0.0003048	1 km = 3280.84 ft	km	ft	3280.84
1 yd = 91.44	yards	cm	91.44	1 cm = 0.01094	cm	yards	0.01094
1 yd = 0.9144	yards	m	0.9144	1 m = 1.0936	m	yards	1.0936
1 yd = 0.000914	yards	km	0.0009144	1 km = 1093.6	km	yards	1093.6
1 mile = 1609.3 m	miles	m	1609.3	1 m = 0.0006214 miles	m	miles	0.0006214
1 mile = 1.6093 km	miles	km	1.6093	1 km = 0.6214 miles	km	miles	0.6214
1 UK nmi = 1.8290 km	UK nmi	km	1.8290	1 km = 0.5467 UK nmi	km	UK nmi	0.5467
1 Intl nmi = 1.8532 km	Intl. nmi	km	1.85320	1 km = 0.5396 intl. nmi	km	Intl nmi	0.5396
1 US nmi = 1.85327 km	US nmi	km	1.85327	1 km = 0.5396 US nmi	km	US nmi	0.5396

Table 9.14 Imperial to metric conversion factors for area.

Conversion factor	To convert	To	Multiply by	Conversion factor	To convert	To	Multiply by
1 sq inch = 645.16 sq mm	sq inch	sq mm	645.16	1 sq mm = 0.0016 sq inch	sq mm	sq inch	0.0016
1 sq inch = 6.4516 sq. cm	sq inch	sq cm	6.4516	1 sq. cm = 0.155 sq inch	sq cm	sq inch	0.155
1 sq inch = 0.0069 sq ft	sq inch	sq ft	0.0069	1 sq ft = 144 sq inch	sq ft	sq inch	144
1 sq inch = 0.000645 sq cm	sq inch	sq m	0.000645	1 sq m = 1550 sq inch	sq m	sq inch	1550.0031
1 sq ft = 929.03 sq m	sq ft	sq cm	929.03	1 sq cm = 0.0011 sq ft	sq cm	sq ft	0.0011
1 sq ft = 0.092903 sq m	sq ft	sq m	0.092903	1 sq m = 10.7639 sq ft	sq m	sq ft	10.7639
1 sq ft = 0.0000000359 sq miles	sq ft	sq miles	0.0000000359	1 sq mile = 27 878 400 sq ft	sq miles	sq ft	27 878 400
1 sq ft = 0.0000000929 sq km	sq ft	sq km	0.0000000929	1 sq km = 10 763 910 sq ft	sq km	sq ft	10 763 910
1 sq yd = 8361.27 sq m	sq yd	sq cm	8361.27	1 sq cm = 0.000120 sq yd	sq cm	sq yds	0.000120
1 sq yd = 0.836127 sq m	sq yd	sq. m	0.836127	1 sq m = 1.1960 sq yd	sq. m	sq yd	1.1960
1 sq yd = 0.0000008361 sq km	sq yd	sq km	0.0000008361	1 sq km = 1 195 990 sq yd	sq km	sq yds	1 195 990
1 sq mile = 2 589 988 sq m	sq miles	sq m	2 589 988	1 sq m = 0.00000039 sq miles	sq m	sq miles	0.00000039
1 sq mile = 2.590 sq km	sq miles	sq km	2.590	1 sq km = 0.3861 sq miles	sq km	sq miles	0.3861
1 acre = 43 560 square feet	acre	sq. ft	43 560	1 sq. ft = 0.00002296 acres	sq. ft	acre	0.00002296
1 acre = 4840 square yd	acre	sq. yd	4840	1 sq. yd = 0.00020661 acres	sq. yd	acre	0.00020661
1 acre = 0.4047 ha*	acres	hectares	0.4047	1 ha = 2.471 acres	hectares	acres	2.471

*1 ha = 10 000 sq m.

Table 9.15 Imperial to metric conversion factors for volume.

Conversion factor	To convert	To	Multiply by	Conversion factor	To convert	To	Multiply by
1 cub inch = 16387.06 cub mm	cub inch	cub mm	16387.06	1 cub mm = 0.000061 cub inch	cub mm	cub inch	0.000061
1 cub inch = 16.39 cub cm	cub inch	cub cm	16.39	1 cub cm = 0.06102 cub inch	cub cm	cub inch	0.06102
1 cub inch = 0.000579 cub ft	cub inch	cub ft	0.000579	1 cub ft = 1728 cub inch	cub ft	cub inch	1728
1 cub inch = 0.00001639 cub cm	cub inch	cub m	0.00001639	1 cub m = 61023.74 cub inch	cub m	cub inch	61023.74
1 cub ft = 28316.85 cub m	cub ft	cub cm	28316.85	1 cub cm = 0.000035 cub ft	cub cm	cub ft	0.0000353
1 cub ft = 0.0283 cub m	cub ft	cub m	0.0283	1 cub m = 35.31 cub ft	cub m	cub ft	35.31
1 cub yd = 764554.86 cub m	cub yd	cub cm	764554.86	1 cub cm = 0.00000131 cub yd	cub cm	cub yd	0.00000131
1 cub yd = 0.765 cub m	cub yd	cub m	0.765	1 cub m = 1.3080 cub yd	cub m	cub yd	1.3080
1 cub mile = 4.168 cub km	cub miles	cub km	4.168	1 cub km = 0.2399 cub miles	cub km	cub miles	0.2399

Table 9.16 Imperial to metric conversion factors for capacity.

Conversion factor	To convert	To	Multiply by	Conversion factor	To convert	To	Multiply by
1 British gallon = 4.546 litres	British gallons	litres	4.546	1 litre = 0.21997 British gallons	litres	British gallons	0.21997
1 US gallon* = 3.785 litres	US gallons	litres	3.785	1 litre = 0.26420 US gallons	litres	US gallons	0.26420
1 British gallon = 1.28 US gallon	British gallons	US gal-lons	1.28	1 US gallon = 0.78125 British gallon	US gallons	British gallons	0.78125

*6.4 US gallons = 5 British Imperial gallons.

Table 9.17 Imperial to metric conversion factors for speed.

Conversion factor	To convert	To	Multiply by	Conversion factor	To convert	To	Multiply by
ft/s = m/s	ft/s	m/s	0.3048	1 m/s = 3.2808 ft per s	m/s	ft per s	3.2808
1 mph = 1.4667 ft per s	mph	ft per s	1.4667	1 ft per s = 0.6818 mph	ft per s	mph	0.6818
1 mph = 0.44704 m/s	mph	m/s	0.44704	1 m/s = 2.2369 mph	m/s	mph	2.2369
1 mph = 1.6093 km/h	mph	km/h	1.6093	1 km per h = 0.6214 mph	km per h	mph	0.6214
1 mph = 0.88 UK knots (nmi per hr)	mph	UK knots	0.88000	1 UK knot = 1.1364 mph	UK knots	mph	1.1364
1 mph = 0.86842 Intl. knots (nmi per hr)	mph	intl. knots	0.86842	1 intl. knot = 1.1515 mph	intl. knots	mph	1.1515
1 mph = 0.86838 US knots (nmi per hr)	mph	US knots	0.86838	1 US knot = 1.1516 mph	US knots	mph	1.1516

Table 9.18 Imperial to metric conversion factors for energy.

Conversion factor	To convert	To	Multiply by	Conversion factor	To convert	To	Multiply by
1 calorie = 4.1868 J	calorie	joule	4.1868	1 joule = 0.238846 calories	Joule	Calories	0.238846
1 BTU = 1055 J	BTU	joule	1055	1 joule = 0.000948 BTU	Joule	BTU	0.000948
1 BTU = 252 calories	BTU	calorie	252	1 calorie = 0.003968 BTU	Calories	BTU	0.003968

Table 9.19 Periodic table in alphabetical order ordered by chemical symbol.

Symbol	Element	Atomic No.	Atomic weight
Ac	Actinium	89	227.0278
Ag	Silver	47	107.868
Al	Aluminium	13	26.981
Am	Americium	95	(243)
Ar	Argon	18	39.948
As	Arsenic	33	74.9216
At	Astatine	85	(210)
Au	Gold	79	196.9665
B	Boron	5	10.81
Ba	Barium	56	137.33
Be	Beryllium	4	9.01218
Bi	Bismuth	83	208.9804
Bk	Berkelium	97	(247)
Br	Bromine	35	79.904
C	Carbon	6	12.011
Ca	Calcium	20	40.08
Cd	Cadmium	48	112.41
Ce	Cerium	58	130.12
Cf	Californium	98	(251)
Cl	Chlorine	17	35.453
Cm	Curium	96	(247)
Co	Cobalt	27	58.9332
Cr	Chromium	24	51.996
Cs	Caesium	55	132.9054
Cu	Copper	29	63.54
Dy	Dysprosium	66	162.5
Ee	Einsteinium	99	(252)
Er	Erbium	68	167.2
Eu	Europium	63	151.96
F	Fluorine	9	18.9984
Fe	Iron	26	55.84
Fm	Fermium	100	(257)
Fr	Francium	87	(223)
Ga	Gallium	31	69.72
Gd	Gadolinium	64	157.25
Ge	Germanium	32	72.5
H	Hydrogen	1	1.00794
He	Helium	2	4.0026
Hf	Hafnium	72	178.4
Hg	Mercury	80	200.5
Ho	Holmium	67	164.9304
I	Iodine	53	126.9045
In	Indium	49	114.82
Ir	Iridium	77	192.2
K	Potassium	19	39.0983
Kr	Krypton	36	83.80

Table 9.19 *(continued)*

Symbol	Element	Atomic No.	Atomic weight
La	Lanthanum	57	138.905
Li	Lithium	3	6.94
Lr	Lawrencium	103	(260)
Lu	Lutetium	71	174.967
Md	Mendelevium	101	(258)
Mg	Magnesium	12	24.305
Mn	Manganese	25	54.938
Mo	Molybdenum	42	95.94
N	Nitrogen	7	14.0067
Na	Sodium	11	22.98977
Nb	Niobium	41	92.9064
Nd	Neodymium	60	144.2
Ne	Neon	10	20.17
Ni	Nickel	28	58.69
No	Nobelium	102	(259)
Np	Neptunium	93	237.0482
O	Oxygen	8	15.999
Os	Osmium	76	190.2
P	Phosphorus	15	30.973
Pa	Protactinium	91	231.0359
Pb	Lead	82	207.2
Pd	Palladium	46	106.42
Pm	Promethium	61	145
Po	Polonium	84	(209)
Pr	Praseodymium	59	140.9077
Pt	Platinum	78	195.0
Pu	Plutonium	94	(244)
Ra	Radium	88	226.0254
Rb	Rubidium	37	85.467
Rb	Rhodium	45	102.99055
Re	Rhenium	75	186.207
Rn	Radon	86	(222)
Ru	Ruthenium	44	101.0
S	Sulphur	16	32.06
Sb	Antimony	51	121.7
Sc	Scandium	21	44.9559
Se	Selenium	34	78.9
Si	Silicon	14	28.085
Sm	Samarium	62	150.3
Sn	Tin	50	118.6
Sr	Strontium	38	87.62
Ta	Tantalum	73	180.9479
Tb	Terbium	65	158.9254
Tc	Technetium	43	98
Te	Tellurium	52	127.60
Th	Thorium	90	232.0381

(continued overleaf)

Table 9.19 *(continued)*

Symbol	Element	Atomic No.	Atomic weight
Ti	Titanium	22	47.8
Tl	Thallium	81	204.383
Tm	Thulium	69	168.9342
U	Uranium	92	238.0289
V	Vanadium	23	50.9415
W	Tungsten	74	183.8
Xe	Xenon	54	131.2
Y	Yttrium	39	88.9059
Yb	Ytterbium	70	173
Zn	Zinc	30	65.38
Zr	Zirconium	40	91.22

9.13 Polarization Matching Matrix

The ability of an antenna to detect an incoming EM wave depends on the polarization of the incident wave relative to the polarization of the receive antenna. If the receive antenna's polarization is the same as that of the incident wave, then the antenna will detect the incident wave without loss. This incident wave is called the co-polar radiation. In the case of a vertical monopole the co-polar radiation is vertically polarized and a horizontally polarized wave is the cross-polar radiation. Ideally a vertical monopole would not receive any of the horizontally polarized wave, but in practice real antennas, especially those that are top loaded, have a cross-polar discrimination (XPD) of the order of 20 dB. Thus the horizontally polarized wave will be detected with a loss equal to this cross-polar level, that is, at a level of −20 dB, below that of a co-polar wave. In the case of circular polarization an RHCP wave will be co-polar to an antenna that receives RHCP, whereas the cross-polarization will be LHCP.

The matrix shown in Table 9.20 depicts only the losses that occur due to receiver and transmitter polarizations.

9.14 Power in dBm and Power in Watts

It should be noted that if power is quoted in decibels, this is a ratio, for example between a transmitter and receiver. In order to work out the actual power, we have either to be given the absolute power in one case or have the power referred to an actual level.

Thus powers are usually referenced to milliwatts and quoted in dBm, that is, dB milliwatts. In this case 1 mW is taken as 0 dBm.

In high-power applications the power is reference to watts and quoted in dBW, and 0 dBW is 1 W. To convert the power in dBm to power in milliwatts, the formula is

$$P_{\mathrm{dBm}} = 10 \log (P_{\mathrm{mW}}).$$

Table 9.20 Polarization matching matrix.

		Transmitter polarization					
		Horizontal →	Vertical ↑	Right slant ↗	Left slant ↖	RHCP ↻	LHCP ↺
Receiver polarization	Horizontal →	Yes no loss	At XP level*	3 dB loss	3 dB loss	3 dB loss	3 dB loss
	Vertical ↑	At XP level*	Yes no loss	3 dB loss	3 dB loss	3 dB loss	3 dB loss
	Right slant ↗	3 dB loss	3 dB loss	Yes no loss	At XP level*	3 dB loss	3 dB loss
	Left slant ↖	3 dB loss	3 dB loss	At XP level*	Yes no loss	3 dB loss	3 dB loss
	RHCP ↻	3 dB loss	3 dB loss	3 dB loss	3 dB loss	Yes no loss	At XP level*
	LHCP ↺	3 dB loss	3 dB loss	3 dB loss	3 dB loss	At XP level*	Yes no loss

*Although cross polarization cannot be detected by perfect antennas, practical antennas have levels of cross polar discrimination of the order of −20 dB. Thus detection is possible at these levels

Thus we can see that if the power is 1 mW, log (1) is zero and thus the power in dBm is also zero. If the power is 10 mW, log (10) is 1, and when we multiply by 10 we get the power in dBm is 10 dBm.

To convert the power in milliwatts to power in dBm, we have to divide the power by 10 and then take the antilog. Thus the following is

$$P_{\mathrm{mW}} = \mathrm{antilog}(P_{\mathrm{dBm}}/10)$$

or alternatively

$$P_{\mathrm{mW}} = 10^{(P_{\mathrm{dBm}}/10)}$$

To convert the power in dBW to power in dBm, we have to add 30. Thus 30 dBm is 0 dBW.

Table 9.21 shows the conversion to linear (absolute) power from the power in dB relative to 1 mW in the case of dBm levels, and the power in dB relative to 1 W in the case of dBW levels. The absolute power levels are given in femtowatts, picowatts, nanowatts, microwatts, milliwatts and watts.

Table 9.21 Conversion table for dBW, dBm and watts.

dBW	−135	−134	−133	−132	−131	−130	−129	−128	−127	−126	−125	−124	−123	−122	−121
dBm	−105	−104	−103	−102	−101	−100	−99	−98	−97	−96	−95	−94	−93	−92	−91
fW	31.62	39.8	50.1	63.1	79.4	100	125.8	158.5	199.5	251.2	316.2	398.1	501.2	631	794.3
dBW	−120	−119	−118	−117	−116	−115	−114	−113	−112	−111	−110	−109	−108	−107	−106
dBm	−90	−89	−88	−87	−86	−85	−84	−83	−82	−81	−80	−79	−78	−77	−76
pW	1	1.26	1.58	2	2.51	3.16	3.98	5.01	6.31	7.94	10	12.59	15.85	19.95	25.12
dBW	−105	−104	−103	−102	−101	−100	−99	−98	−97	−96	−95	−94	−93	−92	−91
dBm	−75	−74	−73	−72	−71	−70	−69	−68	−67	−66	−65	−64	−63	−62	−61
pW	31.62	39.81	50.12	63.1	79.4	100	125.9	158.5	199.5	251.2	316.2	398.1	501.2	630.9	794.3
dBW	−90	−89	−88	−87	−86	−85	−84	−83	−82	−81	−80	−79	−78	−77	−76
dBm	−60	−59	−58	−57	−56	−55	−54	−53	−52	−51	−50	−49	−48	−47	−46
nW	1	1.26	1.58	2	2.51	3.16	3.98	5.01	6.31	7.94	10	12.59	15.85	19.95	25.12
dBW	−75	−74	−73	−72	−71	−70	−69	−68	−67	−66	−65	−64	−63	−62	−61
dBm	−45	−44	−43	−42	−41	−40	−39	−38	−37	−36	−35	−34	−33	−32	−31
nW	31.62	39.81	50.12	63.1	79.4	100	125.9	158.5	199.5	251.2	316.2	398.1	501.2	630.9	794.3
dBW	−60	−59	−58	−57	−56	−55	−54	−53	−52	−51	−50	−49	−48	−47	−46
dBm	−30	−29	−28	−27	−26	−25	−24	−23	−22	−21	−20	−19	−18	−17	−16
μW	1	1.26	1.58	2	2.51	3.16	3.98	5.01	6.31	7.94	10	12.59	15.85	19.95	25.12

dBW	−45	−44	−43	−42	−41	−40	−39	−38	−37	−36	−35	−34	−33	−32	−31
dBm	−15	−14	−13	−12	−11	−10	−9	−8	−7	−6	−5	−4	−3	−2	−1
μW	31.62	39.81	50.12	63.1	79.4	100	125.9	158.5	199.5	251.2	316.2	398.1	501.2	630.9	794.3
dBW	−30	−29	−28	−27	−26	−25	−24	−23	−22	−21	−20	−19	−18	−17	−16
dBm	0	1	2	3	4	5	6	7	8	9	10	11	12	13	14
mW	1	1.26	1.58	2	2.51	3.16	3.98	5.01	6.31	7.94	10	12.6	15.85	19.9	25.12
dBW	−15	−14	−13	−12	−11	−10	−9	−8	−7	−6	−5	−4	−3	−2	−1
dBm	15	16	17	18	19	20	21	22	23	24	25	26	27	28	29
mW	31.62	39.81	50.12	63.1	79.4	100	125.9	158.5	199.5	251.2	316.2	398.1	501.2	630.9	794.3
dBW	0	1	2	3	4	5	6	7	8	9	10	11	12	15	
dBm	30	31	32	33	34	35	36	37	38	39	40	41	42	45	
W	1	1.26	1.58	2	2.51	3.16	3.98	5.01	6.31	7.94	10	12.59	15.85	31.62	
dBW	16	17	18	19	20	21	22	23	24	25	26	27	28	29	
dBm	46	47	48	49	50	51	52	53	54	55	56	57	58	59	
W	39.81	50.12	63.1	79.43	100	125.9	158.5	199.5	251.2	316.2	398.1	501.2	630.9	794.3	

Table 9.22 Preferred SI scientific prefixes.

Prefix	Exa	Peta	Tera	Giga	Mega	Kilo	Milli	Micro	Nano	Pico	Femto	Atto
Abbreviation	E	P	T	G	M	K	m	µ	n	p	f	a
Power of 10	10^{18}	10^{15}	10^{12}	10^{9}	10^{6}	10^{3}	10^{-3}	10^{-6}	10^{-9}	10^{-12}	10^{-15}	10^{-18}

9.15 Preferred SI Scientific Prefixes

The SI is an international system of units. Instead of using decimal fractions and powers of 10 for units, the system uses prefixes that are equivalent to negative and positive powers of 10 (see Table 9.22). Only powers that are multiples of 3 are used and negative powers use lower-case letters, whereas positive powers use upper-case letters. Thus 10^{-6} is denoted by 'μ', whereas 10^6 is denoted by 'M'. An exception to this is 10^3 which is denoted by the lower-case 'k'. This is because the upper case K is used for the Kelvin temperature scale. Upper case K is used in computing for kB (kilobytes). However, kilobytes are strictly speaking not 10^3 but 1024 bytes or 2 raised to the power of 10, that is, 2^{10}. Similarly GB is 2^{20} bytes, that is, 1 048 576 bytes.

Although the prefix 'c' for centi (10^{-2}) is not strictly speaking one of the preferred SI prefixes it is still in common use.

Optical wavelengths are often quoted in angstrom units instead of using nanometres. An angstrom (Å) is 10^{-10} m.

9.16 Terms and Definitions

The terminology used on opposite sides of the Atlantic is different and can sometimes be confusing. Where appropriate, IEEE definitions have been given, from the *IEEE Standard Dictionary of Electrical and Electronics Terms*. However, in some cases these definitions are not easy to understand, so in these cases explanations are given in simpler terms.

Aerial
An aerial is a device radiating or receiving an EM radio signal. It transfers energy between a transmission line and free space. Although the term aerial is synonymous with antenna, it is usually used for wire antennas that are commonly employed at lower frequencies in the RF range, that is, below about 500 MHz.

Antenna
An antenna is a device radiating or receiving an EM radio signal. It transfers energy between a transmission line and free space. Although the term antenna is synonymous with aerial, it is usually used for aerials that are commonly employed at frequencies in the microwave range, that is, above about 500 MHz. Note that the plural is antennas and *not* antennae. Antennae is used for insects.

Antenna correction factor or antenna factor
This is a term or factor which is applied to the reading of the receiver, in order to convert the voltage reading (at the receiver) to field strength at the receiving antenna. This factor takes into account:

1. the effective height/length of the antenna,

2. the loss in the balun matching network between the antenna and the balun,
3. the loss due to the mismatch between the balun matching network and the transmission line connecting the balun to the receiver,
4. the loss due to the total length of the transmission line cables between the antenna and the receiver.

To be more accurate, the antenna factor should also take into account the proximity of the antenna to ground, and the frequency of measurement.

Axial ratio

This is sometimes defined as the reciprocal of ellipticity, i.e. the ratio of the major axis to the minor axis of the polarization ellipse. This ratio varies from infinity for linear polarization, to 1 for circular polarization. However, see the more common definition in Section 1.4.3.

Azimuth plane

The horizontal plane in which the antenna's radiation pattern is measured. Angles in this plane are denoted by the Greek letter ϕ (phi). In the case of aircraft monopole antennas this is the yaw plane of the aircraft.

Balun

This is an abbreviation for 'balanced-unbalanced', and in antenna applications it refers to a device used to connect a balanced two-wire transmission line to an unbalanced antenna, or an unbalanced coaxial line to a balanced antenna. In more general terms, a balun is an impedance transformer designed to couple between balanced and unbalanced circuits.

Bandwidth

This refers to the frequency range of operation of a piece of equipment. It is often quoted as a percentage or fraction of the centre frequency of operation.

Beam efficiency

Beam efficiency is the fraction of total radiated power contained in the main beam, and is thus a measure of the ability of the antenna to concentrate the energy into a narrow angular sector.

Bond

A bond is a low impedance path between two metal surfaces. Bonding straps are used, for instance between the baseplate of a 'blade' and the metal airframe, or between the individual sheets of a ground plane, so that they appear like an electrically continuous ground plane.

Boresight

The boresight of an antenna is defined as the axis along which maximum radiation occurs. This is not necessarily coincident with the main physical axis of the antenna. In the case of a dipole or monopole that is omnidirectional the boresight is sometimes in the direction of the physical antenna, that is, vertical for a vertical monopole.

Circular polarization

This refers to the polarization of a plane wave in which the electric field is circularly polarized, in the plane perpendicular to the direction of propagation of the wave.

Co-polarization
This applies to linear as well as circular or elliptical polarization; it is the polarization of EM radiation that is transmitted or received in the same orientation or sense as that of the receiving antenna.

Cross-polarization
In the case of linear polarization, it is the polarization of the incident EM radiation that is orthogonal (at right angles) to that of the receiving antenna. For circular polarization, it is the opposite hand of polarization to that for which the antenna is designed. Thus if the antenna is designed to receive right hand circular polarization, then left hand circular polarization would be cross-polar radiation and vice versa.

Decibels
When used in relation to powers this is 10 times the logarithm to the base 10 of a ratio of powers or gains of antennas.

In the case of voltages, electric fields or magnetic fields it is 20 times the ratio of magnitude of voltages, or electric or magnetic fields.

The most commonly used units are dBm, dBW, dBμA/m, dBμV and dBV, which are referenced to levels of 1 mW (milliwatt), 1 W, 1 μA/m, 1 μV, and 1 V, respectively.

Directive gain of an antenna
The IEEE dictionary distinguishes between the directive gain at the terminals of an antenna, and the directive gain in a physical medium, such as free space. The directive gain of an antenna is 4π times the ratio of the radiation intensity in a given direction to the total power radiated by the antenna. The term is synonymous with directivity.

Directive gain in a physical medium
In a given direction and at a given point in the far field, this is the ratio of the power flux per unit area from the antenna, to the power flux per unit area from an isotropic radiator delivering the same power to the medium.

Directivity
This term is synonymous with directive gain.

Dominant mode
This is the lowest order or fundamental mode in a waveguide or coaxial line. It is the only mode that can propagate without the other higher-order modes.

Elevation plane
This is the vertical plane in which the antenna's radiation pattern is measured. Angles in the elevation plane are usually denoted by the Greek letter θ (theta).

E plane sectoral horn
A rectangular horn formed by flaring the waveguide walls in the E plane only, that is, the broadside walls of the waveguide are flared.

Emissions
This term is used for radiated as well as conducted EM waves.

Fraunhofer region
This defines the region in the far field of the antenna from infinity to the minimum distance from the antenna, at which a plane wave can be assumed to exist.

Free space waves
A free space wave is one that is transmitted in free space, where the velocity of the wave is approximately 3×10^8 m/s. However, a free space wave is not necessarily a plane wave.

Front-to-back ratio
This is the ratio of the power radiated in the boresight direction, to that in the backward direction. It is usually expressed in dB rather than as a linear quantity.

Gain of an antenna
This is defined as the product of the directivity and the efficiency of the antenna. Since the efficiency has a maximum value of 1 or 100%, the gain is always less than or equal to the directivity.

Guided waves
A guided wave is one that is transmitted by a transmission line, such as a coaxial cable, microstrip track or a waveguide.

Half-power beamwidth
This is defined as the width of the main beam between the points at which the power is half (−3 dB) that of its maximum value. The maximum power value usually, though not exclusively, occurs at boresight.

H plane sectoral horn
A rectangular waveguide horn in which only the H plane or the narrow walls of the waveguide are flared.

Immunity
The IEEE defines immunity as the ability of any equipment or system to reject a radio disturbance. In EMC (Electromagnetic Compatibility) work the ability of equipment to function correctly in EM fields determines its immunity. The higher the level of the interfering EM field, to cause malfunction, the greater the immunity of the equipment. The EM field can be radiated or conducted. This term is usually applied to non-military equipment, susceptibility being used for military equipment.

Isotropic radiator
An isotropic radiator is a hypothetical antenna that radiates equally in all directions, so that surfaces of constant intensity are spheres with the antenna at its centre. This antenna is used as a reference level to express the directive properties of other antennas.

Log periodic dipole antenna
A log periodic dipole antenna is a wideband antenna consisting of a number of dipoles, which are spaced in such a way that the ratio of their lengths is proportional to their distances from the feed point of the array. The impedance and radiation characteristics of the antenna repeat periodically as the logarithm of frequency. This accounts for the name.

Neper
This is the attenuation in dB, which reduces the value of a current or voltage to 1/e (e is the Naperian constant 2.718) of its initial value.

Phase centre
This applies equally to wire and aperture antennas but is used mainly for aperture antennas such as horns and reflectors. It is the centre from which the spherical wave that extends into the far field can be said to emanate.

Plane wave
A plane wave is one in which the wavefront is perpendicular to the direction of propagation. The electric and magnetic field vectors are perpendicular to each other as well as being perpendicular to the direction of propagation.

Polarization
Although this could apply equally to magnetic or electric polarization, it is used almost exclusively to indicate the locus of the extremity of the electric field vector as it varies with time at a fixed point in space. This locus could be a straight line, an ellipse, or a circle.

Pyramid/rectangular horn
This is a horn formed by flaring the walls in both the broad and narrow dimensions of a rectangular waveguide. A pyramid horn is commonly referred to as a rectangular horn.

Resonant dipole
A resonant dipole is usually a dipole which is approximately half a wavelength long at its operating frequency. However the dipole can also be resonant at any multiple of half a wavelength. Although a dipole can be operational at other frequencies, it is most efficient at its resonant frequency.

Shielding/screening effectiveness (SE)
These two terms are used synonymously to denote the ability of a material to exclude or confine EM waves. These materials are used to reduce emission from a device, and/or to prevent malfunction of a device, as a result of external fields.

The shielding effectiveness of a material is defined as the difference in signal loss experienced with and without the material inserted between a transmitting and receiving antenna. This total insertion loss is due to absorption, reflection and multiple internal reflection. It is usually quoted in decibels.

Sidelobe
Any peak, apart from the main beam of the antenna radiation pattern, is a sidelobe. However, it is only the peaks nearest the main beam or the 'near-in' lobes that are sometimes referred to as sidelobes. In order to distinguish between irregularities in the radiation pattern and a genuine sidelobe peak, sidelobes are sometimes defined as the peaks where the difference between the peak and an adjacent trough is at least 3–6 dB.

Skin depth
This usually pertains to waveguides, and is defined as the depth at which the surface current density is reduced by 1 neper.

Susceptibility
This is a measure of the ability of equipment not to be influenced by external EM fields, and is usually used for military equipment. The lower the levels of EM fields that influence the equipment the greater the susceptibility of the equipment.

Travelling wave antenna
An antenna where the fields that produce the EM radiation can be represented by a progressive wave travelling in one direction only. A travelling wave antenna is usually characterized by the fact that it is terminated by a matched load to prevent a reflected wave in the opposite direction to the initial progressive wave.

VSWR (voltage standing wave ratio)
This is the ratio of the maximum to the minimum electric field intensities of the standing wave. The standing wave is made up of two waves propagating in different (usually but not exclusively, opposite) directions.

Wave impedance
The plane wave can be considered as having an impedance which depends on the medium through which it is propagating. For a plane wave this is the square root of the ratio of the permeability to the permittivity of the medium.

Waveguide cut-off frequency
This is the frequency at which the attenuation in the guide approaches infinity, and thus propagation cannot take place below this frequency. The cut-off wavelength is numerically equal to twice the broadside dimension of a rectangular waveguide propagating the dominant mode.

9.17 VSWR to Return Loss

The VSWR is a measure of the degree of matching between the antenna and the RF cable (or waveguide) feeding it.

If the antenna is perfectly matched, there is no reflected wave at the input terminals of the antenna and thus all the power is transferred to the antenna. The VSWR varies between 1.0 for a perfectly matched and ∞ for a very badly matched antenna when all the power is reflected and no power is transmitted.

A measure of the match is sometimes quoted as the return loss R_{dB} and the conversion between the VSWR and the return loss is given by

$$R_{\text{dB}} = 10 \log_{10} \left[\frac{(S-1)^2}{(S+1)^2} \right]$$

where

R_{dB} is the return loss in dB and
S is the VSWR.

A VSWR of 1.4 is considered about average for most wire antennas. This corresponds to a return loss of -15.6 as shown in Table 9.23. Note that the negative sign is often omitted when referring to the return loss.

Table 9.23 Conversion of VSWR to return loss.

VSWR	1	1.01	1.02	1.03	1.04	1.05	1.06	1.07	1.08	1.09
Return loss	$-\infty$	−46.06	−40.09	−36.61	−34.15	−32.26	−30.71	−29.42	−28.3	−27.32
VSWR	1.1	1.11	1.12	1.13	1.14	1.15	1.16	1.17	1.18	1.19
Return loss	−26.44	−25.66	−24.94	−24.29	−23.69	−23.13	−22.61	−22.12	−21.66	−21.23
VSWR	1.2	1.21	1.22	1.23	1.24	1.25	1.26	1.27	1.28	1.29
Return loss	−20.83	−20.44	−20.08	−19.73	−19.4	−19.08	−18.8	−18.5	−18.2	−17.9
VSWR	1.3	1.31	1.32	1.33	1.34	1.35	1.36	1.37	1.38	1.39
Return loss	−17.7	−17.5	−17.2	−16.9	−16.8	−16.5	−16.3	−16.1	−15.94	−15.75
VSWR	1.4	1.41	1.42	1.43	1.44	1.45	1.46	1.47	1.48	1.49
Return loss	−15.56	−15.38	−15.21	−15.04	−14.88	−14.72	−14.56	−14.41	−14.26	−14.12
VSWR	1.5	1.51	1.52	1.53	1.54	1.55	1.56	1.57	1.58	1.59
Return loss	−13.98	−13.84	−13.71	−13.58	−13.45	−13.32	−13.2	−13.08	−12.96	−12.85
VSWR	1.6	1.61	1.62	1.63	1.64	1.65	1.66	1.67	1.68	1.69
Return loss	−12.74	−12.63	−12.52	−12.4	−12.3	−12.2	−12.1	−12.0	−11.9	−11.8
VSWR	1.7	1.71	1.72	1.73	1.74	1.75	1.76	1.77	1.78	1.79
Return loss	−11.7	−11.6	−11.5	−11.5	−11.4	−11.3	−11.2	−11.12	−11.04	−10.96
VSWR	1.8	1.81	1.82	1.83	1.84	1.85	1.86	1.87	1.88	1.89
Return loss	−10.88	−10.8	−10.73	−10.65	−10.58	−10.51	−10.44	−10.37	−10.3	−10.23
VSWR	1.9	1.91	1.92	1.93	1.94	1.95	1.96	1.97	1.98	1.99
Return loss	−10.16	−10.1	−10.03	−9.97	−9.9	−9.84	−9.78	−9.72	−9.66	−9.6
VSWR	2	2.01	2.02	2.03	2.04	2.05	2.06	2.07	2.08	2.09
Return loss	−9.54	−9.48	−9.43	−9.37	−9.32	−9.26	−9.21	−9.16	−9.1	−9.05
VSWR	2.1	2.11	2.12	2.13	2.14	2.15	2.16	2.17	2.18	2.19
Return loss	−9	−8.95	−8.9	−8.85	−8.8	−8.75	−8.7	−8.66	−8.61	−8.56
VSWR	2.2	2.21	2.22	2.23	2.24	2.25	2.26	2.27	2.28	2.29
Return loss	−8.52	−8.47	−8.43	−8.39	−8.34	−8.3	−8.26	−8.21	−8.17	−8.13
VSWR	2.3	2.31	2.32	2.33	2.34	2.35	2.36	2.37	2.38	2.39
Return loss	−8.09	−8.05	−8.01	−7.97	−7.93	−7.89	−7.86	−7.82	−7.78	−7.74
VSWR	2.4	2.41	2.42	2.43	2.44	2.45	2.46	2.47	2.48	2.49
Return loss	−7.71	−7.67	−7.63	−7.6	−7.56	−7.53	−7.49	−7.46	−7.43	−7.39
VSWR	2.5	2.51	2.52	2.53	2.54	2.55	2.56	2.57	2.58	2.59
Return loss	−7.36	−7.33	−7.29	−7.26	−7.23	−7.2	−7.17	−7.14	−7.1	−7.07
VSWR	2.6	2.61	2.62	2.63	2.64	2.65	2.66	2.67	2.68	2.69
Return loss	−7.04	−7.01	−6.98	−6.95	−6.93	−6.9	−6.87	−6.84	−6.81	−6.78
VSWR	2.7	2.71	2.72	2.73	2.74	2.75	2.76	2.77	2.78	2.79
Return loss	−6.76	−6.73	−6.7	−6.67	−6.65	−6.62	−6.59	−6.57	−6.54	−6.52

Table 9.23 (*continued*)

VSWR	2.8	2.81	2.82	2.83	2.84	2.85	2.86	2.87	2.88	2.89
Return loss	−6.49	−6.46	−6.44	−6.41	−6.39	−6.37	−6.34	−6.32	−6.29	−6.27
VSWR	2.9	2.91	2.92	2.93	2.94	2.95	2.96	2.97	2.98	2.99
Return loss	−6.25	−6.22	−6.2	−6.18	−6.15	−6.13	−6.11	−6.09	−6.06	−6.04
VSWR	3	3.01	3.02	3.03	3.04	3.05	3.06	3.07	3.08	3.09
Return loss	−6.02	−6	−5.98	−5.96	−5.94	−5.91	−5.89	−5.87	−5.85	−5.83

Appendix: Abbreviations and Acronyms

The most common abbreviations and acronyms used in the electromagnetic and aircraft industry are listed below. Some symbols are also included, as are combined abbreviations and acronyms such as RadAlt and SatCom.

3D	Three-Dimensional
a	atto (prefix) = 10^{-18}
a	symbol for acceleration
A	NATO frequency band (0–250 MHz)
AAC	Aeronautical/Airline Administrative Communications
AAF	Airway Facilities Service
AAIB	Aircraft Accident Investigation Branch (UK)
AAIU	Air Accident Investigation Unit
AAL	Above Aerodrome Level
AAL	Absolute Assembly Language
AALC	Autonomous Approach Landing Capability
AAMS	Astrium Antenna Measurement Software Package
AANTS	Advanced Antenna Near-field Test System
AAS	Advanced Antenna System
AAS	Advanced Automation System (FAA)
AATC	Automatic Air Traffic Control (System)
AATT	Advanced Aviation Transportation Technology
ABET	Accreditation Board for Engineering and Technology
ABL	Atlas Basic Language
ABM	Asynchronous Balanced Mode
ABV	Absolute Value
AC	Alternating Current
AC	Analog Computer
AC	Automatic Check-Out
AC	Automatic Computer

Introduction to Antenna Placement & Installation Thereza M. Macnamara

AC	Automatic Control
AC	Advisory Circular – not mandatory but derived from FAA regulations
A/C	Aircraft
ACA	American Communications Association
ACA	Automatic Communications Association
ACAA	Air Carrier Association of America
ACARE	Advisory Council for Aeronautical Research in Europe
ACARS	ARINC Communications Addressing Reporting System
ACARS	Airborne Communications Addressing Reporting System
ACAS	Aircraft Collision Avoidance System (FAA)
ACC	Active Clearance Control
ACC	Area Control Centre
ACC	Airspace Control Centre
ACC	Accumulator
ACCESS	Aircraft Communication Electronic Signalling System
ACCESS	Automatic Computer Controlled Electronic Scanning System
ACD	Automatic Call Distributor
ACD	Automatic Conflict Detection
ACDI	Asynchronous Communications Device Interface
ACDO	Air Carrier District Office (FAA)
ACE	Acceptance Check-Out Equipment
ACE	Automatic Check-Out Equipment
ACE	Advanced Control Equipment
ACE	Advanced Certification Equipment
ACE	Association for Cooperation in Engineering
ACE	Aviation Construction Engineers
ACEC	American Consulting Engineers Council
ACES	Applied Computational Electromagnetics Society
ACET	Advisory Committee for Electronics and Telecommunications
ACF	Advanced Communications Function
ACF	Antenna Correction Factor
ACF	Area Control Facility
ACI	Automatic Card Identification
ACIA	Asynchronous Communications Interface Adaptor
ACID	Automatic Classification and Interpretation of Data
ACIL	American Council of Independent Laboratories
ACIS	Association for Computing and Information Sciences
ACK	Acknowledge
ACK	Acknowledge Character
ACL	Application Control Language
ACL	Association for Computational Linguistics
ACL	Atlas Commercial Language
ACLS	American Council of Learned Societies
ACM/GAMM	Association for Computing Machinery/German Association for Applied Mathematics and Mechanics
ACME	Association of Consulting Management Engineers

ACMO	Afloat Communications Management Office (NSEC)
ACMOS	Advanced Complementary Metal Oxide Semiconductor
ACNSS	Advanced Communications, Navigation, Surveillance System
ACOPP	Abbreviated Cobol Preprocessor
ACOS	Advisory Committee on Safety
ACRE	Automatic Check-Out and Readiness
ACS	Advanced Computer Services
ACS	Alternating Current Synchronous
ACS	Automated Communications Set
ACSS	Analogue Computer Subsystem
ACT	Automatic Code Translation
ACTS	All-Channel Television Society
ACTS	Automatic Equipment Telex Services
ACU	Apron Control Unit
ACU	Autopilot Control Unit
AD	Airworthiness Directives
A/D	Analogue to Digital (Conversion)
ADA	Automatic Data Acquisition
ADACC	Automatic Data Acquisition and Computer Complex
ADAM	Advanced Data Management
ADC	Analogue-to-Digital Converter
ADC	Advanced Data Collection
ADC	Air Data Computer
ADCAB	Advanced Cabin, Cabin systems & Multimedia Services for improved Passenger Comfort & better A/C efficiency EU 6th Framework Project
ADCC	Air Defense Control Center
ADF	Automatic Direction Finding
ADI	Attitude Director Indicator
ADIRS	Air Data Inertial Reference System
ADIZ	Air Defense Identification Zone
ADL	Aeronautical Data Link
ADLP	Aircraft Data Link Processor (Mode S)
ADLP	Airborne Data Link Protocol
ADM	Air Data Module
ADMS	Airline Data Management System
ADN	Aircraft Data Network
ADOC	Air Defense Operation Center
ADPCM	Association for Data Processing and Computer Management
ADPS	Automatic Data Processing System
ADR	Alternative Dispute Resolution
ADRAS	Airplane Data Recovery and Analysis System
ADS	Automatic Detection Surveillance
ADS	Air Data System
ADS-B	Automatic Dependent Surveillance-Broadcast
ADT	Air Data Transducer

AE	Auxiliary Equipment
AEA	Aircrew Equipment System
AEA	American Electronics Association
AEA	American Engineering Association
AEA	Association of European Airlines
AEA	Association of Engineers and Associates
AEA	Atomic Energy Authority
AEACII	Atomic Energy Advisory Committee on Industrial Information
AEC	American Engineering Council
AEC	Atomic Energy Commission
AECB	Atomic Energy Control Board (Canada)
AECL	Atomic Energy of Canada (Ltd)
AECMA	European Association for Aerospace Industries
AED	Algol Extended for Design
AEE	Atomic Energy Establishment (UK)
AEEC	Airlines Electronics Engineering Committee (ARINC)
AEI	Associazione Elettrotecnica Italiana
AEM	Association of Electronic Manufacturers
AEPSC	Atomic Energy Plant Safety Committee
AERA	Automated En-route Air Traffic Control
AERE	Atomic Energy Research Establishment (UK)
AEREA	Association of European Research Establishments in Aeronautics
Aero B-Gan	Broadband Global Area Network (Inmarsat)
AESA	Active Electronically Scanned Array
AF	Antenna Factor
AF	Audio Frequency
AFC	Automatic Frequency Control/Compensation
AFD	Airport Facility Directory
AFCCE	Association of Federal Communication Consulting Engineers
AFDX	Avionics Full Duplex Ethernet Switch
AFIS	Airborne/Automatic Flight Information System
AFM	Aircraft Flight Manual
AFNOR	Association Francaise des Normes (France)
AFRL	Air Force Research Laboratory (US)
AFSK	Audio Frequency Shift Keying
AFTN	Aeronautics Fixed Telecommunications Network
A/G	Air to Ground
AGARD	Advisory Group for Aerospace Research and Development (NATO)
AGATE	A General Analysis Tool for EM – BAE Systems software
AGATE	Advanced General Aviation Transport Experiments (NASA)
AGC	Automatic Gain Control
AGED	Advisory Group on Electron Devices
AGEP	Advisory Group on Electronic Parts
AGET	Advisory Group on Electronic Tubes
AGL	Above Ground Level
AGND	Analog Ground

AGSS	ACARS Ground System Standard (AEEC)
AH	Ampere Hours
AHC	Aircraft Heading Computer
AHRS	Attitude Heading Reference System
AI	Artificial Intelligence
AIA	Aerospace Industries Association
AIAA	American Institute of Acronautics & Astronautics
AIC	Aeronautical Information Circular
AID	Aircraft Installation Delay (RadAlt)
AIDS	Airborne/Aircraft Integrated Data System
AIM	Adaptive Integral Method
AIP	Aeronautical Information Publication
AL	Auto Land
AL	Assembly Language
Al	Aluminium
a/l	Airline
ALARM	Air-Launched Anti-Radiation Missile
ALC	Absorber Lined Chamber
ALC	Automatic Levelling Control
ALDAS	Analysis of Low Directivity Antennas on Structures
ALE	Automatic Link Establishment
ALS	Approach Landing System
ALT	Airborne Link Terminal
ALU	Arithmetic and Logic Unit
AM	Ammeter
AM	Amplitude Modulation
A/m	Ampere per Metre
AMASS	Airport Movement Area Safety System
AMRF	Advanced Multifunction Radio Frequency (System)
AMS	Avionics/Acquisition Management System
AMSA	Australian Marine Sciences Association
AMTA	Antenna Measurement Techniques Association
AMU	Avionics Module Unit
A/N	Alphanumeric
ANO	Alphanumeric Output
ANR	Active Noise Reduction
ANSI	American National Standards Institute
ANTF	Airborne Near-field Test Facility – Astrium
ANZAAS	Australian and New Zealand Association for the Advancement of Science
AOA	Angle of Attack
AOA	Angle of Arrival
AOB	Auxiliary Oxygen Bottle
AOC	Automatic Overload Control
AOC	Aeronautical/Aircraft Operational Control
AOC	Airport Obstruction Chart

AOG	Aircraft on Ground
AOM	Aircraft Operating Manual
AOPA	Aircraft Owners and Pilots Association
A/P	Auto Pilot
APC	Automatic Phase Control
APEC	All Purpose Electronic Computer
API	Application Programming Interface
APK	Amplitude and Phase Keyed
APL	Applied Physical Laboratories (US)
APM	Advanced Power Management
APN	ARINC Packet Network
APS	American Physics Society
APU	Auxiliary Power Unit
AQP	Advanced Qualification Programme
AR	Axial Ratio
ARCG	American Research Committee on Grounding
ARINC	Aeronautical Radio Incorporated (affiliated to FCC)
ARMMS	Automated RF and Microwave Measurement Society
ARMS	Aerial Radiological Measurements and survey
ARP	Aerospace Recommended Practice
ARRL	American Radio Relay League
ARS	Amplified Response Spectrum
ARSR	Area/Air Route Surveillance Radar
ART	Automatic Reporting Telephone
ARTCC	Air Route Traffic Control Center
ARTE	Admiralty Research Test Establishment (UK)
ARTE	Aircraft Related R & T
ARTES	Advanced Research in Telecommunications Systems
ARU	Audio Response Unit
AS	Address Sync
A/S	Airspeed
ASA	American Standards Association (now ANSI)
ASAC	Asian Standards Advisory Committee
ASAC	Aviation Security Advisory Committee
ASAI	Adjunct Switch Application Interface
ASAR	Advanced Synthetic Aperture Radar
ASC	Automatic Sensitivity Control
ASC	Automatic System Controller
ASC	Auxiliary Switch Normally Closed
ASCII	American Standard Code of Information Interchange
ASDAR	Aircraft to Satellite Data Relay
ASDL	Aeronautical Satellite Data Link
ASERIS	EADS suite of software
ASET	American Society for Engineering Technology
ASI	American Standards Institute
ASIC	Application Specific Integrated Circuits

ASIK	Application Specific Integrated Keyboard
ASK	Amplitude Shift Keying
ASL	Above Sea Level
ASL	Antenna Software Ltd.
ASM	Airborne Separation Minima
ASM	American Society for Materials
ASP	Applied Signal Processing
ASR	Airport Surveillance Radar
ASTAP	Advanced Statistical Analysis Program
ASTM	American Society for Testing and Materials
ASTOR	Airborne Stand Off Radar (UK)
AT	Ampere-Turn
ATA	Actual Time of Arrival
ATC	Air Traffic Control
ATC	Advanced Technology Centre – BAE Systems
ATCA	Air Traffic Control Association
ATCBS	Air Traffic Control Radar Beacon System
ATD	Antenna Technologies & Design
ATD	Actual Time of Departure
ATE	Actual Time En-route
ATE	Airborne Terminal Equipment
ATE	Automatic Test Equipment
ATE	Avionics Test and Evaluation
ATM	Air Traffic Management
ATM	Asynchronous Transfer Mode
ATME	Automatic Test Measuring Equipment
ATP	Acceptance Test Plan/Procedure
ATP	Advanced Turbo Prop
ATR	Avions de Transport Regional
ATR	Acceptance Test Report
ATSRAC	Aging Transport Systems Rulemaking Advisory Committee (FAA)
AVLC	Aviation VHF Link Control
AVIONICS	Aviation Electronics
AWACS	Airborne Warning and Control System
AWAS	Automated Weather Advisory Station
AWG	American Wire Gauge
Az	Azimuth
AZS	Automatic Zero Set
B	NATO frequency band (250–500 MHz)
BBE	Symbol for magnetic flux density Boundary Element
B/M	Balun and Matching (network)
B/S	Bits per second
BABT	British Approvals Board for Telecommunications
BAES	BAE Systems
BALUN	Balanced To Unbalanced
BAPT	German postal service (for EMC laws)

BASIC	Beginner's All-Purpose Symbolic Instruction Code
BASIC	Basic Algebraic Symbolic Interpretive Compiler
BASIC	Basic Automatic Stored Instruction Computer
BASIS	Bank Automated Service Information System
BB	Baseband
BB	Broadband (Emission)
b-b	Back-To-Back
BBC	Broadband Conducted
BC	Back-Conducted
BC	Binary Code
BCD	Binary Coded Decimal
BCI	Bulk Current Injection
BCI	Binary Coded Information
BCI	Broadcast Interference
BCO	Binary Coded Octal
BCU	Binary Counting Unit
BD	Baud
BD	Binary Decoder
BE	Boundary Element
BEAME	British Electrical and Allied Manufacturers
BECTO	British Electric Cable Testing Organization
BER	Bit Error Rate/Ratio
BFCO	Band Filter Cutoff
BFG	Binary Frequency Generator
B field	Magnetic Flux Density (in tesla)
BFO	Beat Frequency Oscillator
B-GAN	Broadband Global Area Network
BiCMOS	Bipolar CMOS
BiDI	Bi-directional
BiMOS	Bipolar MOS
BISDN	Broadband Integrated Services Digital Network
BIST	Built-In Self Test
BIT	Binary Digit
BIT	Built-In Test
BIT	Bipolar Integrated Technology
BITE	Built-In Test Equipment
BLM	Basic Language Machine
BLOB	Binary Large Object
BN	Binary Number
BO	Blocking Oscillator
BOC	Binary Offset Carrier
BOP	Binary Output Program
BP	Bandpass
BPF	Bandpass Filter
BPI	Bits Per Inch
BPI	Bytes Per Inch

BPN	Breakdown Pulse Noise
BPS	Bits Per Second
BPSK	Binary Phase Shift Keying
BS	British Standard
BTD	Binary to Decimal
BTR	Bit Time Recovery
BVR	Beyond Visual Range
BW	Bandwidth
BW	Beamwidth
BWAN	Backup Wide Area Network
BWG	Birmingham Wire Gauge
BWO	Backward Wave Oscillator
Byte	A grouping of eight bits
C	Aircraft frequency band (3.5–7.5 GHz)
C	ITU frequency band (4–8 GHz)
C	NATO frequency band (500 MHz to 1 GHz)
C	UK (IEE/IET) frequency band (4–8 GHz)
C	Coulomb (SI unit of charge)
C	symbol for capacitance
c	centi (prefix) 10^{-2}
C/N	Carrier-to-Noise ratio
C3I	Command, Control, Communications and Intelligence (system)
C4I	Command, Control, Communications, Computers and Intelligence (system)
C4ISR	Command, Control, Communications, Computers, Intelligence, Surveillance and Reconnaissance
CAA	Civil Aviation Authority/Administration (UK)
C/A code	Coarse Acquisition code (GPS)
CAD	Computer-Aided Design
CAD	Computer-Aided Detection
CAD/CAM	Computer-Aided Design and Computer-Aided Manufacturing
CADAM	Computer Graphics Augmented Design And Manufacturing (System)
CAE	Computer-Aided Education
CAE	Computer-Aided Engineering
CAI	Common Air Interface
CAIS	Common ADA Interface Set
CAM	Central Address Memory
CAM	Computer-Aided Manufacture
CAM	Computer-Aided Management
CAM	Content Addressable Memory
CAMESA	Canadian Military Electronics Standards Agency
CANSO	Civil Air Navigation Services Organization
CAOS	Completely Automatic Operational System
CAP	Capacitor
CAP	Computer-Aided Production
CAP	Combat Air Patrol

CARS	Computer-Aided Routing System
CARS	Community Antenna Relay Service
CASD	Computer-Aided System Design
CASE	Computer Automated Support System
CAT	Computer-Assisted Testing
CATIA	Computer Aided Three-Dimensional Interactive Application
CATM	Collaborative ATM
CATV	Community Antenna Television
CB	Citizen's Band (Radio)
CBEMA	Computer and Business Equipment Manufacturer's Association
CC	Close-Coupled
CC	Coarse Control
CC	Common Collector
CCA	CENELEC Certification Agreement
CCB	Common Carrier Bureau (FCC)
CCD	Charge-Coupled Device
CCI	Co Channel Interference
CCI	Combined form of CCIR and CCIT
CCIR	International Radio Consultative Committee
CCITT	International Telegraph and Telephone Consultative Committee
CCT	Constant Current Transformer
cct	Circuit
CCTV	Closed-Circuit Television
CCW	Counter-Clockwise
CD	Compact Disk
CD	Cable Duct
CD	Current Density
CD	Committee Draft
CDF	Cumulative Density Function
CDL	Common Data Link
CDL	Computer Description Language
CDM	Code-Division Multiplexing
CDMA	Code-Division Multiple Access
CDMS	Control and Data Management System
CDN	Coupling/Decoupling Network
CD-ROM	Compact Disk Read-Only Memory
CDW	Computer Data Word
CE	Conducted Emissions
CE	Common Emitter
CE	Compromising Emanation (TEMPEST)
CEC	Central European Community
CECC	CENELEC Electronics Components Committee
CEI	Computer Extended Instruction
CEI	Council of Engineering Institutions
CEM	Computational Electromagnetics

CEN	Comité Européen de Normalisation (European Standards Coordinating Committee)
CENELEC	Comité Européen de Normalisation Electrotechnique (European Committee for Electrotechnical Standardisation)
CERDIP	Ceramic Dual In-Line Package
CERL	Central Electricity Research Laboratory
CERMET	Ceramic Metal Element
CERMET	Ceramic-to-Metal (Seal)
CF	Conversion Factor
cf	Compared with
CFAR	Constant False Alarm Rate
CFDM	Compounded Frequency Division Multiplex
CG	Centre of Gravity
CGA	Colour Graphics Adapter (640 × 200 pixels)
CGI	Computer Generated Image
CGM	Computer Graphics Metafile
CGS	Centimetre Gram Second (System of Units)
CI	Conducted Interference
CIA	Computer Industry Association
CIDIN	Common ICAO Data Interchange Network
CIMNE	International Centre for Numerical Methods – Spain
CIP	Current Injection Probe
CISPR	Center for International Systems Research (Department of State)
CISPR	Comité International Spécial des Perturbations Radioélectriques
CIT	Computer Integrated Telephony
CIU	Computer Interface Unit
CK	Compressive Receiver
ckt	Circuit
CLASP	Culham Laboratory Advanced Scattering Program
CLS	Closed Loop System
CLU	Central Logic Unit
cm	Centimetre
CMI	Coded Mark Inversion
CMOS	Complementary Metal Oxide Semiconductor/Silicon Or Sapphire
CMR	Common-Mode Rejection
CMRR	Common-Mode Rejection Ratio
CMS	Current-Mode Switching
CMV	Common-Mode Voltage
C/N	Carrier to Noise Ratio
CNC	Computerized Numerical Control
CNES	Centre National d'Etudes Spatiales (France)
CNET	Centre National d'Etudes des Télécommunications (France)
CNI	Communications/Navigation/Identification
CNNA	Council for National Academic Awards
CNR	Carrier-to-Noise Ratio
CNS	Communication and Navigation System

CODEC	Coder-Decoder
COHO	Coherent Oscillator
COL	Computer Orientated Language
COMINT	Communications Intelligence
COMPACT	Compatible Algebraic Compiler and Translator
COMSAT	Communications Satellite
COMSEC	Communications Security, secure communications
CONELRAD	Control of Electromagnetic Radiation
CONOPS	Concept of Operations
CORDIS	Community Research & Development Information Service
CORT	Carrier Operated Relay Threshold
cos	Cosine (Trigonometry)
cosh	Hyperbolic Cosine
COSHH	Control of Substances Hazardous to Health
COTP	Connection-Oriented Transport Protocol
COTS	Commercial off the Shelf
CP	Circularly Polarized
CP	Clock Pulse
CP	Central Processor
CPA	Collision Prediction and Alerting
CPA	Critical Path Analysis
CPE	Circular Position Error
CPEM	Conference on Precision Electromagnetic Measurement
CPI	Characters Per Inch
CPLD	Complex Programmable Logic Device
CPM	Continuous Phase Modulation
CPS	Characters Per Second
cpse	Counterpoise
CPSK	Coherent Phase Shift Keying
CPU	Central Processing Unit
CR	Current Relay
CRA	Conflict Resolution Advisory
CRC	Cyclic Redundancy Check
CRISC	Complex Reduced Instruction Set Computer
CRO	Cathode Ray Oscilloscope
CRPA	Controlled Radiation Pattern Antenna
CRT	Cathode Ray Tube
CS	Conducted Susceptibility
CSA	Canadian Standards Association
CSD	Compass Safe Distance
CSTAC	Commercial Space Transportation Advisory Committee (FAA)
CSV	Comma-Separated Variables
C/T	Carrier to noise Temperature Ratio
CT2	Cordless Telephone (Second Generation)
CTE	Channel Translation Equipment
CUI	Called Unit Identify

CV/DFDR	Cockpit Voice and Digital Flight Data Recorder (black box)
CVR	Crystal Video Receiver
CW	Carrier Wave
CW	Composite Wave
CW	Continuous Wave
CW	Clockwise
CWS	Collision Warner System
CYMB	Cyan Yellow Magenta Black
CYMK	Cyan Yellow Magenta Black
D8PSK	Differential 8-Phase-Shift_Keying
D	NATO frequency band (1–2 GHz)
d	deci (prefix) 1/10
DA	Design Authority
DA	Descent Advisor
D-A	Digital to Analog
DAB	Defense Acquisition Board
DAB	Digital Audio Broadcasting
DACU	Digital to Analogue Converter Unit
DAM	Direct Access Memory
DAP	Download of Aircraft Parameter
DARPA	Defence Advanced Research Projects Agency
DAT	Digital Audio Tape
dB	Decibel
D-B	Digital to Binary
dB(A/m^2)	Decibels referenced to 1 ampere per square metre
dBA	Decibels referenced to 1 ampere
dBa	Decibel Adjusted
dBA	Decibels referenced to 1 microampere
dBA/m	Decibels referenced to 1 microampere per metre
dBa0	noise power on dBa referred to or measured at 0TLP
DBAO	Digital Block AND-OR
dBc	Decibels relative to the carrier
dBi	Decibels referenced to an isotropic (antenna)
dBj	Relative RF signal levels (j = Jerrold Electronics)
dBk	Decibels referred to 1 kilowatt
dBm	Decibels compared with 1 milliwatt
dBm(PSOPH)	noise power in dBm referred measured by a set with psophometric weighting
dBm/m^2	Decibels referred to 1 milliwatt per square metre
dBm/m^2/MHz	Decibels referred to 1 milliwatt per square metre per Megahertz
dBRN	Decibels above Reference Noise
DBS	Direct Broadcast by Satellite
dB μV	Decibels referenced to 1 microvolt
dBV	Decibels referenced to 1 volt
dB μV/m	Decibels referenced to 1 microvolt per metre
dBW	Decibels referenced to 1 watt

dBx	Decibels above the reference coupling
D&C	Design & Clearance
D&C	Displays & Controls
DC	Direct Current
DC	Disconnect Confirm
DCI	Direct Current Injection
DCL	Digital Command Language
DCL	Departure Clearance (European)
DCM	DC Noise Margin
DCPSK	Differential Coherent Phase Shift Keying
DDE	Dynamic Data Exchange
DDS	Digital Data Storage
Deg	Degree
Demod	Demodulator
DEMS	Digital Electronic Message Service
DF	Direction Finding
DFA	Direction Finding Antenna
DFDAMU	Digital Flight Data Acquisition Management Unit
DFS	Deutsche Flugsicherung (Germany)
DFSK	Double Frequency Shift Keying
DFT	Discrete Fourier Transform
DGND	Digital Ground
DGPS	Differential Global Positioning System
D-H	Digital to Hexadecimal
DH	Decision Height
DHS	Department of Homeland Security (USA)
Dibit	a group of four bits
DIF	Data Interchange Format
DiL	Dual in Line
DIN	Deutsches Institut für Normung (German Standards Organization)
DIP	Dual-In-Line Package/PIN
DIS	Draft International Standard
DLC	Duplex Line Control
DLK	Data Link
DLL	Dynamic Link Library
DLR	Deutsches Zentrum für Luft und Raumfahrt (German Aerospace Research Centre)
DLT	Data Line Translator
DM	Data Management
Dm	Decimetre
DM	Delta Modulation
DMA	Direct Memory Access
DMAC	Direct Memory Access Controller
DME	Distance Measuring Equipment
DME/N	Distance Measuring Equipment – Normal
DME/P	Distance Measuring Equipment – Precision

DML	Data Manipulation Language
DMM	Digital Multimeter
DO	Data Output
DO	Digital Output
D-O	Digital to Octal
DOA	Direction of Arrival
DOC	Department of Communications (Canada)
DoD	Department of Defense (USA)
DOD	Direct Outward Dialling
DoE	Department of Energy
DOF	Degree Of Freedom
DOS	Disk Operating System
DOT	Digital Optical Tape
DOT	Department of Transportation
DP	Data Processing
DP	Double Pole
DPCM	Differential Pulse Code Modulation
DPDT	Double Pole Double Throw (switch)
DPI	Dots per Inch
DPMI	DOS protected mode interface
DPNSS	Digital Private Network Signalling System
DPSK	Differential Phase Shift Keying
DPST	Double Pole Single Throw (Switch)
DRA	Defence Research Agency (UK)
DRAM	Dynamic Random Access Memory
DRFM	Digital RF Memory
DRO	Digital Readout
DRO	Doubly Resonant Oscillator
DRO	Dielectric Resonator Oscillator
DROS	Disk Resonant Operator System
DRT	Device Rise Time
DS	Direct Sequence
DSB	Double Sideband
DSB-AM	Double Sideband Full carrier Digital Amplitude Modulation
DSO	Digital Storage Oscilloscope
DSSB	Double Single Sideband
DSSC	Double Sideband Suppressed Carrier (modulation)
Dstl	Defence Science Technical Laboratory
DTE	Data Terminal Equipment
DTI	Department of Trade and Industry (UK)
DTL	Diode-Transistor Logic
DTS	Digital Termination System
DTSC	Digital Termination Shielded Chamber
DUT	Device Under Test
DVM	Digital Voltmeter
DVOM	Digital Volt-Ohm Meter

DVOR	Doppler Very High Frequency Omnidirectional (Radio) Range
DWL	Dominant Wavelength
DZ	Zener Diode
E	NATO frequency band (2–3 GHz)
E	Exa (prefix) 10^{18}
E	symbol for electric field
e	2.71828
EACC	Enroute Area Control Centre
EADS–CCR	European Aeronautic Defence and Space – Common Research Centre in France
EADSIM	Extended Air Defence Simulation
EASA	European Aviation Safety Agency
EATCHIP	European ATC Harmonisation & Integration Program (Eurocontrol)
EATMS	European ATM System (Eurocontrol)
EB	Elastic Buffer
EBU	European Broadcasting Union
EC	Engineering Council
EC	European Community
ECAC	Electromagnetic Analysis Center
ECAC	European Civil Aviation Conference
ECC	Electronic Calibration Center (NBS)
ECC	Error Correction Code
ECE	Economic Commission for Europe
ECM	Electronic Counter Measures
ECMA	Electronic Computer Manufacturers Association
ECMA	European Computer Manufacturers Association
ECMWF	European Centre for Medium Range Weather Forecasting
ECQAC	Electronic Components Quality Assurance Committee
ECR	Electric Cash Register
ECS	Environmental Control System
EDIF	Electronic Data Interchange Format
EDP	Electronic Data Processing
EDPC	Electronic Data Processing Center
EDPE	Electronic Data Processing Equipment
EE&T	Electromagnetic Engineering and Test
EEC	European Economic Community
EECL	Emitter-Emitter Coupled Logic
EED	Electroexplosive Device
EED	Electromagnetic Energy Density
E^3	Electromagnetic Environmental Effects
EEM	Electric Equipment Monitoring
EEPA	Electromagnetic Energy Policy Alliance
EF	Electric Field
EF	Emitter Follower
EFAS	En-route Flight Advisory Service
E-field	Electric Field

EFIS	Electronic Flight Instrument System
EFS	Electric Field Strength
EFT	Electrical Fast Transients
EFT/B	Electrical Fast Transient Burst
EFTA	European Free Trade Association
EGA	Enhanced Graphics Array (640 × 350 pixels)
EGNOS	European Geostationary Navigation Overlay Service – to augment GPS & GLONASS
EHF	Extremely High Frequency (IRR band 30 –300 GHz)
EHP	Electric Horse Power
EIA	Electronic Industries Association
EIAJ	Electronic Industries Association of Japan
EIRP	Equivalent/Effective Isotropically Radiated Power
EIS	Electronic Intelligent System
EISA	Extended Industry Standard Architecture
EITB	Engineering Industry Training Board
EIVE	FDTD code developed by ONERA
EKG	Effective kilogram
El	Elevation
EL	Emergency Locator
ELF	Extremely Low Frequency IRR frequency band (3 to 3 kHz)
ELF	Extremely Low Frequency ITU frequency band (3 Hz to 30 Hz)
ELINT	Electronic Intelligence
ELSEM3D	Method of Moment code developed by ONERA
ELT	Emergency Locator Transmitter
EM	Electromagnetic
EM	Electromagnetics
E/M	Electro-Mechanical
e/m	ratio of electric charge to mass for particles (usually electrons)
EMC	Electromagnetic Compatibility
EMC	Electromagnetically Coupled
EMC	Electromagnetic Coupling
EMCON	Emission Control
EMCS	Electromagnetic Compatibility Standardisation (Program)
EME	Electromagnetic Energy
EME	Electromagnetic Environmental Effects
EMETF	Electromagnetic Environmental Test Facility
EMF	Electromotive Force
EMH	Electromagnetic Health/Hazard
EMI	Electromagnetic Interference
EMINT	Electromagnetic Intelligence
EMM	Electromagnetic Measurement
EMP	Electromagnetic Pulse (radiation)
EMR	Electromagnetic Radiation
EMS	Electromagnetic Susceptibility
EMSA	Electromagnetic Modelling of Satellite Antennas

EMSEC	Emanation Security (TEMPEST)
emu	Electromagnetic Unit
EMW	Electromagnetic Warfare
EN	Europäische Norm (European Standard)
ENAC	Ecole Nationale de l'Aviation Civile (France)
ENR	Equivalent Noise Resistance
ENR	Excess Noise Ratio
EOL	Electrical Overload
EOS	Electrical Overstress
EOTC	European Organisation for Testing and Certification
EPA	Environment Protection Agency
EPC	Electronic Power Conditioner
EPIRB	Emergency Position Indicating Radio Beacon
EPLD	Electronically Programmable Logic Device
EPM	Electronic Protection Measures (Electronic Counter-Countermeasures)
EPM	External Polarization Modulation
EPROM	Electronically/Erasable Programmable Read Only Memory
EPS	Encapsulated Postscript
ERA	European Research Area
ERA	Electrical Research Association
EROM	Erasable Read Only Memory
ERP	Effective Radiated Power
ERP	Equivalent Radiated Power
ERTZ	Equipment Radiation TEMPEST Zone
ES	Earth Station
ES	Electromagnetic Switching
ESA	European Space Agency
ESA	Electronic Scanning Array
ESA	Electrical Surge Arrestor
ESM	Electronic Support Measures
ESR	Equivalent Series Resistance
ESTEC	European Space Technology Centre
ESTLA	Electrically Short Tuned Loop Antenna
ETA	Estimated Time of Arrival
ETD	Estimated Time of Departure
ETE	Estimated Time of En-route
ETI	Elapsed Time Indicator
ETL	Electrotechnical Laboratory (Japan)
ETSI	European Telecommunications Standards Institute
EURATOM	European Atomic Energy Community
EURET	European Research on Transport Systems
EUROCAE	European Organisation for Civil Aviation Equipment
EUROMART	European co-operative Measures for Aeronautical Research & Technology
EUT	Equipment Under Test
eV	Electron Volt

EVM	Electronic Voltmeter
EVOM	Electronic Volt Ohmmeter
EW	Electronic Warfare
EZAP	Enhanced Zonal Analysis Procedures
F	Fahrenheit
F	Farad (unit of capacitance)
F	Fuse
F	NATO frequency band (3–4 GHz)
f	femto (prefix) 10^{-15}
f	symbol for frequency in Hz
fA	femtoampere
FA	Final Approach
FAA	Federal Aviation Administration (USA)
FACTOPO	Multi-domain solver developed by ONERA
FALTRAN	Fortran to Algol Translator
FANS	Future Air Navigation System (ICAO)
FAR	Federal Aviation Regulations
FAT	File Allocation Table
FAX	Facsimile
F/B	Front-to-Back ratio
FB	Feedback
FBR	Feedback Resistance
FBW	Fly By Wire
FC	Ferrite Core
FCC	Federal Communications Commission (USA)
FCC	Flight Control Computer
FCS	Flight Control System
FD	Full Duplex
FDDI	Fibre Distributed Data Interface
FDM	Frequency Division Multiplexer
FDMA	Frequency Division Multiplex Access
FDOS	Floppy Disk Operating System
FDR	Frequency Domain Reflectometry
FDTD	Finite Difference Time Domain
FDTD/VF	Finite Difference Time Domain/Volume Filling
FDX	Full Duplex
FEA	Failure Effect Analysis
FEA	Federal Energy Administration
FEA	Finite Element Analysis
FEANI	Fédération Européenne d'Associations Nationales d'Ingénieurs (France)
FEC	Forward Error Correction
FED	Field Effect Diode
FEKO	Feldberechnung bei Körpern mit beliebiger Oberfläche (field computations involving bodies of arbitrary shape), hybrid solver of MoM and UTD/PO
FEM	Finite Element Method

FEPROM	Flash Electronically/Erasable Programmable Read Only Memory
FERMAT	Asymptotic SBR code developed by ONERA
FERROD	Ferrite-Rod Antenna
FET	Field-Effect Transistor
FF	Far Field
FF	Flip-Flop (multivibrator circuit)
FFT	Fast Fourier Transform
FFT	Final Form Text
FG	Function Generator
FH	Frequency Hopping
FHSS	Frequency Hopping Spread Spectrum
FI	Field Intensity
FI	Frequency Independent
FIC	Frequency Interference Control
FICS	Fully Integrated Communications System
FIM	Field Intensity Meter
FIM	Field Induced Model
FIS	Floating point Instruction Set
FLOPS	Floating Point Operations Per Second
FM	Facilities Management
FM	Frequency Modulation
FMAC	Frequency Management Advisory Council (USA)
FMCW	Frequency Modulated Continuous Wave
FMCW	Frequency Modulated Carrier Wave
FMLC	Fibre Metal Laminates Composites
FMM	Fast Multipole Method
FMSG	Frequency Management Study Group (ICAO)
FMX	FM transmitter
FOB	Field Operations Bureau (FCC)
FOC	Fibre Optic Cable
FOD	Foreign Object Damage
FOL	Fibre Optic Link
FOM	Figure-Of-Merit
FORTRAN	Formula Translation
FOV	Field Of View
FPA	Flight Path Angle
FPGA	Field Programmable Gate Array
FPLA	Field-Programmable Logic Array
FPROM	Programmable Read-only Memory
FPU	Floating-point Unit
FRCS	Flight Recorder Configuration Standard
FRP	Federal Radio navigation Plan (USA)
FRPA	Fixed Radiation Pattern Antenna
FRP	Federal Radionavigation Plan
FS	Field Strength
FSD	Full-Scale Deflection

FSD	Finite State Diagrams
FSDO	Flight Standards District Office (FAA)
FSM	Field Strength Meter
FSS	Frequency Selective Surfaces
FSS	Fixed Satellite Service
FSVM	Frequency Selective Voltmeter
FT	Fourier transform
ft	Feet
FTC	Federal Trade Commission
FTN	Formulated Twisted Nematic
FTZ	Fernmelde Technischer Zentralamt (Germany)
F/V	Frequency To Voltage
fV	femtovolt
fW	femtowatt
FY	Fiscal Year
G	symbol for gravitational force
G	Giga (prefix) 10^9
G	NATO frequency band (4–6 GHz)
G	USA frequency band (150–225 MHz)
G	symbol for conductance/susceptance
GA	General Assembly
G/A	Ground to Air
GALILEO	Future European Navigation Satellite System
GAMA	General Aviation Manufacturers Association
GARTEUR	Group for Aeronautical Research and Technology in Europe
GATM	Global ATM
GBAS	Ground Based Augmentation System
GCAS	Ground Collision Avoidance System
GCHQ	Government Communication Headquarters (UK)
GDI	Graphics Display Interface
GDOP	Geometric Dilution of Precision
GEM	Graphics Environment Manager
GEO	Geostationary/Geosynchronous Earth Orbit
GFLOPS	Giga FLOPS (Billion FLOPS)
GFRP	Glass Fibre Reinforced Plastic
GFSK	Gaussian Frequency Shift Keying
GGS	GPS Ground Station
GHz	Gigahertz (10^9 Hz or 1000 MHz)
GiD-CEM	Proprietary software from CIMNE
GIF	Graphics Interchange Format
GIGO	Garbage In Garbage Out
GINS	GPS Inertial Navigational System
GIS	Geographical Information Systems
GLARE	Glass Reinforced Laminate
GLONASS	Global Navigational Satellite System – Russian GPS
GLS	GPS/GNSS Landing System

GMC	Ground Movement Control
GMR	Ground Mapping Radar 16.7–17 GHz
GMT	Greenwich Mean Time
GND	Ground
GNSS	Global Navigation Satellite System
GO	Geometric Optics
GOA	Generic Open Architecture
GOPS	Giga (Billion) Operations Per Second
GPIB	General Purpose Interface Bus
GPIRS	Global Positioning Inertial Reference System
GPS	Global Positioning System
GPS	General Problem Solver
GSE	Ground Support Equipment
GSM	Group Special Mobile – Global System for Mobile Communications
GSO	Geo Stationary Orbit
GSS	Geostationary Satellite
G/T	Gain-to-noise Temperature Ratio
GTD	Geometric Theory of Diffraction
GUS	Ground Uplink Station
GZ	Ground Zero
H	Henry (unit of inductance)
H	symbol for magnetic field intensity
H	NATO frequency band (5–8 GHz)
HAARP	HF Active Auroral Research Program to study the Earth's Ionosphere
HALE	High Altitude Long Endurance
HBT	Hetrojunction Bipolar Transistor
HCMOS	High speed CMOS
HCP	Horizontal Coupling Plane (For ESD Testing)
HCTMOS	High speed TTL-compatible CMOS
HDL	Hardware Description Language
HDD	Head Down Display
HEMP	High Altitude (>40 km) Electromagnetic Pulse
HEMT	High Electron Mobility Transistor
HERF	Hazards of Electromagnetic Radiation To Flight
HERF	Hazards of Electromagnetic Radiation To Fuel
HERO	Hazards of Electromagnetic Radiation To Ordnance
HERP	Hazards of Electromagnetic Radiation To Personnel
HF	High Frequency (aircraft band 2–30 MHz)
HF	High Frequency (IRR band 3–30 MHz)
HF	High Frequency (ITU band 3–30 MHz)
HF	Human Factors
HFDL	High Frequency Data Link
H-Field	Magnetic Field
HFIM	High Fidelity Intrapulse Modulation
HIRF	High Intensity Radiated Fields
HLA	High Level Architecture

HMD	Head/Helmet Mounted Display
HMSO	Her Majesty's Stationery Office (UK Government)
HP	Holding Pattern
HP	High Power/Pressure
HP	Horizontal Polarization
HP	Horsepower
HPA	High Power Amplifier
HPBW	Half Power Beamwidth
HPD	High Power Density
HPF	Highest Probable Frequency
HPF	High Pass Filter
HPGL	Hewlett-Packard Graphics Language
HPIB	Hewlett-Packard Interface Bus
HPM	High Power Microwaves
HROT3D	Finite Element code developed by ONERA
HSDRFU	High-speed Data Radio Frequency Unit
HSI	Hardness Surveillance Illuminator
HTTL	High-power TTL
HTTP	Hyper Text Transfer Protocol
HUD	Head-Up Display
HUMS	Health and Usage Monitoring System
HV	High Voltage
HVDC	High Voltage Direct Current
HVPS	High Voltage Power Supply
HVR	High Voltage Regulator
HW	Half Wave
HWIL	Hardware in the Loop
Hz	Hertz (cycles per second)
I	NATO frequency band (8–10 GHz)
I	symbol for current
I&C	Instrumentation and Control
I/O	Input-Output (Device)
IAC	Industry Advisory Committee
IADS	Integrated Air Defence System
IAF	Initial Approach Fix
IAP	Installed Antenna Performance
IATA	International Air Transport Association
IBS	Institute for Basic Standards
IBW	Impulse Bandwidth
IC	Integrated Circuit(s)
ICAO	International Civil Aviation Organisation
ICARD	ICAO five letter code and route
ICAS	International Council of the Aeronautical Sciences
ICP	Inherently Conductive Polymers
ID	Inside Diameter
IDC	Insulation Displacement Connector

IDE	Integrated Development Environment
IDM	Improved Data Modem
IDP	Inherently Dissipative Polymers
IDT	Integrated Device Technology
IEC	International Electrotechnical/Engineering Commission
IEFD	Integral Equation Frequency Domain
IEE	Institution of Electrical Engineers UK (now IET)
IEEE	Institute of Electrical and Electronics Engineers (USA)
IEEJ	Institute of Electrical and Engineers (Japan)
IEFD	Integral Equation Frequency-Domain
IEPS	International Electronics Packaging Society
IES	Institute of Environmental Sciences
IET	Institution of Engineering and Technology (formerly IEE)
IF	Intermediate Frequency
IFALPA	International Federation of Airline Pilots Association
IFD	Instantaneous Frequency Discriminator
IFF	Identification Friend and Foe
IFIS	Integrated Flight Information System
IFR	Instrument Flight Rules
IFR	In-flight Refuelling
IFR	Intermediate Frequency Range
IGFET	Insulated-Gate Field-Effect Transistor
IGFET	Isolated-Gate Field-Effect Transistor
IICIT	International Institute of Connector and Interconnection Technology
IKBS	Intelligence Knowledge Based System
IL	Intermediate Level
ILS	Instrument Landing System
IM	Interference Margin
IM	Intermodualtion
IMA	Integrated Modular Avionics
IMINT	Imagery Intelligence
IMPATT	Impact Avalanche And Transit Time (Diode)
IMRS	Integrated Monitoring & Recording System
IMU	Inertial Measurement Unit
IN	Interference To Noise Ratio
INIRC	International Non-ionising Radiation Committee
INMARSAT	International Marine Satellite (Organization)
INS	Inertial Navigation System
INSTANT	INSTalled ANTenna performance prediction software – BAE Systems
INTELSAT	International Telecommunications Satellite (Organization)
Intl.	International
IOC	Integrated Optical Circuits
IOM	Input-Output Multiplexer
ION	Institute of Navigation
IP	Internet Protocol

IPAS	Installed Performance of Antenna on AeroStructures – Sixth Framework EU Project
IPC	Institute for Interconnecting and Packaging Electronics Circuits
IPR	Intellectual Proprietary Rights
IPU	Interface Processing Unit
IR	Infrared
IRIG	Internationally Recognized Interval Generator
IRIG	Inter-Range Instrumentation Group (standard for telemetry on U.S. Government test ranges)
IRPA	International Radiation Protection Association
IRR	International Radio Regulations
IRS	Inertial Reference System
IRT	Institut für Rundfunktechnik GmbH (Germany)
ISA	Industry Standard Architecture
ISB	Independent Side Band
ISBN	International Standard Book Number
ISDN	Integrated Services Digital Network
ISHM	International Society for Hybrid Microelectronics
ISL	Inter-Satellite Links
ISLS	Interrogator Sidelobe Suppression
ISM	Industrial Scientific And Medical
ISO	International Standards Organization
ISR	Intelligence, Surveillance Reconnaissance
IST	Information Systems Technology (NATO)
ISTAR	Intelligence, Surveillance, Targeting and Reconnaissance
ITD	Incremental Theory of Diffraction
ITE	Information Technology Equipment
ITE	Institute of Telecommunications Engineers
ITEM	Interference Technology Engineers' Master
ITT	International Telephone and Telegraph Company
ITU	International Telecommunication Union (UNO)
IU	Interference Unit
IV	Intermediate Voltage
J	NATO frequency band (10–20 GHz)
J	UK (IEE/IET) frequency band (12–18 GHz)
J	Joule – unit of energy
J	symbol for current density
JAA	Joint Aviation Authorities
JAR	Joint Airworthiness Requirements
JEDEC	Joint Electron Device Engineering Council
JFET	Junction Field-Effect Transistor
JIP	Joint Input Processing
JRC	Joint Resource Council (FAA)
JSEP	Joint Services Electronic Programme
JSR	Jam to Signal Ratio
JTIDS	Joint Tactical Information Distribution System

JTRS	Joint Tactical Radio System
K	Aircraft frequency band (12.5–40 GHz)
K	NATO frequency band (20–40 GHz)
K	ITU frequency band (16–27 GHz)
K	UK (IEE/IET) frequency band (12–18 GHz)
K	USA frequency band (10.9–36 GHz)
K	Kelvin (absolute temperature scale)
K	kilo (IT prefix) $2^{10} = 1024$
k	Boltzmann's constant (1.38×10^{-23} J K^{-1})
k	kilo (prefix) 10^3
K_a	Satellite Communications bands 17.7–21.2 GHz and 27.5–31 GHz
K_a	ITU frequency band (27–40 GHz)
kA	kiloAmpere
KB	Kilobyte – 1024 bytes
KBps	Kilobyte per second
Kb	Kilobit – 1024 bits
KCAS	Knots Calibrated Airspeed
KE	Kinetic Energy
keV	kilo-electron-Volt
kg	kilogram (1000 grams)
kg/m^3	kilogram per cubic metre (density)
kgm	kilogram-metre
kHz	kilohertz (1000 Hz)
km	kilometre
KN	Knot
kohm	kilohm (1000 ohm)
KP	Kernel Procedures
KPH	kilometres per hour
KTAS	Knots True Airspeed
kV	kilovolt (1000 volt)
kVA	kilovolt-ampere
kW	kilowatt
kWh	kilowatt-hour
kWhm	kilowatt-hour metre
L	Aircraft frequency band (1–3 GHz)
L	ITU frequency band (1–2 GHz)
L	NATO frequency band (40–60 GHz)
L	symbol for Inductance
L	UK (IEE/IET) frequency band (1–2 GHz)
L	USA frequency band (0.390–1.55 GHz)
L1	GPS band 1575.42 MHz
L2	GPS band 1227.5 MHz
LA	Lighting Arrestor
LAAS	Local Area Augmentation Systems
LAN	Local Area Network
LANTIRN	Low Altitude Navigation and Targeting Infra-red System

Laser	Light Amplification By Stimulated Emission Of Radiation
LASSY	Licence Administration Support System
LATCC	London air-traffic control centre (UK)
lb	pound
LC	Inductance-Capacitance
LC	Level Control
LC	Line Connector
LC	Load Carrier
LCC	Liquid Crystal Cell
LCD	Liquid Crystal Display
LCLV	Liquid-Crystal Light Valve
LCR	Inductance-Capacitance-Resistance
LCS	Large Core Store
LDB	Light Distribution Box
LDE	Linear Differential Equation
LDGPS	Local Area Differential GPS
LDR	Light Dependent Resistor
LDR	Low Data Rate
L_E	Antenna Effective Length (For Electric Field Antennas)
LED	Light Emitting Diode
LEF	Light Emitting Film
LEF	Lighting Effectiveness Factor
LEMP	Electromagnetic Pulse Generated By A Nearby Lightning Strike
LEO	Low Earth Orbit
LF	Load Factor
LF	Low Frequency (IRR 30–300 kHz)
LF	Low Frequency (ITU 30–300 kHz)
LFC	Load Frequency Control
LFSF	Low Frequency Swept Frequency
LGF- LAAS	Local Area Augmentation System Ground Facility
LH	Left-Hand
LHCP	Left Hand Circular Polarization
LIC	Linear Integrated Circuit
LIDAR	Light Detection and Ranging
LIM	Linear Induction Motor
LINLOG	Linear-Logarithmic
LINS	Laser Inertial Navigation System
LISN	Line Impedance Stabilization Networks
LL	Low Level
LLD	Low Level Detector
LLL	Low Level Logic
LLNL	Lawrence Livermore National Laboratory
LLSC	Low Level Sweep Coupling
LLSF	Low Level Swept Frequency
LMS	Least Mean Square
LMT	Local Mean Time

ln	natural or Naperian logarithm
LNA	Low Noise Amplifier
LNB	Low Noise Block Downconverter
LNR	Low Noise Amplifier
LO	Local Oscillator
LOAMP	Logarithmic Amplifier
log	Logarithm to the base 10
LOM	Locator Outer Marker
LORAN	Long Range Navigation
LORAP	Long Range Aerial Photography
LOS	Line of Sight
LOS	Loss Of Signal
LP	Low Pass
LP	Linear Programming
LP	Line Printer
LP	Low Pressure
LPC	Loop-Control
LPD	Linear Power Density
LPD	Low Power Difference
LPDA	Log-Periodic Dipole Antennas Or Array
LPF	Low Pass Filter
LPL	Linear Programming Language
LPO	Low Power Output
LQG	Linear Quadratic Gaussian Control
LR	Load-Resistor (Relay)
LR	Load Ratio
LRI	Line Replaceable Item
LRU	Line Replaceable Unit
LSB	Least Significant Bit
LSB	Lower Sideband
LSC	Least Significant Character
LSD	Least Significant Digit
LSHI	Large-Scale Hybrid Integration
LSI	Large Scale Integration
LSM	Linear Synchronous Motor
LSN	Line Stabilization Network
LSTTL	Large-Scale TTL
LT	Low Tension
LTE	Long Term Evolution (for mobile networks)
LUF	Lowest Usable Frequency
LUHF	Lowest Usable High Frequency
LUT	Local User Terminals (COSPAS-SARSAT)
LV	Low Voltage
LVP	Low Voltage Protection
M	NATO frequency band (60–100 GHz)
M	Mega (prefix) 10^6

M	Mach – Speed of Sound
m	Metre
m	milli (prefix) -10^{-3}
MA	Metric Association
MA	Megampere (10^6 amperes)
mA	milliampere (10^{-3} amperes)
MAC	Module Auxiliary Connector
MAC	Multiple Access Computer
MAC	Multiplexes Analogue Component
MACH	Modular Adaptive Computer Hardware
MADARS	Malfunction Detection, Analysis and Recording System
MADREC	Malfunction And Detection Recording
MAGAMP	Magnetic Amplifier
MAGLOC	Magnetic Logic Computer
MAGMOD	Magnetic Modulator
MAN	Metropolitan Area Network
MASC	Maritime Airborne Surveillance & Control
maser	Microwave Amplification by Stimulated Emission of Radiation
MATE	Multiple-Access Time-Division Equipment
MAWS	Missile Approach Warner System
max	Maximum
MB	Marker Beacon
MBB	Messerschmitt-Bölkow-Blohm
MBDA	Missile systems company controlled by EADS
MBWO	Microwave Backward Wave Oscillator
MCA	Micro Channel Architecture
MCC	Mission Control Centres
MCI	Mode C Intruder
MCM	Monte Carlo Method
MCS	Mobile Communications Systems (IEEE vehicular Technology Technical Committee)
MCS	Modular Computer Systems
MD	Multi Domain
MDAC	Multiplying digital-to-analog convertor
MDIC	Microwave Dielectric Integrated Circuit
MDS	Minimum Detectable (Or Discernible) Signal
MECL	Motorola emitter-coupled logic
MEG	Megohm (10^6 Ωhm)
MEGA	Million (prefix)
MEIS	Multi-Spectral Electro-Optical Imaging System
MEO	Middle Earth Orbit
MESFET	Metal-Semiconductor Field Effect Transistor
MF	Medium Frequency IRR frequency band 300 kHz to 3 MHz
MF	Medium Frequency ITU frequency band 300 kHz to 3 MHz
mF	millifarad
MFD	Multifunction Display

mfd	millifarad
MFS	Magnetic Field Strength (in A/m)
MFSK	Multiple Frequency Shift Keying
mH	millihenry
MHF	Medium High Frequency
Mho	former unit of conductance (siemens now used)
MHRS	Magnetic Heading Reference System
MHz	Megahertz
Mi	mile (statute)
MIC	Micrometer
MIC	Microwave Integrated Circuit
MIC	Minimum Ignition Current
MIC	Mutual Interference Chart
MICAMP	Microammeter
micro	one-millionth (prefix)
Mil	one-thousandth of an inch
MIL-I	Military specification on Interference
MIL-SPEC	Military Specification
MIME	Multi-format/purpose Internet Mail Extensions
MIPS	Million Instructions Per Second
MIS	Management Information System
MKS	Metre-Kilogram-Second (System of units)
MHz	Megahertz
MLS	Microwave Landing System
ML	Main Lobe
ml	millilitre
MLC	Multilayer Ceramic
MLFMA	Multi-Level Fast Multipole Algorithm
MLFMM	Multi-Level Fast Multipole Method
MLPWB	Multilayer Printed-Circuit Wiring Board
MLS	Microwave Landing System
MLV	Multilayer MOV
mm	millimetre
MM	Machine Model (ESD)
MMB	Mass Media Bureau (FCC)
mm-band	millimetric band (ITU 40–300 GHz)
mm-band	millimetric band (common usage 30–300 GHz)
MMF	Magnetomotive Force
MMI	Man Machine Interface
MMIC	Microwave Integrated Circuit
MMR	Multimode Receiver
MoD	Ministry of Defence
MODEM	Modulator-Demodulator
MoE	Measure of Effectiveness
MOF	Maximum Operating Frequency
MoM	Method of Moments

MoP	Measure of Performance
MOPS	Mega (Million) Operations per second
MOS	Metal-Oxide Semiconductor
MOSA	Metal-Oxide Surge Arrestor
MOSAIC	Metal-Oxide Semiconductor Advanced Integrated Circuit
MOSFET	Metallic-Oxide Semiconductor Field Effect Transistor
MOSM	Metal-Oxide Semi-Metal
MOST	Metal Oxide Semiconductor Transistor
MOV	Metal-Oxide Varistor
MP	Main Phase
MP	Multipole
MP-4	An FCC measurement procedure
MPE	Maximum Permissible Exposure
MPH	Miles Per Hour
MPL	Maximum Permissible Level
MPW	Modified Plane Wave
MPX	Multiplexer
MRI	Magnetic Resonance Imaging
MRU	Mobile Radio Unit
MS	Military Standard sheet, prefix to numbered series issued by DoD (USA)
ms	millisecond
MSB	Most Significant Bit
MSC	Mode Stirred Cavity
MSC	Most Significant Character
MSD	Most Significant Digit
MSI	Medium Scale Integration
MSK	Minimum Shift Keying
MSTS	Multisubscriber Time Sharing Systems
MTBF	Mean Time Between Failures
MTBO	Mean Time Between Outages
MTBR	Mean Time Between Repairs/removal
MTF	Mean Time to Failure
MTF	Modulation Transfer Function
MTI	Moving Target Indicator
MTL	Merged-Transistor Logic
MTTR	Mean Time To Repair
MTU	Multiplexer and Terminal Unit
μ	micro (prefix) 10^{-6}
MUF	Maximum Usable Frequency
MUSIC	Multiple Signal Interface Cancellation
MUT	Material Under Test
MUX	Multiplexer
MV	Megavolt
mV	millivolt
MVDS	Microwave Video Distribution System

MW	Megawatt
MW	Microwave
mW	milliwatt
mW/cm^2	milliwatt per square centimetre
MWh	Megawatt-hour
MWRX	Microwave Receiver
MWTX	Microwave Transmitter
MWV	Maximum Working Voltage
N	NATO frequency band (100–200 GHz)
n	nano (prefix) 10^{-9}
N	Refractivity or refractive index
nA	nanoampere
NAB	National Association of Broadcasters
NACSEM	National COMSEC/EMSEC Information Memorandum (TEMPEST)
NACSI	National COMSEC Instruction (TEMPEST)
NACSIM	National COMSEC Instruction Memorandum (TEMPEST)
NAMAS	National Measurement Accreditation Service (UK)
NARTE	National Association of Radio and Telecommunications Engineers
NASA	National Aeronautics Space Administration
NATA	North American Telecommunications Association
NATLAS	National Testing Laboratory Accreditation Scheme (UK)
NATO	North Atlantic Treaty Organisation
NATS	National Air Traffic Services (UK)
NATS	North American Telephony System
NAVAHRS	Navigation Attitude & Heading Reference
NavAid	Navigational Aid
NB	Narrowband
NB	No Bias (relay)
NBC	Narrowband Conducted
NBC	Nuclear Biological And Chemical
NBFM	Narrowband Frequency Modulation
NBR	Narrowband Radiated
NBS	National Bureau of Standards (USA) now NIST
NBSFS	NBS frequency standard
NC	Not Connected
NC	Numerical Control
NDB	Non-Directional Beacon
NDB	Navigation Data Base
NEB	Noise Equivalent Bandwidth
NEC	National Electrical Code
NEC	Nippon Electrical Company (Japan)
NEC	Numerical Electromagnetics Code
neg.	Negative
NEMA	National Electrical Manufacturer's Association
NEMP	Nuclear Electromagnetic Pulse
NEP	Noise Equivalent Power

NEPA	National Environment Policy Act
NF	Near Field
NF	Noise Figure
NF	Noise Frequency
NFP	Near Field Probe
NFPA	National Fire Protection Association
nH	nanohenry
NIH	National Institute of Health
NIST	National Institute of Standards and Technology (USA)
NLD	Non-Licensed Device
NLR	Nationaal Lucht en Ruimtevaart-Laboratorium (National Aerospace Laboratory, The Netherlands)
NM	Nautical miles – usually the international one is used, i.e. 1 nmi = 1.8532 km
NMH	Nickel Metal Hydride
nmi	Nautical miles – usually International one is used, i.e. 1 nmi = 1.8532 km
NMOS	N-type MOS
NMR	Normal-Mode Rejection
NMR	Nuclear Magnetic Resonance
NNI	Network Node Interface
N	Newton (SI unit of force)
NO	Normally Open
NOAA	National Oceanographic and Atmospheric Administration
NONCOHO	Noncoherent Oscillator
NORAD	North American Aerospace Defense Command (USA)
NOx	Nitric & Nitrous Oxide
NPL	National Physical Laboratory (UK)
npn	Negative-Positive-Negative (Transistor)
NPR	Noise Power Ratio
NRC	National Research Council (Canada)
NRC	Nuclear Regulatory Commission
NRL	National Research Laboratory (USA)
NRPB	National Radiological Protection Board (UK)
NRTL	Nationally Recognized Testing Laboratory
NRZ	Non-Return to Zero
NS	National Standard
ns	nanosecond
NSA	Normalized Site Attenuation
NSA	National Security Agency
NTC	Negative Temperature Coefficient
NTG	Nachrichtentechnische Gesellschaft (German Communications Society)
NTIA	National Telecommunications and Information Administration (USA Department of Commerce)
NTISSI	National Telecommunications and Information Systems Security Instruction

NTP	Normal Temperature and Pressure
NTSC	National Television System Committee (USA) 525 lines and 60 fields with a video BW of 4.2 MHz
NTZ	Nachrichtechnische Zeitschrift (Germany)
NUPAC	Nuclear Protection & Control
NURBS	Non-Uniform Rational B Splines
nV	nanovolt = 10^{-9} V
NVLAP	National Voluntary Laboratory Accreditation Program
NWG	National Wire Gauge
0TLP	Zero Transmission Level Point
O	UK (IEE/IET) frequency band (60–90 GHz)
OATS	Open Area Test Site
O-B	Octal To Binary
OBN	Out-of-Band Noise
OBR	Optical Bar Code
OCR	Optical Character Reader
OCR	Optical Character Recognition
OD	Output Data
OD	Outside Diameter
ODR	Overland Downlook Radar
Oe	Oersted (ampere per metre)
OEM	Original Equipment Manufacturer
OET	Office of Engineering and Technology (FCC)
O-H	Octal to Hexadecimal
OHR	Over the Horizon (radar)
OLAN	On-board Local Area Network
OMT	Orthomode Transducer
ONERA	Office National d'Etudes et de Recherches Aérospatiales
ONR	Office of Naval Research
OOK	On-Off Keying
OOP	Object Orientated Programming
OPAS	Overhead Panel ARINC 629 System
OR	Outside Radius
OS	Operating System
OSHA	Occupational Health and Safety Administration
OSI	Open Systems Interconnection
OSPK	Offset Keyed Quadrature Phase Shift Keying
OSR	Output Shift Register
OTDR	Optical Time Domain Reflectometry
OU	Open University (UK)
OV	Overvoltage
OVV	Overvoltage
oz	Ounce
P	USA frequency band (225–390 MHz)
P	Peta (prefix) 10^{15}
p	pico (prefix) 10^{-12}

P/P	Point To Point
P-P	Peak to Peak
PA	Parity
pA	picoampere
PA	Power Amplifier
PA	Pulse Amplifier
PABX	Private Automatic Branch Exchange
PadB	Perceived Noise Level Expressed In Decibels
PAL	Phase Alternating Line (European Television System) 625 lines and 50 fields with a video BW of 5 MHz
PAM	Pole Amplitude Modulation
PAM	Pulse Amplitude Modulation
PAPM	Pulse Amplitude Phase Modulation
PAMR	Public Access Mobile Radio
PARAMP	Parametric Amplifier
PBIT	Parity Bit
PBS	Public Broadcasting Service
PBX	Private Branch Exchange
PC	Personal Computer
PC	Printed Circuit
pC	picocoulomb
pC	picocurie
PCB	Power Circuit Breaker
PCB	Printed Circuit Board
PCB	Polychlorinated Biphenyl
PCL	Printer Control Language
PCM	Pulse-Code Modulation
PCM	Pulse-Count Modulation
PCMA	Paired Carrier Multiple Access
PCMCIA	Personal Computer Memory Card International Association
PCN	Personal Communications Network
P-code	GPS Precision code
PCT	Photon-Coupled Transistor
PCTFE	Polymonochlorotrifluouroethyle
PCU	Power Control Unit
PCU	Power Conversion Unit
PCW	Pulsed Continuous Wave
PCZ	Physical Control Zone (TEMPEST)
pd	Potential Difference
PD	Power Distribution
PDA	Personal Digital Assistant
PDE	Partial Differential Equation
PDF	Probability Density Function
PDF	Portable Document Format
PDIO	Photodiode
PDM	Pulse Delta Modulation

PDM	Pulse Duration Modulation
P-DME	Precision DME
PDQ	Programmed Data Quantizer
PDS	Power Density Spectra
PDS	Power Distribution System
PDW	Pulse Descriptor Word
PE	Parity Enable
PE	Phase Encoded
PE	Polyethylene
PE	Probable Error
PEC	Photoelectric Cell
PEC	Perfect Electric Conductor
PED	Portable Electronic Devices
PEEK	Poly Ether Ether Ketone
PEK	Phase-Exchange Keying
PEL	Picture Element
PEM	Photoelectromagnetics
PEP	Peak Envelope Power
PERT	Program Evaluation and Review Technique
pF	picofarad
PF	Power Factor
PF	Pulse Frequency
pFAST	Passive Final Approach Spacing Tool
PFC	Primary Flight Computer
PFF	Print Format File
PG	Power Gain
PIF	Program Information Files
PIM	Pulse Interval Modulation
pin	Positive-Intrinsic-Negative
PIU	Plug-In Unit
PIV	Peak Inverse Voltage
PIXEL	Picture Element
PL	Plug
PLC	Power-Line Carrier
PLC	Programmable Logic Controller
PLD	Phase Lock Demodulator
PLL	Phase-Locked Loop
pls/s	Pulses Per Second
PLTN	Public Land Telephone Network
PLT	Power-Line Transients
PM	Panel Meter
PM	Permanent Magnet
PM	Phase Modulation
PM	Pulse Modulation
PMC	Pseudo Machine Code
PMFD	Pilots Multi-Function Display

PMOP	Phase Modulation On Pulse
PMOS	P-Channel Metal-Oxide Semiconductor
PN	Pseudonoise
PNDC	Parallel Network Digital Computer
pnp	Positive-Negative-Positive (Transistor)
PO	Physical Optics
PO	Parallel Output
PO	Power Oscillator
POE	Point Of Entry (TEMPEST)
POI	Probability of Intercept
POS	Point Of Sale
POS	Primary Operating System
POST	Power On Self Test
PP	Preprocessor
PP	Push-Pull (amplifier)
P-P	Peak to Peak
PPC	Pulsed Power Circuit
PPD	Penetration Protection Device (TEMPEST)
PPE	Poly Phenyl Ether
PPI	Plan Position Indicator
ppm	Parts Per Million
PPM	Pulse Position Modulation
PPS	Parallel Processing System
PPS	Preferred Power Supply
PPS	Pulses Per Second
PPS	Precision Positioning Service – GPS
PPU	Peripheral Processing Unit
PR	Pseudorandom
PR	Pulse Rate
PR	Power Ratio
PRB	Private Radio Bureau (FCC)
PRC	Pseudo Random Code
PRF	Pulse Repetition Frequency
PRI	Pulse Repetition Interval
PRM	Pulse-Rate Modulation
PRN	Pseudorandom Noise
PROM	Programmable ROM
PRP	Pulse Repetition Period
PRR	Pulse Repetition Rate
PRT	Pulse Repetition Time
PRV	Peak Inverse (Reverse) Voltage
PS	Power Source
PSA	Pressure Sensitive Adhesives
PSD	Phase Sensitive Demodulator
PSD	Phase Shifting Device
PSD	Power Spectral Density

PSD	Pulse Spacing Distribution
PSI	Pound Force Per Square Inch
PSI	Pounds Per Square Inch Absolute
PSOPH	Psophometric Weighting
PSTN	Public Switched Telephone Network
PSU	Power Supply Unit
PT	Pulse Time
PTAN	Precision Terrain Aided Navigation
PTC	Positive Temperature Coefficient
PTD	Physical Theory of Diffraction
PTFE	Poly Tetra Fluouro Ethylene
PTLM	Parallel Transmission Line Modeller
PTM	Pulse Time Multiplex
PU	Power Unit
PVC	Polyvinyl Chloride
PW	Printed Wiring
PW	Pulse Width
PW	Plane Wave
pW	picowatt
PWA	Printed Wiring Assembly
PWB	Printed Wiring Board
PWD	Power Distribution
PWD	Pulse Width Discrimination
PWL	Piecewise Linear
PWM	Pulsewidth Modulation
PWM	Pulsewidth Multiplier
pWp	picowatt Psophometrically Weighted
pwr	Power
PWS	Plane Wave Spectrum
PWV	Peak Working Voltage
PXSTR	phototransistor
PZT	Lead Zirconate Titanate (semiconductor)
Q	Aircraft frequency band (33–50 GHz)
Q	UK (IEE/IET) frequency band (27–40 GHz)
Q	USA frequency band 36–46 GHz
Q	Transistor
QAM	Quadrature Amplitude Modulation
QAM-PAM	Quadrature Amplitude Modulation-Pulse Amplitude Modulation
QARx	Quality Assurance Receiver
QC	Quality Control
QFE	Field Effect Transistor
QNH	No Height Barometric Pressure At Sea Level
QP	Quasi Peak
Q-P	Quasi Peak
QPA	Quasi-Peak Adapter
QPR	Quadrature Partial Response

QPSK	Quadratic Phase Shift Keying
QRM	Man-Made Interference
QU	Unijunction Transistor
QUAM	Quadrature Amplitude Modulation
QVR	Varistor
QWD	Quantum Well Diode
RA	Radio Altitude
RA	Resolution Advisory
R&D	Research and Development
R&T	Research and Technology
R/T	Real Time
R/T Net	Radio Telephone Network
R/W	Read-Write (Head)
RAC	Representative Advisory Committee
RACE	Random Access Computer Equipment
RACE	Rapid Automatic Checkout Equipment
RACS	Remote Access Calibration System
RAD	Random Access Disk
RAD	Rapid Access Disk
rad	Radian (plane angle)
RadAlt	Radar/Radio Altimeter
radar	Radio Assisted Detection and Ranging
RADEM	Random Access Delta Modulation
RADFAC	Radiation Facility
RADHAZ	Radiation Hazards
Radiac	Radioactivity Detection, Identification, and Computation
RADIMT	Radiation Intelligence
radome	radar dome
RAE	Royal Aeronautical Establishment (UK – now Dstl)
RAF	Royal Air Force (UK)
RAM	Radar Absorbing Material
RAM	Random Access Memory
RASP	Real-Time Surface Picture
RB	Reset Bus
RB	Relative Bearing
RB	Return To Bias
RBE	Radiation Biological Effectiveness
RC	Remote Control
RC	Resistance Capacitance
RCTL	Resistor-Capacitance Transistor Logic
RCVR	Receiver
RD	Radiation Detector
RD	Radiation Detection
RD	Received Data
Rd	Rutherford
RDF	Radio Direction Finding

RE	Radiated Emission
READ	Remote Electronic Alphanumeric Display
REC	Radiant Energy Conversion
REC	Rectifier
RECSTA	Receiving Station
REDAC	Research Engineering and Development Advisory Committee (FAA)
REG	Radio Frequency Environment Generator
REG	Radar Environment Generator
RES	Grounding Resistor
RES	Resistor
RF	Radio Frequency
RFA	Recticulated Radio Absorber
RFC	Radio Frequency Choke
RFC	Radio-Frequency Compatibility Program
RFI	Radio-Frequency Interference
RFPG	Radio Frequency Protection Guidelines
RFS	Radio-Frequency Shift
RGCSP	Review of the General Concept of Separation Panel (ICAO)
RH	Radiological Health
RH	Rheostat
RH	Relative Humidity
RHCP	Right Hand Circular Polarization
RHE	Radiation Hazard Effects
RHS	Right Hand Side
RI	Radio Interference
RI	Resistance Induction
RIN	Royal Institute of Navigation (UK)
RISC	Reduced Instruction Set Computer
RL	Resistance-Inductance
RLC	Resistance-Inductance-Capacitance
RM	Radio Monitoring
RMI	Radio Magnetic Indicator
RMOS	Refractory MOS
RMS	Radiation Monitoring System
RMS	Root Mean Square
RN	Reference Noise
RNAV	Area Navigation
RO	Receive Only
ROBIN	Remote On-Line Business Information Network
ROC	Receiver Operating Characteristics
ROM	Read-Only Memory
ROM	Rough Order of Magnitude
ROS	Read-Only Storage
RPM	Revolutions Per Minute
RPN	Reverse Polish Notation
RRSWG	Random Repetitive Square Wave Pulse Generator

RS	Response Spectrum
RS	Radiated Susceptibility
RSL	Received Signal Level
RSS	Root Sum Square
RT	Receiver Transmitter
RT	Remote Terminal
RTCA	Radio Technical Commission For Aeronautics
RTD	Research And Technology Development
RTI	Referred To Input
RTL	Resistor-Transistor Logic
RTMA	Radio Television Manufacturers' Association
RTO	Referred To Output
RVR	Runway Visual Range
R-W	Read-Write (Head)
RWG	Rao–Wilton–Glisson basis function used in CEM
RWR	Radar Warning Receiver
Rx	Receiver/Receive
RX	Receive Serial Line
RZL	Return to Zero Level
RZM	Return to Zero Mark
S	Aircraft frequency band 2.5–4 GHz
S	ITU frequency band 2–4 GHz
S	UK (IEE/IET) frequency band 2–4 GHz
S	USA frequency band 1.55–5.2 GHz
s	Second
SA	Spectrum Analyser
SA	Successive Approximation
SAA	System Applications Architecture
SAD	Silicon Avalanche Diode
SAE	Society of Automotive Engineers
SAL	Symbolic Assembly Language
SAM	Surface-to-Air Missile
SAM	Sequential Access Memory
SAMPE	Society for the Advancement of Material and Process Engineering
SANTANA	Smart Antenna Terminal
SAR	Search And Rescue
SAR	Synthetic Aperture Radar
SAR	Single Axis Reference
SAR	Specific Absorption Rate
SAR	Storage Address Register
SAR	Successive Applications Register
SAR	Synthetic Aperture RADAR
SARSAT	Search And Rescue Satellite
SAT	Satellite Communications
SatCom	Satellite Communications
SAW	Surface Acoustic Wave

SB	Sideband
SB	Synchronization Bit
SBAS	Satellite/Space Based Augmentation Systems
SBD	Schottky-barrier diode
SBR	Shooting and Bouncing Ray techniques (PO, UTD, PTD)
SBX	S-band transmitter
SC	Semiconductor
SC	Single Contact
SC	Switching Coil
S/C	Signal-to-Clutter Ratio
SCC	Single Cotton Covered (Wire)
SCC	Single-Conductor Cable
SCIF	Sensitive Compartmented Information Facility
SCODL	Scan Conversion Object Description Language
SCSI	Small Computer System Interface
SDH	Synchronous Digital Hierarchy
SDIS	System Design And Integration Service
SDRAM	Synchronous Dynamic Random Access Memory
SE	Shielding or Screening Effectiveness
SEAD	Suppression of Enemy Air Defences
sec	Second
SECO	Sequential Coding
SELDS	Shielded Enclosure Leak Detection System
SET	Science Engineering & Technology
SET	Sensors & Electronics Technology
SEV	Schweizerische Elektrotechnischer Verein (Swiss Electrotechnical Association)
SF	Single Frequency
SFD	Sudden Frequency Deviation
SFDR	Spurious-Free Dynamic Range
S/H	Sample and Hold
SHF	Super High Frequency (ITU band 3–30 GHz)
SHG	Second-Harmonic Generation
SI	Systeme Internationale d'Unites (International system of units)
SIGINT	Signals Intelligence
sin	sine (Trigonometry)
SINCGARS	Single Channel Ground and Airborne Radio System
sinh	hyperbolic sine
SIMMOD	Airspace and Airport Simulation Model
SINAD	Signal Plus Noise And Distortion or Signal To Noise And Distortion
SIP	Single-In-Line Package
SIR	Signal-to-Interference Ratio
SLB	Sidelobe Blanking
SLC	Sidelobe Cancellation
SLF	Super Low Frequency ITU band 30–300 Hz
SLL	Sidelobe Level

SLR	Sideways Looking Radar
SLS	Sidelobe Suppression
SLS	Satellite Landing System
SLWL	Straight Line Wavelength
SMA	Sub Miniature Amphenol
SMART	Standard Modular Avionics Repair and Test
SMC	Stirred Mode Cavity
SMD	Surface Mounted Device
SME	Small & Medium size Enterprises
SMM	Standard Method Of Measurement
SMPS	Switch-Mode Power Supply
SMT	Surface Mounted Technology
SMTP	Simple Mail Transfer Protocol
SN	sine (of the amplitude)
S/N	Serial Number
S/N	Signal-to-Noise (Ratio)
SNA	System Network Architecture
SNF	System Noise Figure
SNI	Signal-to-Noise Improvement
SNR	Signal-to-Noise Ratio
SOIC	Small Outline IC
SOL	Solenoid
SOLV	Solenoid valve
SOM	Space Oblique Mercator
SONET	Synchronous Optical network (USA)
SOP	Small Outline Plastic
SP	Single Pole
SP3T	Single Pole Triple Throw (switch)
SP4T	Single Pole Quadruple Throw (switch)
SPDT	Single Pole Double-Throw (switch)
SPEHS	Single Point Excitation For Hardness Surveillance
SPFP	Single Point Failure Potential
SPG	Single Point Ground
SPICE	Simulation Program For Integrated Circuit Emphasis
SPM	Signal Processing Modem
SPS	Standard Positioning Service – GPS
SPS	Standby Power System
SPST	Single Pole Single Throw
SQW	Square Wave
Sr	Steradian (solid angle)
SRD	Step-Recovery Diode
SRSS	Square Root of the Sum of Squares
SS	Signal Strength
SS	Swedish Standard
SSB	Single Sideband
SSBFM	Single Sideband Frequency Modulation

SSBSC	Single Sideband Suppressed Carrier
SSG	Small Signal Gain
SSI	Small Signal Integration
SSIG	Single Signal
SSPA	Solid State Power Amplifier
SSPM	Single-Sideband Pulse Modulation
SSR	Secondary Surveillance Radar
SSRA	Spread Spectrum Random Access
SSSC	Single Switched Suppressed Carrier
ST	Sawtooth
ST	Single Throw
STABLO	Stabilized Local Oscillator
STAMO	Stabilized Master Oscillator
STAP	Space Time Adaptive Processing
Stat	Electrostatic Units in CGS System (Prefix)
Stby	Standby
STC	Short Time Constant
STDM	Statistical Time-Division Multiplexing
STDM	Synchronous Time-Division Multiplexing
STN	Super Twisted Nematic
STP	Standard Temperature and Pressure
STTE	Special to Type Equipment
SUN	Stamford University Network
SUT	System Under Test
SUT	Site Under Test
SW	Short Wave
SWG	Standard Wire Gauge (British)
SWG	Stubs Wire Gage (USA)
SWR	Standing Wave Ratio
SWTL	Surface-Wave Transmission Line
SX	Simplex Signalling
SYCOM	Synchronous Communications
T	Tera (prefix) 10^{12}
T	Tesla (SI unit of magnetic flux density)
TACAN	Tactical Air Navigation
TACAS	Traffic Alert and Collision Avoidance System
TAM	Telephone Answering Machine
tan	tangent (Trigonometry)
tanh	hyperbolic tangent
TAS	True Aircraft Speed
TATC	Transatlantic Telephone Cable
TB	Time Base
TC	Time Constant
TC	Tantalum Capacitor
TC	Temperature Coefficient
TC	Technical Committee

TCAS	Traffic Collision Avoidance System
TCC	Temperature Coefficient Of Capacitance
TCDL	Tactical Common Data Link
TCE	Thermal Coefficient Of Expansion
TCG	Tune-Controlled Gain
TCR	Temperature Coefficient Of Resistance
TCXO	Temperature-Controlled Crystal Oscillator
TCXO	Temperature-Compensated Crystal Oscillator
TD	Time Delay
TD	Tunnel Diodes
TDDL	Time Division Data Link
TDF	Two-Degrees-Of-Freedom
TDM	Time Division Multiplexer
TDMA	Time Division Multiple Access
TDR	Time Delay Relay
TDR	Time Domain Reflectometry
TE	Transverse Electric (field)
TEA	Tri Ethyl Aluminium
TEM	Transverse Electromagnetic
TEMPEST	USA government radiated emissions security program
TFT	Thin-Film Technology
TFT	Thin-Film Transistor
THD	Total Harmonic Distortion
THz	Terahertz 10^{12} Hz
TI	Transfer Impedance
TIEC	Tactical Information Exchange Capability
TIFF	Tagged Image Format File
TL	Transmission Level
TM	Tone Modulation
TM	Transmission Matrix
TM	Transverse Magnetic (field)
TOD	Time Of Day
TOP	Technical Office Protocol
TP	Test Procedure
TPON	Telecommunications Passive Optical Network
TQC	Total Quality Control
TQM	Total Quality Management
TR	Transmit-Receive
TR	Transmitter-Receiver
TRF	Tuned Radio Frequency
TRU	Transformer Rectifier Unit
TSB	Twin Sideband
TSC	Technical Sub-Committee
TSI	Threshold Signal-To-Interference Ratio
TSR	Terminate But Stay Resident
TSS	Tangential System Sensitivity

TTE	Telecommunications Terminal Equipment
TTL	Transistor-Transistor Logic
TTY	Teletypewriter
TVAR	TEMPEST Vulnerability Assessment Request
TVI	Television Interference
TVS	Transient Voltage Suppressor
TW	Travelling Wave
TWA	Travelling-Wave-Amplifier
TWCRT	Travelling-Wave Cathode-Ray Tube
TWS	Track While Scan
TWSB	Twin Sideband
TWT	Travelling-Wave Tube
TWTA	Travelling-Wave Tube Amplifier
TX	Transmit
TX	Transmit Serial Line
UAE	Unexpected Application Error
UCAV	Unmanned Combat Air Vehicle
UHF	Ultra High Frequency (aircraft band 225–400 MHz)
UHF	Ultra High Frequency (IRR band 300 MHz to 3 GHz)
UHF	Ultra High Frequency (ITU band 300 MHz to 1 GHz)
UL	Underwriter's Laboratory
ULF	Ultra Low Frequency (ITU band 300–3000 Hz)
UML	Unified Modelling Language
UNICOM	Universal Communications
UNO	United Nations Organisation
UPS	Uninterruptible Power Supply
URAV	Uninhabited/Unmanned Reconnaissance Air Vehicle
URSI	International Union of Radio Science
US	Ultrasonic
USB	Upper Sideband
UTC	Universal Time Coordinate
UTD	Uniform/Unified Theory of Diffraction
UTM	Universal Transverse Mercator
UUT	Unit Under Test
UV	Ultra Violet
UW	Unique Word
UWB	Ultra Wideband
V	UK (IEE/IET) frequency band (40–60 GHz)
V	USA frequency band (46–56 GHz)
V	Volt
V	Voltmeter
V/Hz	Volts per Hertz
V/m	Volt per metre (unit of electric field strength)
VA	Voltampere
VAC	Volts Alternating Current
VAM	Voltammeter

VAR	Volt-Ampere Reactive
VARISTOR	Variable resistor
VCCI	Voluntary Control Commission for Interference (Japan)
VCCS	Voltage Controlled Current Source
VCD	Variable Capacitance Diode
VCF	Voltage-Controlled Frequency
VCO	Voltage-Controlled Oscillator
VCP	Vertical Coupling Plane (For ESD Testing)
VCXO	Voltage-Controlled Crystal Oscillator
VD	Voltage Drop
VDC	Volts Direct Current
Vdc	Volts DC
VDE	Verband Deutscher Electrotechniker – Society of German Electrical Engineers
VDR	Voltage Dependent Resistor
VDT	Video Display Terminal
VDU	Video Display Unit
VF	Variable Frequency
VFC	Voltage-To-Frequency Convertor
Vfg	Verfügung (FTZ regulations that affect EDP equipment)
VFM	Very Fast Method
VFM	Very Fast Method
VFR	Visual Flight Rules
VGA	Variable Gain Amplifier
VGA	Video Graphics Array (640 × 480 pixels)
VHDL	VHSIC Hardware Description Language
VHF	Very High Frequency (aircraft band 30–225 MHz)
VHF	Very High frequency (IRR band 30–300 MHz)
VHF	Very High frequency (ITU band 30–300 MHz)
VHSIC	Very High Speed Integrated Circuit
VLF	Very Low Frequency (IRR band 3–30 kHz)
VLF	Very Low Frequency (ITU band 3–30 kHz)
VLSI	Very Large Scale Integration
VM	Velocity Modulation
VM	Voltmeter
VME	Versa Module European (IEEE 1014)
VNE	Velocity Never Exceed
VoIP	Voice Over IP (Internet Protocol)
VOM	Volt-Ohm Meter
VOR	VHF Omnidirectional (Radio) Range
VP	Vertical Polarization
VPN	Virtual Private Network
VSB	Vestigial Sideband
VSWR	Voltage Standing Wave Ratio
VTO	Voltage Tuned Oscillator
VWL	Variable Word Length

W	USA frequency band (56–100 GHz)
WAAS	Wide Area Augmentation Systems
WAN	Wide Area Network
WARC	World Administrative Radio Conference
WB	Wide Band
Wb	Weber (unit of magnetic flux)
Wb/m^2	Weber per square metre (unit of magnetic flux density same as Tesla)
WBCO	Waveguide Below Cutoff
WBIF	Wide-Band Intermediate Frequency
WBL	Wide-Band Limiting
WD	Working Draft
WDB	Wide Band
WG	Waveguide
WG	Wire Gauge
WG	Working Group (CISPR)
WGBC	Waveguide Below Cutoff
WL	Wavelength
WM	Wattmeter
WMF	Windows Metafile Format
WOD	Word Of Day
WT	Wait
WV	Working Voltage
WVDC	Working Voltage, Direct Current
X	Aircraft frequency band (6–12.5 GHz)
X	ITU frequency band (8–12 GHz)
X	UK (IEE/IET) frequency band (8–12 GHz)
X	USA frequency band (5.2–10.9 GHz)
XCO	Crystal Controlled Oscillator
XO	Crystal Oscillator
XPD	Cross Polar Discrimination
XPI	Cross Polar Interference
XSECT	Cross Section
XT	Cross Talk
XTAL	Crystal
YAG	Yttrium Aluminium Garnet
YIG	Yttrium Iron Garnet
ZA	Zero Adjusted
zkW	Zero kilowatt
ZZF	Zentralamt für Zulassungen im Fernmeldewesen (a branch of the Deutsche Bundespost that issues permits to market EMC compliant equipment in Germany)

Index